Endophyte Biotechnology

Potential for Agriculture and Pharmacology

CABI Biotechnology Series

Biotechnology, in particular the use of transgenic organisms, has a wide range of applications including agriculture, forestry, food and health. There is evidence that it could make a major impact in producing plants and animals that are able to resist stresses and diseases, thereby increasing food security. There is also potential to produce pharmaceuticals in plants through biotechnology, and provide foods that are nutritionally enhanced. Genetically modified organisms can also be used in cleaning up pollution and contamination. However, the application of biotechnology has raised concerns about biosafety, and it is vital to ensure that genetically modified organisms do not pose new risks to the environment or health. To understand the full potential of biotechnology and the issues that relate to it, scientists need access to information that not only provides an overview of and background to the field, but also keeps them up to date with the latest research findings.

This series, which extends the scope of CABI's successful 'Biotechnology in Agriculture' series, addresses all topics relating to biotechnology including transgenic organisms, molecular analysis techniques, molecular pharming, *in vitro* culture, public opinion, economics, development and biosafety. Aimed at researchers, upper-level students and policy makers, titles in the series provide international coverage of topics related to biotechnology, including both a synthesis of facts and discussions of future research perspectives and possible solutions.

Titles available

1. **Animal Nutrition with Transgenic Plants**
   Edited by G. Flachowsky
2. **Plant-derived Pharmaceuticals: Principles and Applications for Developing Countries**
   Edited by K.L. Hefferon
3. **Transgenic Insects: Techniques and Applications**
   Edited by M.Q. Benedict
4. **Bt Resistance: Characterization and Strategies for GM Crops Producing *Bacillus thuringiensis* Toxins**
   Edited by Mario Soberón, Yulin Gao and Alejandra Bravo
5. **Plant Gene Silencing: Mechanisms and Applications**
   Edited by Tamas Dalmay
6. **Ethical Tensions from New Technology: The Case of Agricultural Biotechnology**
   Edited by Harvey James
7. **GM Food Systems and Their Economic Impact**
   Tatjana Brankov and Koviljko Lovre
8. **Endophyte Biotechnology: Potential for Agriculture and Pharmacology**
   Edited by Alexander Schouten

Endophyte Biotechnology

Potential for Agriculture and Pharmacology

Edited by

Alexander Schouten

Laboratory of Phytopathology, Wageningen University & Research, Wageningen, The Netherlands

CABI

CABI is a trading name of CAB International

CABI
Nosworthy Way
Wallingford
Oxfordshire OX10 8DE
UK

Tel: +44 (0)1491 832111
Fax: +44 (0)1491 833508
E-mail: info@cabi.org
Website: www.cabi.org

CABI
745 Atlantic Avenue
8th Floor
Boston, MA 02111
USA

T: +1 (617)682-9015
E-mail: cabi-nao@cabi.org

A catalogue record for this book is available from the British Library, London, UK.

Library of Congress Cataloging-in-Publication Data

Names: Schouten, Alexander, editor. | C.A.B. International, issuing body.
Title: Endophyte biotechnology : Potential for agriculture and pharmacology / edited by Alexander Schouten.
Description: Wallingford, Oxfordshire, UK ; Boston, MA : CABI, [2019] | Includes bibliographical references and index.
Identifiers: LCCN 2019016880| ISBN 9781786399427 (hbk : alk. paper) | ISBN 9781786399441 (epub) | ISBN 9781786399434 (epdf)
Subjects: | MESH: Endophytes | Biotechnology | Symbiosis | Bacterial Physiological Phenomena | Fungi--physiology | Plants--metabolism
Classification: LCC TP248.27.F86 | NLM TP248.27.M53 | DDC 660.6--dc23 LC record available at https://lccn.loc.gov/2019016880

ISBN-13: 978 1 78639 942 7 (hardback)
978 1 78639 943 4 (ePDF)
978 1 78639 944 1 (ePub)

Commissioning Editor: David Hemming
Editorial Assistant: Emma McCann
Production Editor: Shankari Wilford

Typeset by SPi, Pondicherry, India
Printed and bound in the UK by Severn, Gloucester

Contents

Contributors

Alec Baird, University of California, Los Angeles, California, USA
Gabriele Berg, Institute of Environmental Biotechnology, Graz University of Technology, Graz, Austria
Alessandro Bergna, Austrian Centre of Industrial Biotechnology, Graz, Austria
John R. Caradus, Grasslanz Technology Limited, Palmerston North, New Zealand
Tomislav Cernava, Institute of Environmental Biotechnology, Graz University of Technology, Graz, Austria
Sang Cho, Adaptive Symbiotic Technologies, Seattle, Washington, USA
Stéphane Compant, Bioresources Unit, Center for Health & Bioresources, AIT Austrian Institute of Technology GmbH, Tulln, Austria
Amanda F. Currie, School of Biological Sciences, Royal Holloway, University of London, Egham, Surrey, UK
Sherif S. Ebada, Department of Pharmacognosy, Faculty of Pharmacy, Ain Shams University, Cairo; Department of Pharmaceutical Chemistry, Faculty of Pharmacy, Mu'tah University, Al-Karak, Jordan
Weaam Ebrahim, Department of Pharmacognosy, Faculty of Pharmacy, Mansoura University, Mansoura, Egypt
Alan C. Gange, School of Biological Sciences, Royal Holloway, University of London, Egham, Surrey, UK
Pablo A. García-Parisi, IFEVA, Facultad de Agronomía, Universidad de Buenos Aires, CONICET, Buenos Aires, Argentina; Cátedra de Forrajicultura, Departamento de Producción Animal, Facultad de Agronomía, Universidad de Buenos Aires, Buenos Aires
Zachery Gray, Adaptive Symbiotic Technologies, Seattle, Washington, USA
Evan Groover, University of California, Berkeley, California, USA
Roman Harto, Adaptive Symbiotic Technologies, Seattle, Washington, USA
Marian Hsieh, Adaptive Symbiotic Technologies, Seattle, Washington, USA
Linda J. Johnson, AgResearch Limited, Grasslands Research Centre, Palmerston North, New Zealand
Jutta Ludwig-Müller, Institut für Botanik, Technische Universität Dresden, Dresden, Germany
Katie Malmberg, Adaptive Symbiotic Technologies, Seattle, Washington, USA
Ryan Manglona, Adaptive Symbiotic Technologies, Seattle, Washington, USA
Malia Mercer, Adaptive Symbiotic Technologies, Seattle, Washington, USA

Christina A. Müller Bogotá, Institute of Environmental Biotechnology, Graz University of Technology, Graz, Austria
Natalie Nasman, Adaptive Symbiotic Technologies, Seattle, Washington, USA
Tatiana Nicklason, Adaptive Symbiotic Technologies, Seattle, Washington, USA
Melanie-Maria Obermeier, Institute of Environmental Biotechnology, Graz University of Technology, Graz, Austria; Austrian Centre of Industrial Biotechnology, Graz, Austria
Marina Omacini, IFEVA, Facultad de Agronomía, Universidad de Buenos Aires, CONICET, Buenos Aires; Cátedra de Ecología, Departamento de Recursos Naturales y Ambiente, Facultad de Agronomía, Universidad de Buenos Aires, Buenos Aires
Peter Proksch, Institute of Pharmaceutical Biology and Biotechnology, Heinrich-Heine-University Düsseldorf, Düsseldorf, Germany
Regina Redman, Adaptive Symbiotic Technologies, Seattle, Washington, USA; Symbiogenics, Seattle, Washington, USA
Melissa Rienstra, Adaptive Symbiotic Technologies, Seattle, Washington, USA
Rusty Rodriguez, Adaptive Symbiotic Technologies, Seattle, Washington, USA; Symbiogenics, Seattle, Washington, USA; University of Washington, Seattle, Washington, USA
Alexander Schouten, Laboratory of Nematology, Wageningen University & Research, Wageningen, The Netherlands. (Current address: Laboratory of Phytopathology, Wageningen University & Research, Wageningen, The Netherlands)
Jessica-Rose Spong, School of Biological Sciences, Royal Holloway, University of London, Egham, Surrey, UK
Corinne Vacher, BIOGECO, INRA, University of Bordeaux, Pessac, France
Alex Van Inwegen, Adaptive Symbiotic Technologies, Seattle, Washington, USA
Andy VanHooser, Adaptive Symbiotic Technologies, Seattle, Washington, USA
Christin Zachow, Institute of Environmental Biotechnology, Graz University of Technology, Graz, Austria

Preface

Virtually all plants in nature build intimate associations with microbes. Some microbes not only reside on the plant surface but are also capable of migrating into the plant without showing a distinct phenotype and eliciting disease responses. These rather elusive endophytic microbes are often closely related to plant pathogenic species and can even display the endophytic behaviour in one plant species while being pathogenic in another. Plants that have a persisting endophytic association with microbes often have a significant advantage over those that have not, because their performance and survival under stressful biotic and abiotic conditions, such as herbivory, disease, drought, extreme temperatures, or a combination of these, is positively affected by this association. Expanding the knowledge on plant–endophyte interactions is of crucial importance for future developments in plant breeding and sustainable agricultural practices. What is more, as they are involved in biotrophic interactions, endophytes and, by combining their biosynthetic pathways, the plant–endophyte association may hold new peptides and metabolites with valuable properties for pharmacological and biotechnological purposes.

In the past ten years, the fundamental and applied research on endophytes has significantly accelerated by using state-of-the-art molecular, biochemical, microscopical and biological techniques. This book aims at appreciating the added value of the current accumulated knowledge on endophytes by elaborating on the latest insights regarding microorganisms, their mesmerizing diversity and distribution, their intriguing interactions with plants, their ecological functions, and their benefits and applications in agriculture, biotechnology and medicine.

I sincerely acknowledge all the colleagues who contributed, thus making this book possible: David Hemming at CABI for inviting me to edit this book in this series and both David Hemming and Emma McCann at CABI for their friendly support, advice and patience.

Alexander Schouten
Wageningen
February 2019

1 Introduction

Alexander Schouten*

Laboratory of Phytopathology, Wageningen University & Research, Wageningen, The Netherlands

1.1 Microbes: Ancient Allies in Sustaining Plant Life

Fossil records revealing the presence of arbuscular and fungal structures in 400-million-year-old plants indicate that intimate relationships between plants and microorganisms are very ancient (Remy *et al.*, 1994; Taylor *et al.*, 1995; Redecker *et al.*, 2000; Taylor *et al.*, 2005; Taylor and Krings, 2005; Krings *et al.*, 2007; Labandeira and Prevec, 2014). Due to the sometimes devastating plant diseases in crops caused by microorganisms, such as late blight in potato, Panama disease in banana, rusts in cereals and brown spot in rice (Klinkowski, 1970; Marquardt, 2001), most microorganisms were initially mistrusted and plants were considered to be rather vulnerable to microbial invasion. Based on research over the past six decades, this view has gradually changed, and the current view is that in nature plants are not all that vulnerable to microbial diseases and can cope perfectly with both biotic and abiotic stress conditions. In this concept, the microbial community in the rhizosphere, phyllosphere and endosphere is even considered a true asset for plant survival (Rodriguez *et al.*, 2008; Rodriguez and Redman, 2008), and may even be deliberately recruited and manipulated by the plant to maximize growth and development. Microbes can protect plants against biotic (pests and pathogens) and abiotic (extreme temperatures, drought, chemical contaminants) stress conditions and facilitate nutrient uptake. A typical and very practical illustration that particular microorganisms can benefit the plant is the presence of specific antibiotic-producing pseudomonads in wheat and barley, which significantly reduce root disease caused by the soilborne fungus *Gaeumannomyces graminis* var. *tritici*. Although this disease can be significantly detrimental in the first three to five growing seasons, by sustaining a strict monoculture approach, the bacterial population is capable of accumulating to effective antagonistic levels in the plant's rhizosphere in several important growing areas, such as the Inland Pacific Northwest of the USA, The Netherlands and the UK. In this way, these crops have been successfully cultivated for many decades without showing significant detrimental effects caused by the pathogen (Gerlagh, 1968; Shipton, 1972; Baker and Cook, 1974; Gurusiddaiah *et al.*, 1986; Cook, 2003; Weller *et al.*, 2007). This counterintuitive monoculture approach proved to be crucial in suppressing the take-all disease because it could reappear when this growing strategy was interrupted, e.g. by fallow or crop rotation (Baker and Cook, 1974; Cook *et al.*, 1995; Cook, 2007).

The ability to establish a beneficial association with particular microbes is most likely not an easy task, considering the vast numbers of microorganisms that can be found in both the rhizosphere and phyllosphere. It has been calculated that the rhizosphere, which is the thin zone of soil around the root in which the microbial life is affected through root exudates (Curl and Truelove, 1991), can harbour up to 10^{11} colony forming units (cfu) of prokaryotic cells (Shafer and Blum, 1991), comprising more than 30,000 different species

* Corresponding author e-mail: sander.schouten@wur.nl

(Mendes *et al.*, 2011), and more than 10^7 cfu of fungi per gram of fresh root (Shafer and Blum, 1991). And for the rhizoplane itself, prokaryotic population densities of 10^7 cfu per gram of fresh material were calculated (Benizri *et al.*, 2001; Bais *et al.*, 2006). The microbial community structures within the rhizosphere environments are nevertheless tremendously inconsistent in place and time (Sasse *et al.*, 2018). This inconsistency is regulated not only by abiotic factors, such as soil type and geographical location, but also by the plant species, its genotype and its developmental stage (Micallef *et al.*, 2009a,b; Weinert *et al.*, 2011; Inceoglu *et al.*, 2013). Root cap, border and other root cells release an array of constituents such as insoluble mucilage and soluble (antibiotic) exudates as well as volatile organic carbons (Walker *et al.*, 2003; Bais *et al.*, 2006; Hartmann *et al.*, 2009; Jones *et al.*, 2009). Within root exudates, sugars, amino compounds, organic acids, fatty acids, sterols, growth factors, nucleotides, flavones, enzymes, together with an array of miscellaneous compounds, such as auxins, scopoletin, hydrocyanic acid and microbial growth stimulants and inhibitors, were detected (Curl and Truelove, 1991), thus illustrating the chemical complexity of the rhizosphere.

1.2 The Plant Endosphere as Habitat for Microorganisms

Primarily from the rhizosphere, a selected number of (beneficial) microorganisms are allowed access to the endosphere of the plant (Bulgarelli *et al.*, 2013). This is nevertheless a rather oversimplified view, as some microorganisms are obligate endophytic and not only transferred horizontally but also vertically, i.e. through seeds, such as the grass endophytes, and are therefore not encountered in the bulk soil or rhizosphere (Schardl *et al.*, 2004). Nevertheless, the microbial community within the endosphere is significantly less complex than that of the rhizosphere (Compant *et al.*, 2010; Edwards *et al.*, 2015; Vandenkoornhuyse *et al.*, 2015). At first, only arbuscular mycorrhizae (AMs) and rhizobia (Denison and Kiers, 2011) were considered and extensively scrutinized (Parniske, 2008; Denison and Kiers, 2011). But it is now evident that virtually every plant can allow a much broader assortment of microorganisms, particularly bacteria and fungi, to reside in its endosphere without really exhibiting its presence (Rodriguez and Redman, 2008). As for arbuscular AMs and rhizobia, the endosphere is regarded as a protective environment and serving as an important carbon source for the microbe, whereas the benefit for the host plant is often more difficult to define. This is because these benefits may be more indirect and multifaceted, not only facilitating nutrient uptake (García-Garrido and Ocampo, 2002), as described for AMs and rhizobia, but also providing other means to increase plant vigour, growth and development (Sikora, 1992; Rodriguez and Redman, 2008; Aly *et al.*, 2011; Bakker *et al.*, 2013; Ludwig-Müller, 2015), which may only be determined by considering the environmental or ecological context (Schardl, 2001; Müller and Krauss, 2005; Rodriguez *et al.*, 2008; Redman *et al.*, 2011). In all, these endophytes can be mutualistic, serving the host plant in ways AMs and rhizobia may not.

Similar to AMs and rhizobia, it is believed that chemical queues released by the roots are involved in the recruiting of microorganisms from the bulk soil and subsequently manipulating the evolved microbial community, all aiming at allowing beneficial microorganisms to enter the rhizosphere and endosphere, while simultaneously repressing or repelling unwelcome, pathogenic or parasitic microorganisms. However, the occurrence of diseases indicates that the selection for beneficial microorganisms is error prone. As discussed in Chapter 2, this volume, this may not be without reason because the difference between a pathogenic and beneficial microorganism can in some cases be subtle. The underlying mechanisms for the change in microbial behaviour are still poorly understood. Nevertheless, biological, genetic and molecular studies suggest that both plant and endophytes are responsible. Knowledge on these issues is elemental when endophytes are to be exploited for agricultural practices.

1.3 Exploiting Endophytes

Endophytes can be exploited in several ways. Firstly, they can be used in agricultural practices to support plant vigour, growth and development, even making it possible to reduce the usage of fertilizers and pesticides and to grow plants under less ideal conditions, such as water deficiency, increased soil salinity and high temperatures. Secondly, endophytes are known to synthesize a plethora of chemical constituents. This is most likely because, over time, the intimate association between plants and endophytes led to complex chemical interactions, not only with the host plant but also with competitors. These constituents may benefit not only the plant but also humans. From many (medicinal) plants, endophytes are being characterized that are by themselves able to synthesize compounds relevant for pharmacological (Aly *et al.*, 2011) and agronomical reasons.

1.4 Aim of This Book

The aim of this book is to give an overview on the current knowledge about endophytic fungi and bacteria, their diversity, their relationships with pests and pathogens, their distribution and activities inside the plant and their (potential) applications in developing more sustainable agricultural practices. Furthermore, the identification of chemical constituents synthesized by endophytes or by the endophyte–host plant association is discussed, as they can be most relevant for identifying novel compounds relevant for medicine, such as antibiotics and anticancer drugs, and for agriculture, such as biologically sound pesticides. It demonstrates that the current research on endophytes is highly technology-based on every level, relying on state-of-the-art molecular, biochemical, microscopical, computational and biological methods.

References

Aly, A.H., Debbab, A. and Proksch, P. (2011) Fungal endophytes: Unique plant inhabitants with great promises. *Applied Microbiology and Biotechnology* 90, 1829–1845. doi:10.1007/s00253-011-3270-y

Bais, H.P., Weir, T.L., Perry, L.G., Gilroy, S. and Vivanco, J.M. (2006) The role of root exudates in rhizosphere interactions with plants and other organisms. *Annual Review of Plant Biology* 57, 233–266. doi:10.1146/annurev.arplant.57.032905.105159

Baker, K.F. and Cook, R.J. (1974) *Biological Control of Plant Pathogens*. W.H. Freeman and Company, San Francisco, California.

Bakker, P.A.H.M., Berendsen, R.L., Doornbos, R.F., Wintermans, P.C. and Pieterse, C.M. (2013) The rhizosphere revisited: Root microbiomics. *Frontiers in Plant Science* 4, 165. doi:10.3389/fpls.2013.00165

Benizri, E., Baudoin, E. and Guckert, A. (2001) Root colonization by inoculated plant growth-promoting rhizobacteria. *Biocontrol Science and Technology* 11, 557–574. doi:10.1080/09583150120076120

Bulgarelli, D., Schlaeppi, K., Spaepen, S., Ver Loren van Themaat, E. and Schulze-Lefert, P. (2013) Structure and functions of the bacterial microbiota of plants. *Annual Review of Plant Biology* 64, 807–838. doi:10.1146/annurev-arplant-050312-120106

Compant, S., Clement, C. and Sessitsch, A. (2010) Plant growth-promoting bacteria in the rhizo- and endosphere of plants: Their role, colonization, mechanisms involved and prospects for utilization. *Soil Biology and Biochemistry* 42, 669–678. doi:10.1016/j.soilbio.2009.11.024

Cook, R.J. (2003) Take-all of wheat. *Physiological and Molecular Plant Pathology* 62, 73–86. doi:10.1016/S0885-5765(03)00042-0

Cook, R.J. (2007) Take-all decline: Model system in the science of biological control and clue to the success of intensive cropping. In: Vincent, C., Goettel, M. and Lazarovits, G. (eds) *Biological Control: A Global Perspective*. CAB International, Wallingford, UK. doi:10.1079/9781845932657.0000

Cook, R.J., Thomashow, L.S., Weller, D.M., Fujimoto, D., Mazzola, M., Bangera, G. and Kim, D.S. (1995) Molecular mechanisms of defense by rhizobacteria against root disease. *Proceedings of the National Academy of Sciences of the United States of America* 92, 4197–4201.

Curl, E.A. and Truelove, B. (1991) *The Rhizosphere*. Springer, Berlin/Heidelberg.

Denison, R.F. and Kiers, E.T. (2011) Life histories of symbiotic rhizobia and mycorrhizal fungi. *Current Biology* 21, R775–R785. doi:10.1016/j.cub.2011.06.018

Edwards, J., Johnson, C., Santos-Medellin, C., Lurie, E., Podishetty, N.K., *et al.* (2015) Structure, variation, and assembly of the root-associated microbiomes of rice. *Proceedings of the National Academy of Sciences of the United States of

America 112, E911–E920. doi:10.1073/pnas.1414592112

García-Garrido, J.M. and Ocampo, J.A. (2002) Regulation of the plant defence response in arbuscular mycorrhizal symbiosis. *Journal of Experimental Botany* 53, 1377–1386.

Gerlagh, M. (1968) Introduction of *Ophiobolus graminis* into new polders and its decline. *Netherlands Journal of Plant Pathology* 74(Suppl. 2), 97.

Gurusiddaiah, S., Weller, D.M., Sarkar, A. and Cook, R.J. (1986) Characterization of an antibiotic produced by a strain of *Pseudomonas fluorescens* inhibitory to *Gaeumannomyces graminis* var. *tritici* and *Pythium* spp. *Antimicrobial Agents and Chemotherapy* 29, 488–495.

Hartmann, A., Schmid, M., van Tuinen, D. and Berg, G. (2009) Plant-driven selection of microbes. *Plant and Soil* 321, 235–257. doi:10.1007/s11104-008-9814-y

Inceoglu, O., van Overbeek, L.S., Falcao Salles, J. and van Elsas, J.D. (2013) Normal operating range of bacterial communities in soil used for potato cropping. *Applied and Environmental Microbiology* 79, 1160–1170. doi:10.1128/AEM.02811-12

Jones, D.L., Nguyen, C. and Finlay, R.D. (2009) Carbon flow in the rhizosphere: Carbon trading at the soil-root interface. *Plant and Soil* 321, 5–33. doi:10.1007/s11104-009-9925-0

Klinkowski, M. (1970) Catastrophic plant diseases. *Annual Review of Phytopathology* 8, 37–60. doi:10.1146/annurev.py.08.090170.000345

Krings, M., Taylor, T.N., Hass, H., Kerp, H., Dotzler, N. and Hermsen, E.J. (2007) Fungal endophytes in a 400-million-yr-old land plant: Infection pathways, spatial distribution, and host responses. *New Phytologist* 174, 648–657. doi:10.1111/j.1469-8137.2007.02008.x

Labandeira, C.C. and Prevec, R. (2014) Plant paleopathology and the roles of pathogens and insects. *International Journal of Paleopathology* 4, 1–16. doi:10.1016/j.ijpp.2013.10.002

Ludwig-Müller, J. (2015) Plants and endophytes: Equal partners in secondary metabolite production? *Biotechnology Letters* 37, 1325–1334. doi:10.1007/s10529-015-1814-4

Marquardt, S. (2001) "Green havoc": Panama disease, environmental change, and labor process in the Central American banana industry. *The American Historical Review* 106, 49–80.

Mendes, R., Kruijt, M., de Bruijn, I., Dekkers, E., van der Voort, M., *et al.* (2011) Deciphering the rhizosphere microbiome for disease-suppressive bacteria. *Science* 332, 1097–1100. doi:10.1126/science.1203980

Micallef, S.A., Channer, S., Shiaris, M.P. and Colon-Carmona, A. (2009a) Plant age and genotype impact the progression of bacterial community succession in the Arabidopsis rhizosphere. *Plant Signaling & Behavior* 4, 777–780. doi:10.1093/jxb/erp053

Micallef, S.A., Shiaris, M.P. and Colon-Carmona, A. (2009b) Influence of *Arabidopsis thaliana* accessions on rhizobacterial communities and natural variation in root exudates. *Journal of Experimental Botany* 60, 1729–1742. doi:10.1093/jxb/erp053

Müller, C.B. and Krauss, J. (2005) Symbiosis between grasses and asexual fungal endophytes. *Current Opinion in Plant Biology* 8, 450–456.

Parniske, M. (2008) Arbuscular mycorrhiza: The mother of plant root endosymbioses. *Nature Reviews Microbiology* 6, 763–775.

Redecker, D., Kodner, R. and Graham, L.E. (2000) Glomalean fungi from the Ordovician. *Science* 289, 1920–1921.

Redman, R.S., Kim, Y.O., Woodward, C.J.D.A., Greer, C., Espino, L., Doty, S.L. and Rodriguez, R.J. (2011) Increased fitness of rice plants to abiotic stress via habitat adapted symbiosis: A strategy for mitigating impacts of climate change. *PLOS ONE* 6, e14823. doi:10.1371/journal.pone.0014823

Remy, W., Taylor, T.N., Hass, H. and Kerp, H. (1994) Four hundred-million-year-old vesicular-arbuscular mycorrhizae. *Proceedings of the National Academy of Sciences of the United States of America* 91, 11841–11843. doi:10.1073/pnas.91.25.11841

Rodriguez, R. and Redman, R. (2008) More than 400 million years of evolution and some plants still can't make it on their own: Plant stress tolerance via fungal symbiosis. *Journal of Experimental Botany* 59, 1109–1114.

Rodriguez, R.J., Henson, J., van Volkenburgh, E., Hoy, M., Wright, L. *et al.* (2008) Stress tolerance in plants via habitat-adapted symbiosis. *ISME Journal* 2, 404–416.

Sasse, J., Martinoia, E. and Northen, T. (2018) Feed your friends: Do plant exudates shape the root microbiome? *Trends in Plant Science* 23, 25–41. doi:10.1016/j.tplants.2017.09.003

Schardl, C.L. (2001) *Epichloë festucae* and related mutualistic symbionts of grasses. *Frontiers in Plant Science* 33, 69–82. doi:10.1006/fgbi.2001.1275

Schardl, C.L., Leuchtmann, A. and Spiering, M.J. (2004) Symbioses of grasses with seedborne fungal endophytes. *Annual Review of Plant Biology* 55, 315–340.

Shafer, S.R. and Blum, U. (1991) Influence of phenolic acids on microbial populations in the rhizosphere of cucumber. *Journal of Chemical Ecology* 17, 369–389. doi:10.1007/bf00994339

Shipton, P.J. (1972) Take-all in spring-sown cereals under continuous cultivation: Disease progress and decline in relation to crop succession and nitrogen. *Annals of Applied Biology* 71, 33–46. doi:10.1111/j.1744-7348.1972.tb04714.x

Sikora, R.A. (1992) Management of the antagonistic potential in agricultural ecosystems for the biological control of plant parasitic nematodes. *Annual Review of Phytopathology* 30, 245–270. doi:10.1146/annurev.py.30.090192.001333

Taylor, T.N. and Krings, M. (2005) Fossil microorganisms and land plants: Associations and interactions. *Symbiosis* 40, 119–135.

Taylor, T.N., Remy, W., Hass, H. and Kerp, H. (1995) Fossil arbuscular mycorrhizae from the early Devonian. *Mycologia* 87, 560–573. doi:10.2307/3760776

Taylor, T.N., Hass, H., Kerp, H., Krings, M. and Hanlin, R.T. (2005) Perithecial ascomycetes from the 400 million year old Rhynie chert: An example of ancestral polymorphism. *Mycologia* 97, 269–285.

Vandenkoornhuyse, P., Quaiser, A., Duhamel, M., Le Van, A. and Dufresne, A. (2015) The importance of the microbiome of the plant holobiont. *New Phytologist* 206, 1196–206. doi:10.1111/nph.13312

Walker, T.S., Bais, H.P., Grotewold, E. and Vivanco, J.M. (2003) Root exudation and rhizosphere biology. *Plant Physiology* 132, 44–51. doi:10.1104/pp.102.019661

Weinert, N., Piceno, Y., Ding, G.C., Meincke, R., Heuer, H., *et al.* (2011) PhyloChip hybridization uncovered an enormous bacterial diversity in the rhizosphere of different potato cultivars: Many common and few cultivar-dependent taxa. *FEMS Microbiology Ecology* 75, 497–506. doi:10.1111/j.1574-6941.2010.01025.x

Weller, D.M., Landa, B.B., Mavrodi, O.V., Schroeder, K.L., de la Fuente, L., *et al.* (2007) Role of 2,4-diacetylphloroglucinol-producing fluorescent *Pseudomonas* spp. in the defense of plant roots. *Plant Biology* 9, 4–20. doi:10.1055/s-2006-924473

2 Endophytic Fungi: Definitions, Diversity, Distribution and Their Significance in Plant Life

Alexander Schouten*

Laboratory of Phytopathology, Wageningen University & Research, Wageningen, The Netherlands

Abstract

Endophytes are set opposite to pathogens and therefore should colonize plants asymptomatically. However, as will be illustrated, endophytic fungi may behave differently under various biotic and abiotic circumstances, in which the host plant can play a defining role as well. The genetic differences between an endophytic fungus and a phylogenetically related pathogenic fungus may vary significantly. Nevertheless, over the years endophytic fungi have frequently been isolated and never elicit disease symptoms in various host plants. Such true endophytes are considered mutually beneficial; the endophyte, embedded in the stable, protective and resource-rich environment of the host plant, supports the host plant to sustain biotic and abiotic stress conditions. The mechanisms by which endophytic fungi protect the host plant against biotic stress factors are generally diverse because they can directly antagonize pests or pathogens, trigger plant defence mechanisms or do both simultaneously.

2.1 Endophytes Defined

The term endophyte indicates a heterogeneous group of microorganisms, primarily consisting of bacteria and fungi. As it means 'inside the plant', an endophyte can essentially be any microorganism that resides for a certain period inside a plant at a certain point during its lifetime, regardless of its beneficial, detrimental or neutral impact on the host plant during this period. Over the years, the definition has nevertheless evolved, indicating not only the endospheric environment in which this organism can be encountered but also the particular relationship it has with the host plant, which is considered as being neutral or beneficial (Petrini, 1991; Wilson, 1995; Stone *et al.*, 2004). The term endophyte has thus become more meaningful, standing opposite to the term pathogen. However, microorganisms can be very dynamic in their behaviour, and for several endophytes, depending on the host plant species or even genotype and physiological and developmental stages of both host plant and endophyte, disease symptoms may still be elicited (Wilson, 1995; Kuldau and Yates, 2000; Schulz and Boyle, 2005). Thus, the definition is not as solid as envisaged. Arbuscular mycorrhizae (AMs) are often set aside from endophytic fungi because they do form arbuscules, which are specialized fungal structures responsible for nutrient transfer between fungus and plant, when proliferating into the roots (Wilson, 1995; Brundrett, 2004; Rodriguez *et al.*, 2009).

The association between microorganisms and plants can be considered a continuum, in which pathogens can be found at one end and true endophytes (Wilson, 1995; Mostert *et al.*,

* Corresponding author e-mail: sander.schouten@wur.nl

2000) at the other end (Wilson, 1995; Schulz and Boyle, 2005). The relationship between host plants and true endophytes has evolved to a level that no visual symptoms during colonization of and subsequent proliferation inside the host plant are being provoked at any time. True endophytes are therefore considered as being capable of maintaining a continuous balanced association that is often mutualistic (Brundrett, 2004; Schulz and Boyle, 2005). For some, the definition of a true endophyte is still too loose because it can apply to both transient and obligate endophytes. In the more stringent definition, true endophytes, which are also called systemic endophytes, would be interpreted as being obligate endophytes, which can only survive by their endophytic association with the host plant and are therefore transmitted vertically through seeds and/or vegetative structures over time (Wani *et al.*, 2015). In this chapter the more relaxed definition for a true endophyte is used.

2.2 How to Obtain and Analyse Fungal Endophytes from a Plant

Whatever the exact definition, the continuous balanced association with the host plant means that true endophytes in particular are elusive, meaning that these organisms are often overlooked in practice and their potential in sustaining plant life ignored. And finding fungal endophytes in itself can be difficult, particularly when they are obligate endophytic (Schulz and Boyle, 2005). One can use histological methods, but that makes further studies, like biological studies, difficult, if not impossible. Immunological or metagenomic analysis may give insight in the presence of endophytic isolates residing inside the plant but as these generally are destructive methods, further biological assays cannot be done. The best approach for obtaining live endophytes that can be used in subsequent biological studies is the screening of surface-sterilized plant tissue, such as shoots or roots, placed on growth media suitable for fungal growth. An antibiotic may be added to prevent endophytic bacteria from proliferating as well. Fungal colonies that emerge from the plant tissue are then purified by transferring them individually to fresh media, after which they can be identified. When sporulating, individual spores can be subcultured to exclude any contamination and to obtain pure isolates. Such a screening approach is thus highly biased towards fungi that can be cultured *in vitro*, but has the advantage that the encountered endophytes can be easily maintained and studied and used in biological assays for evaluating their beneficial effects on plants, which may not necessarily be the plants they were isolated from. Over the years thousands of endophytic fungal isolates have been characterized (Petrini, 1986; Schulz *et al.*, 1993, 1995, 1998; Schulz and Boyle, 2005; Yan *et al.*, 2011; Miles *et al.*, 2012; Sánchez Márquez *et al.*, 2012).

2.3 Diversity of Fungal Endophytes

On the basis of their ecology, a differentiation between balanciaceous endophytes and the nonbalanciaceous endophytes is often made (Schulz and Boyle, 2005). The balanciaceous endophytes, also referred to as grass endophytes, are phylogenetically related and comprise the ascomycete genera *Epichloë* and *Balansia* (anamorphs *Neotyphodium* and *Ephelis*, respectively) within the Clavicipitaceae family (Schulz and Boyle, 2005). Balanciaceous endophytes are unique in the sense that they have probably evolved from insect-parasitic fungi rather than plant-parasitic fungi (White *et al.*, 2002). They are obligate endophytic, transmitted both horizontally and vertically, i.e. through seeds and vegetative parts, and growing intercellularly in a strictly controlled manner in the above-ground plant tissue. Balanciaceous endophytes generate specific mycelial structures by which the uptake of nutrients is facilitated. Some grass endophyte species manifest themselves at a certain stage as being antagonistic to their host, suppressing seed production (choke disease), thereby preventing vertical transmission, whereas others always remain elusive, staying mutualistic by enhancing growth, development and desiccation tolerance of the host plant and reducing herbivory (Schardl *et al.*, 2004; see also Chapter 7, this volume).

The majority of the thousands of nonbalanciaceous endophytes isolated from various plant species are acsomycetes, with *Alternaria*, *Colletotrichum*, *Fusarium*, *Trichoderma*, *Chaetomium* and *Acremonium* being the dominating genera (Petrini, 1986; Schulz *et al.*, 1993, 1995, 1998; Yan *et al.*, 2011; Miles *et al.*, 2012; Sánchez Márquez *et al.*, 2012). Among those, the genera *Fusarium*, particularly the species *Fusarium oxysporum*, and *Trichoderma* are generally the most prominent (Kuldau and Yates, 2000; Bacon and Yates, 2006; Maciá-Vicente *et al.*, 2008; Yan *et al.*, 2011). To a lesser extent, several basidiomycetes can also be found among the endophytic isolates, with *Piriformospora indica* as the best studied example (Qiang *et al.*, 2012). Based on their ecology, the classification of endophytes may be refined by considering host range, transmission mechanisms, their simultaneous occurrence (biodiversity) inside a single host plant and ecological role (Rodriguez *et al.*, 2009). In that case, class 1 endophytes are the balansiaceous endophytes, having a narrow host range. Class 2, 3 and 4 comprise the nonbalanciaceous endophytes, in which class 2 endophytes can be distributed throughout the plant, root aerial plant parts and rhizome, with, like class 1, low species abundancy (biodiversity) within a single host plant and potential horizontal and vertical (seed coats, seeds and rhizomes) transmission. Class 3 endophytes are only locally distributed within the aerial plant parts and show a high species abundancy within a single host, which can reach more than 20 species in a single leaf (Arnold and Herre, 2003). Class 4 comprises dark septate fungi (Jumpponen and Trappe, 1998; Andrade-Linares *et al.*, 2011), which have only been observed in the roots and have a broad host range.

2.4 How Different Are Endophytic Fungi from Pathogenic Fungi?

As indicated in Chapter 1, the current view is that the distinction between a microorganism being pathogenic or endophytic may not be very discrete. It may depend on the host plant species or, sometimes, host plant variety or cultivar. Also, many fungal species harbour individual isolates that have an endophytic or pathogenic lifestyle or both, emphasizing that plant pathogens and endophytes may in fact be genetically quite similar. This can be illustrated by looking at the genus *Fusarium*. Isolates within the *Fusarium graminearum* species complex (FGSC) cause diseases in cultivated grasses, known as Fusarium head blight in wheat and barley and Fusarium ear rot in maize. They produce mycotoxins, which can also be found in the harvested grains and therefore is a major health issue for humans and animals. It was shown that 25 native North American grass species harbour FGSC isolates, which all grow asymptomatically with little or no trichothecene accumulation, although they were capable of producing these mycotoxins in wheat. There are indications that the coexistence of North American grasses and isolates from FGSC is very ancient, suggesting that evolutionary processes have shaped the host–fungus relationship into a benign and possibly mutualistic interaction (Lofgren *et al.*, 2018). *F. oxysporum* is a cosmopolite, always saprophytically competent and notorious for the various pathogenic isolates that have been characterized, which have been grouped in *formae speciales* on the basis of their host plant species (Lievens *et al.*, 2008). The majority of the *F. oxysporum* isolates are, however, harmless and can even perform as true endophytes by supporting the host plant in antagonizing fungal pathogens, plant-parasitic nematodes and insects (Alabouvette and Couteaudier, 1992; Hallmann and Sikora, 1994; Griesbach, 1999; Schouten, 2016). Several isolates were found to be restricted to the endosphere of the roots. Not all endophytic *F. oxysporum* isolates, capable of colonizing banana, could be distinguished from pathogenic *F. oxysporum* f. sp. *cubense* isolates in a phylogenetic analysis, which was based on sequences of the ribosomal intergenic spacers (Kurtz *et al.*, 2008). Becoming pathogenic may be the result of a mutual interaction going astray, in which the host plant cannot properly manipulate and contain the endophyte, resulting in the proliferation of the fungus into critical areas of the plant, like particular cells or the vascular tissue. An endophytic isolate of *Fusarium verticillioides* systemically propagated only intercellularly, whereas a pathogenic strain also invaded intracellularly (Bacon and Hinton, 1996). The endophyte *P. indica* shows all the features of a mutualist by promoting plant growth and supporting the plant in resisting biotic and abiotic stress elements, although this fungus is notably aggressive when colonizing and proliferating, because it induces cell death inside the roots. However, only the cortex is affected in this way and vascular tissue remains intact, thus not harming the development and functioning of the root (Deshmukh *et al.*, 2006; Jacobs *et al.*, 2011; Qiang *et al.*, 2012). Then again, xylem-associated fungal endophytes have frequently been identified (Stone *et al.*, 2000; Martín *et al.*, 2015; Pérez-Martinez *et al.*, 2018; Win *et al.*, 2018), suggesting that such type of invasive growth can be handled by the host plant.

Comparsion of whole genomes of *F. verticillioides*, *F. graminearum*, both pathogenic on cereals, and *F. oxysporum* f. sp. *lycopersici* (Fol) revealed lineage-specific (LS) genomic regions in the latter, covering more than one-quarter of its genome, including four entire chromosomes, 3, 6, 14 and 15, and parts of chromosomes 1 and 2 (Ma *et al.*, 2010). LS chromosome 14 harbours genes coding for the unrelated small proteins Six1 (Avr3) and Six3 (Avr2), which are involved in virulence on tomato (Rep *et al.*, 2004; Houterman *et al.*, 2009) and secreted by the pathogen during proliferation into the xylem system (Houterman *et al.*, 2007; van der Does *et al.*, 2008a), together with a gene coding for an *in planta*-secreted oxidoreductase (ORX1) (Houterman *et al.*, 2007). These genes were initially thought to be exclusively present in *F. oxysporum* strains causing tomato wilt (van der Does *et al.*, 2008b). Transfer of two LS chromosomes into the non-pathogenic *F. oxysporum* strain Fo47 (Fo47), which was capable of colonizing the outer cortex of flax roots (Olivain *et al.*, 2003) and lacks LS chromosomes, resulted in mutant strains with varying levels of virulence on tomato. In those mutants, chromosome 14 was present and the most virulent mutant additionally contained a smaller chromosome comprising a fragment present in two LS chromosomes, 3 and 6 (Ma *et al.*, 2010). Recently, *F. oxysporum* f. sp. *cubense* tropical race 4 (TR4), which is currently the most threatening banana pathogen for the dominant Cavendish cultivars grown worldwide, was shown to possess three *SIX1* homologues, *SIX1a*, *b* and *c*, with a sequence similarity of 74%, 63% and 73%, respectively, when compared to Fol *SIX1*. A TR4 *SIX1a* gene deletion mutant was severely reduced with respect to its virulence and the subsequent ectopic reintegration of the Focub-*SIX1a* gene into this deletion mutant fully reestablished virulence to wild-type levels again (Widinugraheni *et al.*, 2018). By using PCR analysis only *SIX5* and *SIX6* could be detected in the endophytic *F. oxysporum* strain Fo162, suggesting that the genome of this isolate, like Fo47, lacks one or more LS regions (Eschweiler and Schouten, unpublished). Although it does colonize, *F. oxysporum* Fo162 is non-pathogenic on various plant species, such as tomato, where it was originally isolated from (Hallmann and Sikora, 1994), banana, squash (*Cucurbita pepo* L.), melon (*Cucumis melo* L.) and *Arabidopsis* (Vu *et al.*, 2006; Menjivar *et al.*, 2011; Martinuz *et al.*, 2015).

Supernumerary chromosomes, also known as conditionally dispensable chromosomes, dispensable chromosomes, accessory chromosomes or minichromosomes have been frequently encountered in fungal genomes (Covert, 1998; Bertazzoni *et al.*, 2018). They are dispensable for basic, saprophytic growth but can be imperative for colonizing certain ecological niches (Covert, 1998; Bertazzoni *et al.*, 2018). In *Nectria haematococca* (anamorph *Fusarium solani*), a pea pathogen, 1.6-Mb supernumerary chromosomes were identified, which contain functional genes encoding proteins that detoxify the pea phytoalexin pisatin (PDA genes) and the chickpea phytoalexins maackiain and medicarpin (MAK genes), together with genes associated with pathogenicity on pea (PEP genes) (Miao *et al.*, 1991; Covert *et al.*, 1996; Kistler *et al.*, 1996; Wasmann and VanEtten, 1996). Loss of this chromosome resulted in loss of the ability to cause disease (Covert, 1998; VanEtten *et al.*, 1998). Loss of a 1.1–1.8 Mb chromosome in *Alternaria alternata* resulted in loss of

AM-toxin production and loss in pathogenicity on apple. Among the many genes encoded on this chromosome, genes encoding proteins for AM-toxin synthesis were identified (Harimoto *et al.*, 2007). There are, however, no data on whether the loss of these supernumerary chromosomes resulted in losing the ability to colonize the host asymptomatically.

A single allelic mutation in *Colletotrichum magna* can turn the pathogen into a fungus displaying a completely endophytic lifestyle in susceptible watermelon cultivars (Freeman and Rodriguez, 1993). The presence of this path-1 mutant also prevented the pathogenic wild-type *C. magna*, *Colletotrichum orbiculare* and unrelated *F. oxysporum* f. sp. *niveum* from eliciting disease in watermelon. Remarkably, the path-1 mutant could extend its host range, as it also proliferated endophytically in wild-type *C. magna*-resistant cucurbit cultivars (Freeman and Rodriguez, 1993; Redman *et al.*, 2001). The host range for *C. magna* could even be expanded further to at least four plant families, Solanaceae, Fabaceae, Poaceae and Rosaseae, in which both wild-type and path-1 mutant proliferated asymptomatically in various species (Redman *et al.*, 2001).

Overall, endophytes may thus be latent pathogens, in which virulence traits are successfully repressed, even for extended periods, by the pathogen, by the host plant or both (Schulz and Boyle, 2005), and true endophytes have gradually evolved from plant pathogenic fungi (Saikkonen *et al.*, 1998), by losing traits that initiate a disease phenotype without losing the ability to invade the plant. For these latent pathogens, the change from beneficial lifestyle into a pathogenic one may be triggered by environmental factors (Junker *et al.*, 2012). Typical examples are found within the balanciaceous endophytes causing choke disease in grasses. Epichloë species generally proliferate in the intercellular space of stems, leaves, inflorescences and seeds of the grass plant without eliciting disease symptoms. However, especially under nutrient-poor conditions in the soil, they can turn pathogenic, developing fungal stroma or sclerotia on tillers and suppressing the development of their host plant's inflorescence (Malinowski and Belesky, 2000; see also Chapter 7, this volume). *Iriartea deltoidea*, the dominating palm tree in many wet lowland Neotropical forests, is frequently colonized asymptomatically by the endophytic ascomycete *Diplodia mutila*. This fungus may cause disease in seedlings, although this occurred only under intense light conditions. It was suggested that light-induced hydrogen peroxide, generated by the fungus, initiated hypersensitivity and necrosis of palm tissue (Álvarez-Loayza *et al.*, 2011). From two guarana (*Paullinia cupana*) genotypes, the second most abundant endophytic species encountered was *Phomopsis asparagi*, which is known to be pathogenic on asparagus (*Asparagus officinalis*), causing stem blight (Yin *et al.*, 2012). In *in vitro* tests, *P. asparagi* and another isolated endophyte, *Peyronellaea pinodella*, developed microsclerotia on roots, suggesting that these species may become pathogenic in nature as well, under yet unidentified conditions (Álvarez-Loayza *et al.*, 2011). The ascomycete genus *Botrytis* (teleomorph *Botryotinia*) harbours many aggressive necrotrophic plant pathogenic species. Nevertheless, the interior of multiple tissues of apparently healthy host plant species, belonging to the Asteraceae, Brassicacae, Primulacae and Rosaceae families, could frequently harbour *Botrytis* species, reaching tissue sample infection frequencies of 50% or more. *Botrytis cinerea* and *Botrytis pseudocinerea* were predominant, although novel species were encountered as well. The asymptomatic infection only advanced into tissue damage and fungal sporulation upon tissue stress or when the host tissue reached maturity or became senescent by itself (Shaw *et al.*, 2016). Such observations should, however, be considered with care because, as discussed later in this chapter, other endophytes may have pacified *P. asparagi*, *P. pinodella* and the *Botrytis* species.

2.5 The Influence of the Host Plant on Asymptomatic Proliferation

Host plant species or genotype may have effective tools in properly negotiating asymptomatic interactions as well. Similar to the wild-type, the *C. magna* path-1 mutant formed appressoria and systemically infected the stem, but it proliferated slower within the host tissue. This was attributed to host defence responses mounted by the host plant (Freeman and Rodriguez, 1993; Redman *et al.*, 1999). When the stress-activated mitogen-activated protein kinase *sakA* in the fungus *Epichloë festucae* was disrupted, this grass endophyte turned into a more aggressive mutant, proliferating more vigorously inside the host plant and provoking disease responses, such as early senescence and disturbance of the normal plant phenotype and growth (Eaton *et al.*, 2010, 2011). At the molecular level, the host plants accumulated transcripts typical for defence against pathogens upon infection with the *sakA* mutant. The involvement of *sakA* in repressing a disease phenotype suggests that the host plant exerts some sort of pressure on the fungus, thereby forcing it into submission. In maize, a high total antioxidant activity and free phenolic acid concentrations at the beginning of kernel development significantly reduced deoxynivalenol contamination in the mature seeds at harvest (Giordano *et al.*, 2017). Wheat genotypes most resistant towards FHB disease in a field situation displayed the highest ferulic acid contents in wheat spikes (Martin *et al.*, 2017). Mycotoxin production by *F. oxysporum* f. sp. *niveum* was increased more in the presence of exudates from susceptible rather than resistant watermelon (*Citrullus lanatus*) cultivars and after flowering rather than before flowering. Both before and after flowering, a higher ratio of antifungal phenolic acids – p-hydroxybenzoic, phthalic, gallic, coumaric, cinnamic, ferulic, salicylic and sinamic acids – to fungus-stimulating phenolic acids – vanillic and syringic acids – was observed in the root exudates of resistant cultivars when compared to susceptible ones (Wu *et al.*, 2009). In *in vitro* assays, ferulic acid was shown to be an efficient inhibitor of the mycotoxin type B trichothecene biosynthesis in a dose-dependent manner, in *Fusarium culmorum* and *F. graminearum* strains, through reducing Tri-gene transcription (Boutigny *et al.*, 2009).

2.6 The Diversity and Distribution of Fungal Endophytes Inside a Plant

There are multiple factors that influence the endophytic fungal population structure, such as the host species or cultivar, host plant tissue, geographical location, soil type, physiological condition of the plant and season. Fungal endophytes can be found throughout the plant, such as roots, stem, leaves, flowers and seeds, in which individual species or isolates may proliferate systemically or remain restricted to certain plant parts or even areas within plant parts. As already mentioned, grass endophytes colonize intercellular space of stems, leaves, inflorescences and seeds but not roots (Schardl *et al.*, 2004). In four rice (*Oryza sativa*) cultivars grown in the Guangdong province, South China, endophytic actinomycetes, particularly *Streptomyces* species, dominated in the Panyu district, whereas endophytic fungi, particularly *Fusarium* species, dominated in the Wushan district (Tian *et al.*, 2004). This was associated with the soil features, being acidic in the Wushan district, favouring the growth and colonization of fungi, and alkalic in the Panyu district, favouring the growth and colonization of actinomycetes. In the same study, endophytic fungi were more or less evenly distributed between roots and leaves, whereas the actinomycetes were more frequently encountered in the roots (Tian *et al.*, 2004). The dominating endophytes, collected from rice in the Bhadra River Project Area in Southern India, were bacterial *Streptomyces* species and the fungal species *Chaetomium globosum*, *Penicillium chrysogenum*, *F. oxysporum* and *Cladosporium* sp., although the colonization rates differed, being 40% in roots and 26% in leaves during the winter season of 2005 and 20% in roots and 9% in leaves during the summer season of 2006 (Naik *et al.*, 2009). It was assessed that leaves of tropical trees are hotspots of fungal endophyte diversity, harbouring numerous species that had never before been recovered from other biomes (Arnold and Lutzoni, 2007). A study of three healthy tea (*Camellia sinensis*) cultivars, Hokumei, Sayamakaori and Yabukita, resulted in 520 endophytic isolates from new leaf, old leaf, bark and xylem, which could be classified into 44 taxa, of which 93% belonged to the ascomycetes. The endophytes in stem tissue were more diverse than in leaf tissue and the xylem tissue was the least colonized. Tissue type was more important than cultivar with respect to the endophytic community structure. The dominating species encountered was *Colletotrichum gloeosporioides* f. sp. *camelliae*, which favoured bark and old leaf tissue, and *Pleosporales* sp., which favoured new leaf tissue. *C. gloeosporioides* f. sp. *camelliae* colonized the Yabukita cultivar significantly less frequently (Win *et al.*, 2018). In oilseed rape (*Brassica napus*), the fungal endophyte composition was diverse as well, resulting in 40 species, with 80% belonging to the ascomycetes, 17.5% to the basidiomycetes and 2.5%, i.e. only *Rhizopus oryzae*, to the zygomycetes (Zhang *et al.*, 2014). The most dominant species encountered was *A. alternata*, followed by *C. globosum*, *Clonostachys rosea*, *F. oxysporum*, *Fusarium proliferatum* and *Periconia* sp. The various species were not evenly distributed among roots, stem and leaves. Only two species, *A. alternata* and *C. globosum*, were isolated from all three parts; six species, such as *F. oxysporum* and *Fusarium tricinctum*, were only isolated from the roots; 20 species, including all basiodiomycetes, were only isolated from the stem; and three species were only isolated from leaves. In a previous study (Chen *et al.*, 2004), the endophytic content of oilseed rape differed regarding several genera, although *Acremonium*, *Alternaria*, *Aspergillus*, *Chaetomium*, *Epicoccum*, *Fusarium*, *Nigrospora*, *Penicillium* and *Rhizoctonia* were isolated as well. The differences were attributed to the geographical locations where the oilseed rape plants were grown (Zhang *et al.*, 2014).

From the bark, branches and leaves of *Taxus chinensis* var. *mairei* growing in the Jiangxi, Zhejiang and Chongqing regions of China, the isolated fungal endophytes comprised 145 genera, of which 125 belonged to the ascomycetes, 14 to the basidiomycetes, 5 to zygomycetes, and 1 could not be defined. Species richness was higher in the branches and bark than in the leaves. Locality and tissue both affected the encountered dominant genera (Wu *et al.*, 2013). From seven- to ten-day-old cucumber (*C. sativus*) seedlings, cultivated in 87 different soil samples gathered from tropical, subtropical and temperate climatic regions, 514 endophytic fungal isolates from roots (275 isolates), stems (88 isolates) and leaves (151 isolates) were collected and grown on potato dextrose agar (PDA) (Yan *et al.*, 2011). Of the 294 endophytic fungi that could be properly cultivated on PDA and produced spores, 189 isolates could be characterized and belonged to 16 genera, primarily *Fusarium* (91 isolates) and, to a lesser extent, *Trichoderma* (25 isolates). *Fusarium* was found in all plant parts, but this was not the case for the other genera, such as *Actinomucor*, *Paeacilomyces*, *Curvularia* and *Trichoderma*, which were only recovered from roots, roots and leaves and roots and stems, respectively.

From two guarana (*P. cupana*) genotypes, growing in two sites of the Brazilian Amazon basin, 256 endophytic fungal strains, comprising 8 species in seeds, 23 in roots and 3 in both organs, were identified (de Azevedo Silva *et al.*, 2018). The ascomycete *Xylogone ganodermophthora* was the most abundant species in the roots and one of the eight species never encountered before as endophyte, although the genus *Xylogone* has been identified as being endophytic in *T. chinensis* var. *mairei* in China (Wu *et al.*, 2013). The endophytic community structure in *P. cupana* depended on the geographic location in which the plant was growing, the clonal type and plant organ, with the latter being the most relevant (de Azevedo Silva *et al.*, 2018). These examples are to illustrate the rich diversity of fungal endophytes that can be encountered. By no means is it intented to describe them all. Many more studies regarding the screening for and characterization of endophytes from an array of plant species can be found in the literature.

2.7 Ecological Roles of Endophytes in Plant Life

It is assumed that the overall outcome of the endophyte–host plant interaction generally is beneficial for both parties. The endophyte is protected against soil-bound competitors by the plant while having a direct access to nutrient supplies. At the same time, the endophyte can protect its niche, i.e. the host plant, against other competing organisms, such as insects, nematodes, bacteria and fungi, and support the host plant with respect to growth and development, the uptake of nutrients from the soil and to sustain adverse abiotic conditions, such as drought, extreme temperatures, soil salinity and contaminated soils (Sikora, 1992; Schardl *et al.*, 2004; Müller and Krauss, 2005; Trillas and Segarra, 2009; Weyens *et al.*, 2009; Aly *et al.*, 2011; Bakker *et al.*, 2013; Ludwig-Müller, 2015; see also Chapters 5, 6, 7, 8 and 9, this volume). Because of the potential benefits endophytes may provide, it is often difficult to determine what the actual benefits for the host plant are. The study on the beneficial activity of endophytes is therefore generally trial-and-error based and, as a consequence, highly biased by the type of research in which the endophyte–plant interaction is embedded. Nevertheless, the ecological context in which plants grow may provide clues. Such habitat-adapted symbiosis (Rodriguez *et al.*, 2008; Redman *et al.*, 2011) with endophytes was observed when plants were exposed to abiotic stress conditions. In coastal regions, dune grass (*Leymus mollis*) was associated with particular endophytic *F. culmorum* isolates, which supported the plant in tolerating salt stress (Rodriguez *et al.*, 2008). The association with the endophytic *Curvularia protuberate* conferred heat tolerance of panic grass (*Dichanthelium lanuginosum*), allowing its growth on geothermal soils of Yellowstone National Park (Redman *et al.*, 2002). Inoculation of broad bean (*Vicia faba*) with an endophytic root-associated *Acremonium strictum* isolate resulted in a significant lower performance of the aphid *Aphis fabae* with respect to fecundity (Jaber and Vidal, 2009). Inoculation of tomato (*Solanum lycopersicum*) with *A. strictum* revealed an even more severe effect on the unspecialized cotton bollworm (*Helicoverpa armigera*), which was significantly reduced in growth rate, displayed an extended development time and suppressed moulting, and produced smaller and less viable pupae. The emerging adults had a lower fecundity. These effects could not be related to the amount of foliage consumed, indicating that *A. strictum* inoculation did not have a deterring effect on feeding (Jallow *et al.*, 2004). Similar effects were observed when *H. armigera* was reared on *A. strictum*-inoculated broad bean (Jaber and Vidal, 2010).

What is more, endophytes may be beneficial in various ways simultaneously. Grass endophytes are typical in this matter. As mentioned before, they can benefit the host plant by enhancing growth, development and desiccation tolerance and reducing herbivory by insects, nematodes and vertebrates (Elmi *et al.*, 2000; Kauppinen *et al.*, 2016; see also Chapter 7, this volume). *Beauveria bassiana* may be known for protecting plants against insect pests (Arnold and Lewis, 2005; Jaber and Vidal, 2009, 2010; Reay *et al.*, 2010; Akello and Sikora, 2012; Biswas *et al.*, 2013), but can protect the host plant against diseases (Ownley *et al.*, 2008, 2010; Jaber and Alananbeh, 2018) and enhance plant growth as well (Jaber and Enkerli, 2017; Jaber and Araj, 2018; Jaber and Ownley, 2018). Although primarily free living, some isolates of *Trichoderma* species can asymptomatically colonize plants and promote plant growth. In cucumber, *Trichoderma asperellum* T-203 (also known as *Trichoderma harzianum*) was shown coiling its hyphe around the root, forming appessorium-like structures on the root surface, and subsequently migrating intercellularly into the cortex, most likely by using lytic enzymes to degrade cell walls (Yedidia *et al.*, 1999, 2000). However, *T. asperellum* T-203 was also shown to parasitize the extraradical structures of the AM *Glomus* sp. MUCL 41833 and, through this, to extend into the intraradical AM structures. AMs may thus serve as a gateway for endophytic colonization and proliferation of *Trichoderma* species (De Jaeger *et al.*, 2010). The endophytic *F. oxysporum* Fo162 stimulate root growth and, as determined by split-root assays, systemically reduce the infection of sedentary nematodes in *Arabidopsis* (Martinuz *et al.*, 2015; see also Chapter 9, this volume).

The impact of endophytes may be complex in the sense that it not only affects plant herbivory but also completes food webs. Aphid density was three times higher on

Italian ryegrass (*Lolium multiflorum*) grown in the absence of *Neotyphodium* colonization, although the aphid species *Rhopalosiphum padi* and *Metopolophium festucae* were differently affected (Omacini *et al.*, 2001). However, the absence of the endophyte resulted in an eightfold increase of parasitized aphids comprising primary parasitoids, which attack the live aphids, and secondary parasitoids, which attack either parasitized live or mummified aphids. The primary parasitoids were nevertheless relatively less successful on endophyte-free plants, because of an unequal increase in secondary parasitism. The body size of secondary parasitoids emerging from *R. padi* mummies was larger on endophyte-free plants, an effect which was not observed for secondary parasitoids emerging from *M. festucae*. The overall outcome was that the aphid and secondary parasite reproduction benefited from the absence of the endophyte, whereas this benefit was not observed for the primary parasites, regardless of their increased attack rate (Omacini *et al.*, 2001).

Mortality of the fall armyworm, *Spodoptera frugiperda*, caused by the entomopathogenic nematode, *Steinernema carpocapsae*, was reduced when the insect fed on *Lolium perenne*, infected with *Neotyphodium lolii* (Richmond *et al.*, 2004). *S. carpocapsae* was less fatal to the fourth and fifth instars of the black cutworm *Agrotis ipsilon* when feeding on *N. lolii* infected ryegrass as well (Kunkel and Grewal, 2003). The nematodes seem to be affected by toxins produced by grass endophytes. Ergonovine malate increased and ergocristine decreased the rates of nematode infection of black cutworm and *in vitro* experiments demonstrated a repressed growth of *Xenorhabdus nematophila*, which is the symbiotic bacterium of *S. carpocapsae*, in the presence of ergocristine (Kunkel *et al.*, 2004). Although the larvae had a significantly lower biomass when feeding on endophyte-infected ryegrass, it seems that the herbivore can tolerate these ergot alkaloids, while simultaneously exploiting them for its own gain against the entomopathogenic nematode's symbiotic bacterium that is essential for the parasitism on insects (Richmond *et al.*, 2004). Cheatgrass (*Bromus tectorum*) can host many different fungal endophytes, although individual plants generally contain a subset. One of these endophytes is *Fusarium* cf. *torulosum*, which is particularly preferred by the fungivorous nematode *Paraphelenchus acontioides*. Studies indicated a mutualistic interaction between nematode and endophyte, in which the nematode supports the proliferation of this endophyte in the host plant, without affecting the host plant (Baynes *et al.*, 2012).

2.8 Competition between Endophytes and Pathogens

The *C. magna* path-1 mutant conferred resistance not only towards the *C. magna* wild-type but also towards *C. orbiculare* and *F. oxysporum* f. sp. *niveum* in watermelon and towards the oomycete *Phytophthora capsica* in a squash (*C. pepo*) variety (Freeman and Rodriguez, 1993; Redman *et al.*, 2001). However, other than established for *F. oxysporum* f. sp. *niveum*, both *C. magna* wild type and *P. capsica* could never be reisolated from the path-1-colonized plant tissue suggesting that, although endophyte-conferred resistance may be common (Redman *et al.*, 2001), the level is different at which the pathogen is controlled. This suggests that different mechanisms are responsible for the observed resistance. The studies by Zhang *et al.* (2014) indicated that oilseed rape harbours various endophytic fungi that can protect it against disease development caused by *Sclerotinia sclerotiorum* and *B. cinerea*. In the presence of endophytes, *B. cinerea* was apparently living asymptomatially, as it could even be recovered from endophyte-colonized plants. From *Espeletia grandiflora* and *Espeletia corymbosa* (Asteraceae), plant species that are endemic for the Paramo region in the Andean mountains, two endophytes were collected, identified as *Aureobasidium pullulans* and *Paraconiothyrium sporulosum*, which individually protected tomato plants against *Rhizoctonia solani* in a co-inoculation experiment (Miles *et al.*, 2012). The endophytic *Phoma eupatorii* isolate 8082 nearly eliminated the infection of potato by *Phytophthora infestans* (de Vries *et al.*, 2018).

2.9 Competition among Endophytes

Just as plant-colonizing microorganisms can compete with fungal and bacterial pathogens, they can compete with each other. A metagenomic analysis revealed that grass endophytes reduced root biomass and mycorrhizal colonization but significantly stimulated the production of root exudates (Omacini *et al.*, 2012). The overall consequences of these endophyte-imposed changes are discussed further in Chapter 8. As mentioned before, *T. harzianum* can also parasitize the structures of the AM *Glomus* sp. MUCL 41833 both intraradically and extraradically. This caused a decrease in the overall viability of the AM mycelial structures (De Jaeger *et al.*, 2010). As determined in a split-root experiment, the endophytic antagonistic bacterium *Rhizobium etli*, strain R12, systemically reduced the colonization of tomato by the fungal endophyte *F. oxysporum* Fo162 (Martinuz *et al.*, 2012a). The endophytic fungus *Alternaria tenuissima* increased the synthesis of several polyketides, including the antifungal stemphyperylenol when co-cultured *in vitro* with the endophytic fungus *Nigrospora sphaerica*. Stemphyperylenol was toxic for *N. sphaerica* but not for the host plant yacón (*Smallanthus sonchifolius*), even at relatively high concentrations (Chagas *et al.*, 2013).

2.10 Mechanisms Involved in Endophyte-mediated Antagonism

The actual mechanisms by which endophyte-mediated antagonism towards a pest or pathogen works can be diverse. These mechanisms can be operating simultaneously and be both direct, immediately directed towards the targeted organism, and indirect, through triggering local or systemic plant defences that are aimed at the targeted organism. All these options complicate the dissecting of endophyte-mediated antagonism. In the following section the general mechanisms are discussed.

2.10.1 Attacking and trapping

Trichoderma species are rather opportunistic fungi as they can attack and devour many different (plant pathogenic) fungi, such as *R. solani*, *A. alternata*, *B. cinerea* (teleomorph *Botryotinia fuckeliana*) and *S. sclerotiorum*, and nematodes, such as *Meloidogyne*, *Heterodera* and *Globodera* species, either in the soil or in the roots (Chet and Baker, 1981; Bordallo *et al.*, 2002; Singh and Mathur, 2010; Druzhinina *et al.*, 2011; Escudero and Lopez-Llorca, 2012). Endophytic *Metarhizium* and *Beauveria* species are known to kill and consume a wide range of insect species (McKinnon *et al.*, 2017). Several endophytic fungal isolates are known to be nematophagous. Like *T. asperellum*, isolates from *Acremonium implicatum*, *Paecilomyces lilacinus* and *Arthrobotrys oligospora* can trap or attack, kill and devour nematodes at various developmental stages (Bordallo *et al.*, 2002; Rumbos and Kiewnick, 2006; Singh and Mathur, 2010; Escudero and Lopez-Llorca, 2012). For penetrating, mechanical forces through appressorial structures and lytic enzymes are employed to enter nematode eggs, larvae and adults (Curtis *et al.*, 2011; Lin *et al.*, 2013; Yao *et al.*, 2015). *A. implicatum* has been shown to produce chitinases, which play a role in penetrating nematode eggs, in *in vitro* experiments. Endophytic strains of the trapping fungus *A. oligospora* perform better in controlling nematodes than the soilborne strains. This quality was attributed to the support by the host plant when the endophyte has to proliferate into the rhizosphere and produce its trapping structures (Bordallo *et al.*, 2002).

2.10.2 Competition for space and resources

The endophytes *Monographella nivalis* var. *neglecta* strain 114 and *Pyrenochaeta cava*, both isolated from elm (*Ulmus* spp.), are potential antagonists towards the causal pathogen of the Dutch elm disease (DED), *Ophiostoma novo-ulmi* (Martin *et al.*, 2013). *P. cava* (Schulzer) Gruyter, Aveskamp & Verkley was isolated from xylem samples, and *M. nivalis* var. *neglecta* strain 114 was isolated from bark. *M. nivalis* var. *neglecta* is the teleomorph of *Microdochium majus* (Wollenw.) Glynn & S.G. Edwards, previously known as *Fusarium nivale* var. *majus* Wollenw. (Glynn *et al.*, 2005), and is primarily known as a snow mold and an endophyte of grasses and cereals (Dahl, 1934; Sieber *et al.*, 1988), although it can also cause symptomless endophytic infections in elms (Martín *et al.*, 2013). By using phenotype microarrays, the carbon utilization profiles of the highly virulent DED pathogen and the asymptomatic endophyte isolates were determined (Blumenstein *et al.*, 2015). This showed that the endophytes displayed an extensive substrate utilization overlap with the pathogen, which was especially the case for substrates that are principal to carbon metabolism, such as sugar alcohols, tri- and tetra-saccharides and monosaccharides, and for fatty acids. *P. cava* showed the highest utilization rates for the various sugars and sugar alcohols, whereas *M. nivalis* var. *neglecta* 114 isolate showed the broadest niche with respect to C-substrate utilization. It was postulated that, next to the release of bioactive constituents by *M. nivalis* var. *neglecta* 114, the protection of elm by both endophytes could be based on the competition for substrates.

Sedentary plant-parasitic nematodes actively establish an intricate biotrophic relationship with their host by altering particular root cells into nurse cells, also called syncytia (for cyst nematodes), or giant cells (for root knot nematodes, RKNs), for the duration of their development and reproduction, from which they actively acquire their nourishment by means of their stylet (Gheysen and Mitchum, 2011; Goverse and Smant, 2014). The root-associated endophyte *F. oxysporum* Fo162 impaired or suspended the development of the RKN (*Meloidogyne incognita*) larvae into adult females and decreased reproduction rates (Martinuz *et al.*, 2012b). An impaired or suspended development, together with an increase in the male–female ratio was also found for another RKN species, *Meloidogyne graminicola*, in rice, after inoculation with the endophytic *Fusarium moniliforme* strain Fe14 (Le *et al.*, 2016). As sex determination among RKNs is primarily epigenetically driven (Papadopoulou and Triantaphyllou, 1982; Chan *et al.*, 2010), the observed postinfection effects collectively suggest that particular endophytes negatively affect the initiation of the giant cell to such a level that the extraction of sufficient nourishment by the nematode to perform optimally is compromised. The intense transport of sugars from phloem into the giant cell (Hofmann *et al.*, 2009a, b) may also serve as an important energy source for endophytes (Martinuz *et al.*, 2012b). A particular class of plant sugar transporters (SWEET) take part in the loading of sugars from the phloem parenchyma, via the apoplast, into the phloem companion cell (Chen *et al.*, 2010; Chen, 2014). Bacterial symbionts and fungal and bacterial pathogens were shown to be capable of initiating SWEET gene expression, suggesting that these microorganisms drive sugar translocation into the apoplast, thus enabling immediate access to sugars for their own gain (Chen, 2014).

2.10.3 Chemical antibiosis

Similar to most fungi, fungal endophytes are able to produce an assortment of secondary metabolites, of which the biological function is in most cases not or only partially known (Tan and Zou, 2001; Zhang *et al.*, 2006; Aly *et al.*, 2011). These constituents may affect the host plant, such as auxins (plant pathogenic), microorganisms, herbivores or a combination of these (Schulz *et al.*, 2015; see also Chapter 9, this volume). For example, antibiosis was, at least in part, the suggested mode of action by which endophytes prevented *R. solani* and *P. infestans* infection in tomato and potato, respectively (Miles *et al.*, 2012; de Vries *et al.*, 2018). The grass endophytes are known for their antagonistic activity towards insect and vertebrate herbivores based on the production of toxic constituents (Bush *et al.*, 1997; Lane *et al.*, 2000; see also Chapter 7, this volume). The presence of the endophyte can make a host less attractive or more repulsive for herbivores. Italian ryegrass (*L. multiflorum*) infected by *Epichloë occultans* reduced the feeding preference of young Aberdeen Angus steers (*Bos taurus*) in a choice experiment, which was associated with the presence of alkaloids and changes on grass metabolome (Hernández-Agramonte *et al.*, 2018). The RKN *M. incognita* preferred root exudates extracted from tomato over those extracted from tomato colonized by the endophytic *F. oxysporum* Fo162 (Kunkel and Grewal, 2003; Dababat and Sikora, 2007). The same preference was observed for *M. graminicola* in rice in the absence and presence of *F. moniliforme* strain Fe14 (Le *et al.*, 2016).

Loline, ergovaline and α-ergocryptine have nematocidal activity, whereas ergocornine and ergonovine are generally nematostatic (Bacetty *et al.*, 2009a, b). A nematode seems capable of sensing particular compounds because ergovaline had a repelling effect on *Pratylenchus scribneri* at both high and low concentrations (Bacetty *et al.*, 2009a). *N*-formylloline served as an attractant at concentrations lower than 20 μg/ml, whereas it served as a repellent at higher concentrations. Nevertheless, when the ergot alkaloid biosynthesis was disrupted, the negative effect of the ryegrass-endophyte association with *P. scribneri* was not affected. In the absence of nematode infection, ergot alkaloid accumulation in the pseudostem was undetectable or not high enough, which also questioned the meaning of these toxins in resistance towards nematodes (Panaccione *et al.*, 2006). In other studies, loline accumulation could nonetheless reach nematocidal levels inside roots (Bush *et al.*, 1993; Bacetty *et al.*, 2009b).

Nonbalanciaceous endophytes, such as *Fusarium*, *Acremonium*, *Trichoderma*, *Chaetomium* and *Paecilomyces* species, can all produce various constituents when grown *in vitro*. Volatile organic compounds of the *F. oxysporum* CanR-46 endophyte from oilseed rape could protect tomatoes against *B. cinerea* tomato fruit rot in a postharvest situation (Zhang *et al.*, 2014). Culture filtrates of *A. strictum*, *A. implicatum*, *P. lilacinus*, *T. harzianum* and *F. oxysporum* were lethal to *M. incognita* second-stage preparasitic juveniles (Goswami *et al.*, 2008; Tian *et al.*, 2014; Bogner *et al.*, 2017). Culture filtrates of *F. oxysporum* Fo162 contained constituents that notably affected several, but not all, tested nematode species (Amin, 1994; Hallmann and Sikora, 1996; Athman *et al.*, 2006). Preparasitic juveniles of *Heterodera schachtii*, *M. incognita*, *Meloidogyne javanica* and *Meloidogyne arenaria* were particularly sensitive, being fully immobilized. Only 60% of mixed stages of the migratory endoparasites *Radopholus similis* and *Pratylenchus zeae* were immobilized, and none of the the mycophagous species *Aphelenchoides composticola* and microphagous species *Panagrellus redivivus* were affected, collectively suggesting that larval stages were more sensitive to the tested culture filtrates (Hallmann and Sikora, 2011). From the endophytic basidiomycete *P. indica*, exudates as well as cell wall extracts had an effect on vigour, infectivity, development and fecundity of the sedentary cyst nematode *H. schachtii* (Daneshkhah *et al.*, 2013).

Except for one example (Bogner *et al.*, 2017), culture filtrates have thus far not been assessed for the particular nematocidal constituents present. Amazingly, of the various nematode constituents that were produced by the endophyte *F. oxysporum* Fo162, idole-3-acetic acid (IAA) and 4-hydroxybenzoic acid were most toxic for *M. incognita* preparasitic juveniles (Bogner *et al.*, 2017; see also Chapter 9, this volume).

Because two biosynthetic pathways are to a certain extent combined in the intimate association between plant and endophyte, supplementary constituents (derivatives) may be produced, which cannot be produced by the individual organisms alone (Ludwig-Müller, 2015). Furtheremore, the two organisms may reciprocally trigger or enhance particular synthesis pathways (Ludwig-Müller, 2015; see also Chapter 12, this volume).

2.10.4 Induced plant defences and tolerance

The order of arrival may sometimes determine the success of disease control, suggesting that an endophytic interaction with the plant must first be initiated in order to obtain a successful control of pathogen. Endophytic *F. verticillioides* strains, which could suppress *Ustilago maydis*, the causal agent of smut disease in maize, were only effective in reducing disease severity in seedlings upon a simultaneous inoculation with the pathogen (Lee *et al.*, 2009). A non-pathogenic *Verticillium dahliae* strain could prevent infection by a pathogenic strain, when it was inoculated earlier or simultaneously (Shittu *et al.*, 2009). In wild lima bean (*Phaseolus lunatus*), endophytes, identified as *Cochliobolus cynodontis*, *Hyphozyma variabilis*, *Cochliobolus australiensis*, *Keissleriella genistae* and *Fusarium* sp., could individually control *Pseudomonas syringae* best when they were preinoculated (Adame-Álvarez *et al.*, 2014).

Although no apparent disease symptoms are observed, it is now generally recognized that plants do respond to invading and proliferating endophytes (Rodriguez *et al.*, 2009) both locally and systemically. This may not be without reason because the entering of the plant may be sometimes as aggressive as found for pathogens as already mentioned for the non-pathogenic *C. magna* path-1 mutant, which still formed appressoria (Freeman and Rodriguez, 1993), *T. asperellum* T-203, which coiled its hyphe around the root, formed appessorium-like structures on the root surface and most likely used lytic enzymes to degrade cell walls for intercellular migration into the cortex (Yedidia *et al.*, 1999, 2000), and *P. indica*, which killed root cortex cells (Deshmukh *et al.*, 2006; Jacobs *et al.*, 2011; Qiang *et al.*, 2012). Initially the plant may respond to endophytes with immune responses that are similar to those found for invading pathogenic microorganisms, but which are subsequently modulated, thus allowing an asymptomatic colonization while simultaneously priming for defence against other microorganisms (Zamioudis and Pieterse, 2012). Priming is a latent form of induced resistance, in which there is no apparent accumulation of defence-related transcripts, such as those coding for phenyl ammonia lyase (PAL) and pathogenesis-related (PR) proteins. Upon additional biotic or abiotic stresses, such as root-associated microorganisms, particular synthetic compounds, pathogens and herbivores, defence responses are quickly activated, which can be observed molecularly by an often dramatic increase in the defence-related transcripts (Conrath *et al.*, 2006; Heil and Silva Bueno, 2007; Frost *et al.*, 2008; Pineda *et al.*, 2010). Next to priming, immediate defences may be activated as well. The endophytic nematophagous *A. oligospora* and *Pochonia chlamydosporia* both induced papillae formation and other cell wall appositions in barley and tomato, respectively (Escudero and Lopez-Llorca, 2012; Larriba *et al.*, 2015). Similar appositions were observed when cucumber (*C. sativus*) was inoculated with *T. asperellum* T-203 (Yedidia *et al.*, 1999, 2000). Although they are generally associated with plant resistance, these appositions can be elicited by both pathogenic and non-pathogenic fungi (Beswetherick and Bishop, 1993; Heitefuss, 1997; Bao and Lazarovits, 2001). The colonization and proliferation of both *A. oligospora* and *P. chlamydosporia* was not obstructed, and the deposition of phenolics (including lignin), proteins and callose, and the triggering of particular plant defence responses was implied as one of the mechanisms to antagonize plant-parasitic nematodes (Larriba *et al.*, 2015). Transcriptome analysis of barley root colonization by *P. chlamydosporia* showed an enrichment of transcripts involved in abiotic stress responses, primarily those coding for heat shock proteins, transcripts associated with plant hormone biosynthesis, such as auxin, ethylene and jasmonic acid (JA), and transcripts related to effector-triggered immunity (ETI) and pattern-triggered immunity (PTI) (Larriba *et al.*, 2015). Inoculation of cucumber plants with the nematode antagonistic root-endophyte *T. asperellum* T-203 resulted in an increase in lipoxygenase (Lox1), ethylene receptor 1 (ETR1) and constitutive triple response 1 gene B (CTR1) transcript accumulation in both roots and leaves, suggesting a triggering of the JA/ethylene-mediated induced defences (Shoresh *et al.*, 2005). A strong correlation was found between the abilities of *Trichoderma virens* strains to trigger terpenoid phytoalexin defence compounds in cotton seedlings, and the control of *R. solani*. *T. virens* harbours genes coding for non-ribosomal peptide synthetases (NRPSs) and polyketide synthase/NRPS (PKS/NRPS) hybrid enzymes. When one of PKS/NRPS hybrid genes was mutated, the ability of *T.* virens to induce the defence responses through PAL (phenylalanine ammonia lyase) was impaired, whereas the induction of another defence response, i.e. allene oxide synthese, was not (Viterbo *et al.*, 2005; Djonović *et al.*, 2007; Shoresh and Harman, 2008; Vargas *et al.*, 2008; Mastouri *et al.*, 2010; Shoresh *et al.*, 2010; Mukherjee *et al.*, 2012). When *Trichoderma*-preinoculated cucumber plants were challenged with the bacterial leaf pathogen *P. syringae* pv. *lachrymans*, a higher systemic expression of the pathogenesis-related genes encoding for chitinase 1, β-1,3-glucanase and peroxidase was increased when compared to the challenged non-inoculated control plants, indicating that the endophyte enhanced plant defences. Transcripts coding for the stress-related glutathione-*S*-transferase and glutathione-dependent FALDH proteins had also accumulated (Shoresh *et al.*, 2005). The glutathione-*S*-transferase substrate, the thiol GSH, can serve as an antioxidant and performs in the detoxification of xenobiotics and in abiotic and biotic stress tolerance (Xiang *et al.*, 2001; Rouhier *et al.*, 2008; Foyer and Noctor, 2009). Sufficient concentrations of (homo)GSH are a prerequisite in the root nodulation process and therefore seem relevant for the proper formation of the symbiotic interaction between *Rhizobium* spp. and legumes (Frendo *et al.*, 2005). Conversely, because it can regulate related stress defence genes, GSH is involved in the resistance towards oomycetes, bacterial pathogens and insect herbivores (Ball *et al.*, 2004, Parisy *et al.*, 2007, Schlaeppi *et al.*, 2008). In barley roots, an infection with the pathogenic *F. culmorum* lowered ascorbate and GSH accumulation levels (Harrach et al., 2013). In *Arabidopsis*, a mutation of the γ-glutamylcysteine synthetase coding gene (*PAD2*) led to a mutant that contained only about 20% of the GSH found in wild-type plants (Parisy *et al.*, 2007), accumulated lower levels of camalexin (Glazebrook and Ausubel, 1994) and glucosinolates (Schlaeppi *et al.*, 2008), and turned out to be susceptible to the generalist insect *Spodoptera littoralis* (Schlaeppi *et al.*, 2008) and hypersusceptible to the oomycete *Phytophthora porri* (Roetschi *et al.*, 2001). *P. porri* resistance was impartial of SA and JA/ethylene signalling. *P. indica* colonization in barley resulted, at particular stages, in the accumulation of transcripts involved in phytohormone metabolism, primarily gibberellin, auxin and abscisic acid (ABA), although salicylic acid (SA)-associated transcript accumulation was reduced (Schäfer *et al.*, 2009). Although these studies were all done in the absence of an additional biotic stressor, such as a pathogen, it still seems that for *T. asperellum* T-203, *P. chlamydosporia* and *P. indica* trigger JA/ethylene-mediated defence pathways, whereas SA-mediated pathways are not involved (Schouten, 2016).

The conclusion that the JA/ethylene pathway is the general pathway triggered in endophyte-mediated induced resistance should, however, be considered with care (Schouten, 2016). In the aforementioned example of wild lima bean, in which *P. syringae* could be controlled by individual preinoculation with various endophytic isolates (Adame-Álvarez *et al.*, 2014), the control of the biotrophic *P. syringae* was assumed to be mediated by the salicylic acid-dependent pathways (Cameron *et al.*, 1999; Cui *et al.*, 2005), which usually inhibits JA signalling (Caarls *et al.*, 2015). The root-associated endophytic *A. oligospora* increased the activity of plant defence-related enzymes, such as PAL, polyphenol oxidase, chitinase, glucanase, superoxide dismutase, catalase and peroxidases in tomato leaves after infection with the RKN *M. incognita*, indicating that these responses were systemic (Singh *et al.*, 2013). In the presence of the root-associated endophytic *F. oxysporum* Fo162, PAL, PR1 and PR5 transcript accumulation in tomato leaves was significantly increased within 24 h after being challenged with the whitefly *Trialeurodes vaporariorum*, whereas LOX transcript accumulation had not changed (Eschweiler *et al.*, 2014). And another non-pathogenic root-associated *F. oxysporum* strain Fo47, able to induce systemic resistance to pathogenic *F. oxysporum* isolates in tomato, caused an accumulation of PR1 transcripts and increased chitinase, β-1,3-glucanase and β-1,4-glucosidase activity (Fuchs *et al.*, 1997; Duijff *et al.*, 1998). In flax, Fo47 caused a biphasic H_2O_2 accumulation, typical for a hypersensitive response (HR), together with Ca^{2+} spiking in the host plant (Olivain *et al.*, 2003). All this would indicate that an endophyte can initiate the SA-dependent systemic acquired resistance (SAR) (van Loon *et al.*, 1998). HR responses were proposed to cause the observed delay in postinfection development of another RKN *M. arenaria* on susceptible and resistant peanut cultivars (Proite *et al.*, 2008). SAR may also lead to the accumulation of particular secondary plant metabolites that are detrimental towards nematodes. In *N. coenophialum*-inoculated tall fescue plants, metabolites were generated, which reduced the motility of the burrowing nematode *P. scribneri* (Bacetty *et al.*, 2009a). Several plant-derived phenolics were reported that could lead to nematode paralysis (Wuyts *et al.*, 2006) and a phytoalexin, the phenylphenalenone anigorufone, found in a *R. similis*-resistant banana cultivar, showed nematocidal activity (Hölscher *et al.*, 2014). Also, PAL, the initial enzyme of the phenylpropanoid biosynthesis pathway, involved in phenylphenalenone synthesis, was significantly induced in *R. similis*-infected roots of the resistant banana cultivar Ykm5 (Wuyts *et al.*, 2006). *F. oxysporum* Fo162 may as well elicit anigorufone accumulation in the banana roots (Schouten, 2016), similar to what is found for the pathogenic *F. oxysporum* f. sp. *cubense* (Luis *et al.*, 1994), and possibly other phytoalexins with, currently, unknown nematocidal activity in other plant species.

However, to complicate things further, split-root experiments in tomato revealed that both SA and methyl jasmonate, when applied to the inducer side, can reduce *M. incognita* colonization at the responder side (Selim, 2010), and systemic resistance induced against nematodes was reported to be independent of SA accumulation inside the roots in the case of soilborne antagonistic fluorescent pseudomonads (Siddiqui and Shaukat, 2004). For the endophyte-induced defence pathways against pests and pathogens, cross-talk between SA- and JA/ethylene-mediated defence pathways (Caarls *et al.*, 2015; Vos *et al.*, 2015) must therefore still be taken into account. Studies in rice suggests that both pathways are in fact part of a much more complex defence-regulating network, in which ABA and brassinosteroids act as repressors and the ABA pathway is activated by the sedentary nematode itself to induce rice susceptibility (Kyndt *et al.*, 2014, 2017). Whether endophytes can affect this overarching network has so far not been studied.

Nevertheless, *in vitro* studies have shown that endophytic fungi can, like many bacteria and fungi, produce plant hormones, such as auxins, cytokinins and gibberellins (Tan and Zou, 2001; Zhang *et al.*, 2006; Aly *et al.*, 2011; Redman *et al.*, 2011; see also Chapter 9, this volume). When *Arabidopsis* seedlings were inoculated with either *T. virens* or *Trichoderma atroviride*, the typical auxin-related phenotypes, e.g. increased biomass or accelerated root development, were observed. Mutant analysis showed that disruption of auxin transport or signalling reduced growth promotion and increased root proliferation after *T. virens* inoculation (Contreras-Cornejo *et al.*, 2009). *In vitro* experiments with *Arabidopsis* revealed an increase in root proliferation in the presence of *F. oxysporum* Fo162, an endophytic isolate that is also capable of synthesizing indole acetic acid (IAA) *in vitro* (Bogner *et al.*, 2017). Such an increased root growth may increase tolerance to nematode infections because this will compensate for the restricted water transport caused by the location of the nurse cells, initiated by sedentary nematodes, or the damaged root system, caused by burrowing nematodes (Haverkort *et al.*, 1992; Hoveland, 1993; Haverkort *et al.*, 1996). Hormones or functional analogues also may have a more direct impact on the infecting nematode. First, as was mentioned before, IAA was shown to be toxic for the nematode *M. incognita* (Bogner *et al.*, 2017; see also Chapter 9, this volume). Second, sedentary plant-parasitic nematodes delicately control pathways responsive to auxins, ethylene and, possibly, other phytohormones, which are crucial for properly establishing and maintaining the nurse cells (Gheysen and Mitchum, 2011). By altering the IAA content, the proper initiation and maintenance of the nurse cell could be disturbed. Both *H. schachtii* and *M. incognita* could produce cytokine-comprising exudates, with benzyladenine and zeatin-type cytokinins being the most dominant, at sufficient levels to modify the physiological processes inside a host plant (De Meutter *et al.*, 2003). In *H. schachtii*, the silencing of a cytokinin-synthesizing isopentenyltransferase gene resulted in a reduced nurse cell expansion (Siddique *et al.*, 2015).

2.11 Conclusions and Outlook

Fungal endophytes are omnipresent, and a screening generally results in many isolates belonging to different species, apparently forming a small ecosystem within the plant. The encountered endophytic species generally do not differ from plant pathogenic species. Genetically, however, they may be significantly different, lacking whole chromosomes or sections thereof. Small mutations as well may convert a pathogenic fungus into an endophytic fungus and *vice versa*. The reason for such discrepancies is still poorly understood, particularly because plant species or genotype, too, can influence whether a fungus can proliferate asymptomatically inside the host. Nevertheless, there are fungal endophytes that proliferate inside host plants and never elicit disease symptoms.

The association with such true endophytes can affect whole food webs and be beneficial for the host plant against biotic and abiotic stress conditions, although the exact beneficial role(s) may be difficult to determine. The ecological setting in which the association occurs may give clues. Endophytic fungi can directly and indirectly improve the quality of plant life by competing with pests and pathogens, ameliorating abiotic stress and/or triggering plant resistance mechanisms. Hence, endophytes are particularly interesting from an agronomical point of view, as they can reduce disease incidence or ameliorate the effect of high temperatures, drought or salinity in crop plants (see Chapters 5–9, this volume, for further reading). Although there are already examples of successful applications, an integrated biological, molecular and biochemical research on aspects of plant–endophyte associations is still necessary to further expand the applications for endophytes in agricultural practices.

As indicated, endophytes can synthesize many biological active compounds, all of which have the potential to be used for pharmacological purposes (see Chapters 10–12, this volume, for further reading). The currently accessible state-of-the-art technologies, such as next-generation sequencing, integrated microscopical and biochemical techniques, and site-directed mutagenesis, will help to accelerate the research on host–endophyte interactions, bringing a broader application of endophytes in agricultural practices closer than ever before.

References

Adame-Álvarez, R.M., Mendiola-Soto, J. and Heil, M. (2014) Order of arrival shifts endophyte-pathogen interactions in bean from resistance induction to disease facilitation. *FEMS Microbiology Letters* 355, 100–107. doi:10.1111/1574-6968.12454

Akello, J. and Sikora, R.A. (2012) Systemic acropedal influence of endophyte seed treatment on *Acyrthosiphon pisum* and *Aphis fabae* offspring development and reproductive fitness. *Biological Control* 61, 215–221. doi:10.1016/j.biocontrol.2012.02.007

Alabouvette, C. and Couteaudier, Y. (1992) Biological control of *Fusarium* wilts with non-pathogenic fusaria. In: Tjamos, E.C., Papavizas, G.C. and Cook, R.J. (eds) *Biological Control of Plant Diseases: Progress and Challenges for the Future*. Plenum Press, New York.

Álvarez-Loayza, P., White, J.F., Torres, M.S., Balslev, H., Kristiansen, T., Svenning, J.C. and Gil, N. (2011) Light converts endosymbiotic fungus to pathogen, influencing seedling survival and niche-space filling of a common tropical tree, *Iriartea deltoidea*. *PLOS ONE* 6, e16386. doi:10.1371/journal.pone.0016386

Aly, A.H., Debbab, A. and Proksch, P. (2011) Fungal endophytes: Unique plant inhabitants with great promises. *Applied Microbiology and Biotechnology* 90, 1829–1845. doi:10.1007/s00253-011-3270-y

Amin, N. (1994) Untersuchung über die Bedeutung endophytischer Pilze für die biologosche Bekämpfung des wanderenden Endoparasiten *Radopholus similis* (Cobb) Thorne an Bananen. PhD. thesis, University of Bonn, Bonn, Germany.

Andrade-Linares, D.R., Grosch, R., Restrepo, S., Krumbein, A. and Franken, P. (2011) Effects of dark septate endophytes on tomato plant performance. *Mycorrhiza* 21, 413–422. doi:10.1007/s00572-010-0351-1

Arnold, A.E. and Herre, E.A. (2003) Canopy cover and leaf age affect colonization by tropical fungal endophytes: Ecological pattern and process in

Theobroma cacao (Malvaceae). *Mycologia* 95, 388–398.
Arnold, A.E. and Lewis, L.C. (2005) Ecology and evolution of fungal endophytes and their roles against insects. In: Vega, F.E. and Blackwell, M. (eds) *Insect-Fungal Associations: Ecology and Evolution*. Oxford University Press, Inc., New York.
Arnold, A.E. and Lutzoni, F. (2007) Diversity and host range of foliar fungal endophytes: Are tropical leaves biodiversity hotspots? *Ecology* 88, 541–549. doi:10.1890/05-1459
Athman, S.Y., Dubois, T., Viljoen, A., Labuschagne, N., Coyne, D., *et al.* (2006) *In vitro* antagonism of endophytic *Fusarium oxysporum* isolates against the burrowing nematode *Radopholus similis*. *Nematology* 8, 627–636.
Bacetty, A.A., Snook, M.E., Glenn, A.E., Noe, J.P., Hill, N., *et al.* (2009a) Toxicity of endophyte-infected tall fescue alkaloids and grass metabolites on *Pratylenchus scribneri*. *Phytopathology* 99, 1336–1345. doi:10.1094/PHYTO-99-12-1336
Bacetty, A.A., Snook, M.E., Glenn, A.E., Noe, J.P., Nagabhyru, P. and Bacon, C.W. (2009b) Chemotaxis disruption in *Pratylenchus scribneri* by tall fescue root extracts and alkaloids. *Journal of Chemical Ecology* 35, 844–850. doi:10.1007/s10886-009-9657-x
Bacon, C.W. and Hinton, D.M. (1996) Symptomless endophytic colonization of maize by *Fusarium moniliforme*. *Canadian Journal of Botany* 74, 1195–1202. doi:10.1139/b96-144
Bacon, C.W. and Yates, I.E. (2006) Endophytic root colonization by *Fusarium* species: Histology, plant interactions, and toxicity. In: Schulz, B., Boyle, C. and Sieber, T.N. (eds) *Microbial Root Endophytes*. Springer, Berlin.
Bakker, P.A.H.M., Berendsen, R.L., Doornbos, R.F., Wintermans, P.C. and Pieterse, C.M. (2013) The rhizosphere revisited: Root microbiomics. *Frontiers in Plant Science* 4, 165. doi:10.3389/fpls.2013.00165
Ball, L., Accotto, G.P., Bechtold, U., Creissen, G., Funck, D., *et al.* (2004) Evidence for a direct link between glutathione biosynthesis and stress fefense gene expression in *Arabidopsis*. *The Plant Cell* 16, 2448–2462. doi:10.1105/tpc.104.022608
Bao, J.R. and Lazarovits, G. (2001) Differential colonization of tomato roots by nonpathogenic and pathogenic *Fusarium oxysporum* strains may influence *Fusarium* wilt control. *Phytopathology* 91, 449–456.
Baynes, M.A., Russell, D.M., Newcombe, G., Carta, L.K., Rossman, A.Y. and Ismaiel, A. (2012) A mutualistic interaction between a fungivorous nematode and a fungus within the endophytic community of *Bromus tectorum*. *Fungal Ecology* 5, 610–623. doi:10.1016/j.funeco.2012.03.004
Bertazzoni, S., Williams, A.H., Jones, D.A., Syme, R.A., Tan, K.C. and Hane, J.K. (2018) Accessories make the outfit: Accessory chromosomes and other dispensable dna regions in plant-pathogenic fungi. *Molecular Plant-Microbe Interactions* 31, 779–788. doi:10.1094/mpmi-06-17-0135-fi
Beswetherick, J.T. and Bishop, C.D. (1993) An ultrastructural-study of tomato roots inoculated with pathogenic and nonpathogenic necrotrophic fungi and a saprophytic fungus. *Plant Pathology* 42, 577–588.
Biswas, C., Dey, P., Satpathy, S., Satya, P. and Mahapatra, B.S. (2013) Endophytic colonization of white jute (*Corchorus capsularis*) plants by different *Beauveria bassiana* strains for managing stem weevil (*Apion corchori*). *Phytoparasitica* 41, 17–21. doi:10.1007/s12600-012-0257-x
Blumenstein, K., Albrectsen, B.R., Martín, J.A., Hultberg, M., Sieber, T.N., Helander, M. and Witzell, J. (2015) Nutritional niche overlap potentiates the use of endophytes in biocontrol of a tree disease. *BioControl* 60, 655–667. doi:10.1007/s10526-015-9668-1
Bogner, C.W., Kamdem, R.S.T., Sichtermann, G., Matthäus, C., Hölscher, D., *et al.* (2017) Bioactive secondary metabolites with multiple activities from a fungal endophyte. *Microbial Biotechnology* 10, 175–188. doi:10.1111/1751-7915.12467
Bordallo, J.J., Lopez-Llorca, L.V., Jansson, H.B., Salinas, J., Persmark, L. and Asensio, L. (2002) Colonization of plant roots by egg-parasitic and nematode-trapping fungi. *New Phytologist* 154, 491–499. doi:10.1046/j.1469-8137.2002.00399.x
Boutigny, A.L., Barreau, C., Atanasova-Penichon, V., Verdal-Bonnin, M.N., Pinson-Gadais, L. and Richard-Forget, F. (2009) Ferulic acid, an efficient inhibitor of type B trichothecene biosynthesis and *Tri* gene expression in *Fusarium* liquid cultures. *Mycological Research* 113, 746–753. doi:10.1016/j.mycres.2009.02.010
Brundrett, M. (2004) Diversity and classification of mycorrhizal associations. *Biological Reviews of the Cambridge Philosophical Society* 79, 473–495. doi:10.1017/S1464793103006316
Bush, L.P., Fannin, F.F., Siegel, M.R., Dahlman, D.L. and Burton, H.R. (1993) Chemistry, occurrence and biological effects of saturated pyrrolizidine alkaloids associated with endophyte grass interactions. *Agriculture, Ecosystems & Environment* 44, 81–102. doi:10.1016/0167-8809(93)90040-V
Bush, L.P., Wilkinson, H.H. and SchardL, C.L. (1997) Bioprotective alkaloids of grass-fungal endophyte symbioses. *Plant Physiology* 114, 1–7. doi:10.1104/pp.114.1.1
Caarls, L., Pieterse, C.M.J. and Van Wees, S.C.M. (2015) How salicylic acid takes transcriptional

control over jasmonic acid signaling. *Frontiers in Plant Science* 6, 170. doi:10.3389/fpls.2015.00170

Cameron, R.K., Paiva, N.L., Lamb, C.J. and Dixon, R.A. (1999) Accumulation of salicylic acid and PR-1 gene transcripts in relation to the systemic acquired resistance (SAR) response induced by *Pseudomonas syringae* pv. *tomato* in *Arabidopsis*. *Physiological and Molecular Plant Pathology* 55, 121–130. doi:10.1006/pmpp.1999.0214

Chagas, F.O., Dias, L.G. and Pupo, M.T. (2013) A mixed culture of endophytic fungi increases production of antifungal polyketides. *Journal of Chemical Ecology* 39, 1335–1342. doi:10.1007/s10886-013-0351-7

Chan, Y.L., Yang, A.H., Chen, J.T., Yeh, K.W. and Chan, M.T. (2010) Heterologous expression of taro cystatin protects transgenic tomato against *Meloidogyne incognita* infection by means of interfering sex determination and suppressing gall formation. *Plant Cell Reports* 29, 231–238. doi:10.1007/s00299-009-0815-y

Chen, L.J., Sun, G.Y., Zhang, R. and Guo, J.Q. (2004) Isolation and identification of endophytic fungi on *Brassica napus*. *Journal of Shihezi University (Natural Science Edition)* 22(Suppl. 1), 66–68.

Chen, L.Q. (2014) SWEET sugar transporters for phloem transport and pathogen nutrition. *New Phytologist* 201, 1150–1155. doi:10.1111/nph.12445

Chen, L.Q., Hou, B.H., Lalonde, S., Takanaga, H., Hartung, M.L., *et al.* (2010) Sugar transporters for intercellular exchange and nutrition of pathogens. *Nature* 468, 527–532.

Chet, I. and Baker, R. (1981) Isolation and biocontrol potential of *Trichoderma hamatum* from soil naturally suppressive to *Rhizoctonia solani*. *Phytopathology* 71, 286–290. doi:10.1094/Phyto-71-286

Conrath, U., Beckers, G.J.M., Flors, V., Garcia-Agustin, P., Jakab, G., *et al.* (2006) Priming: Getting ready for battle. *Molecular Plant-Microbe Interactions* 19, 1062–1071.

Contreras-Cornejo, H.A., Macías-Rodríguez, L., Cortés-Penagos, C. and Lopéz-Bucio, J. (2009) *Trichoderma virens*, a plant beneficial fungus, enhances biomass production and promotes lateral root growth through an auxin-dependent mechanism in *Arabidopsis*. *Plant Physiology* 149, 1579–1592.

Covert, S.F. (1998) Supernumerary chromosomes in filamentous fungi. *Current Genetics* 33, 311–319. doi:10.1007/s002940050342

Covert, S.F., Enkerli, J., Miao, V.P.W. and VanEtten, H.D. (1996) A gene for maackiain detoxification from a dispensable chromosome of *Nectria haematococca*. *Molecular and General Genetics* 251, 397–406.

Cui, J., Bahrami, A.K., Pringle, E.G., Hernandez-Guzman, G., Bender, C.L., Pierce, N.E. and Ausubel, F.M. (2005) *Pseudomonas syringae* manipulates systemic plant defenses against pathogens and herbivores. *Proceedings of the National Academy of Sciences of the United States of America* 102, 1791–1796. doi:10.1073/pnas.0409450102

Curtis, R.H.C., Jones, J.T., Davies, K.G., Sharon, E. and Spiegel, Y. (2011) Plant nematode surfaces. In: Davies, K.G. and Spiegel, Y. (eds) *Biological Control of Plant-Parasitic Nematodes: Building Coherence Between Microbial Ecology and Molecular Mechanisms*. Springer, Heidelberg, Germany.

Dababat, A.E.F.A. and Sikora, R.A. (2007) Influence of the mutualistic endophyte *Fusarium oxysporum* 162 on *Meloidogyne incognita* attraction and invasion. *Nematology* 9, 771–776. doi:10.1163/156854107782331225

Dahl, A.S. (1934) Snowmold of turf grasses as caused by *Fusarium nivale*. *Phytopathology* 24, 197–214.

Daneshkhah, R., Cabello, S., Rozanska, E., Sobczak, M., Grundler, F.M., Wieczorek, K. and Hofmann, J. (2013) *Piriformospora indica* antagonizes cyst nematode infection and development in *Arabidopsis* roots. *Journal of Experimental Botany* 64, 3763–3774. doi:10.1093/jxb/ert213

Deshmukh, S., Hueckelhoven, R., Schaefer, P., Imani, J., Sharma, M., *et al.* (2006) The root endophytic fungus *Piriformospora indica* requires host cell death for proliferation during mutualistic symbiosis with barley. *Proceedings of the National Academy of Sciences of the United States of America* 103, 18450–18457.

de Azevedo Silva, F., Liotti, R.G., de Araújo Boleti, A.P.D., de Melo Reis, E., Passos, M.B.S., *et al.* (2018) Diversity of cultivable fungal endophytes in *Paullinia cupana* (Mart.) Ducke and bioactivity of their secondary metabolites. *PLOS ONE* 13, 34.e0195874. doi:10.1371/journal.pone.0195874

De Jaeger, N., Declerck, S. and de La Providencia, I.E. (2010) Mycoparasitism of arbuscular mycorrhizal fungi: a pathway for the entry of saprotrophic fungi into roots. *FEMS Microbiology Ecology* 73, 312–322. doi:10.1111/j.1574-6941.2010.00903.x

De Meutter, J., Tytgat, T., Witters, E., Gheysen, G., Van Onckelen, H. and Gheysen, G. (2003) Identification of cytokinins produced by the plant parasitic nematodes *Heterodera schachtii* and *Meloidogyne incognita*. *Molecular Plant Pathology* 4, 271–277. doi:10.1046/j.1364-3703.2003.00176.x

de Vries, S., von Dahlen, J.K., Schnake, A., Ginschel, S., Schulz, B. and Rose, L.E. (2018)

Broad-spectrum inhibition of *Phytophthora infestans* by fungal endophytes. *FEMS Microbiology Ecology* 94(4). doi:10.1093/femsec/fiy037

Djonović, S., Vargas, W.A., Kolomiets, M.V., Horndeski, M., Wiest, A. and Kenerley, C.M. (2007) A proteinaceous elicitor Sm1 from the beneficial fungus *Trichoderma virens* is required for induced systemic resistance in maize. *Plant Physiology* 145, 875–889. doi:10.1104/pp.107.103689

Druzhinina, I.S., Seidl-Seiboth, V., Herrera-Estrella, A., Horwitz, B.A., Kenerley, C.M., *et al.* (2011) *Trichoderma*: The genomics of opportunistic success. *Nature Reviews Microbiology* 9, 749–759. doi:10.1038/nrmicro2637

Duijff, B.J., Pouhair, D., Olivain, C., Alabouvette, C. and Lemanceau, P. (1998) Implication of systemic induced resistance in the suppression of *Fusarium* wilt of tomato by *Pseudomonas fluorescens* WCS417r and by nonpathogenic *Fusarium oxysporum* Fo47. *European Journal of Plant Pathology* 104, 903–910.

Eaton, C.J., Cox, M.P., Ambrose, B., Becker, M., Hesse, U., Schardl, C.L. and Scott, B. (2010) Disruption of signaling in a fungal-grass symbiosis leads to pathogenesis. *Plant Physiology* 153, 1780–1794. doi:10.1104/pp.110.158451

Eaton, C.J., Cox, M.P. and Scott, B. (2011) What triggers grass endophytes to switch from mutualism to pathogenism? *Plant Science* 180, 190–195. doi:10.1016/j.plantsci.2010.10.002

Elmi, A.A., West, C.P., Robbins, R.T. and Kirkpatrick, T.L. (2000) Endophyte effects on reproduction of a root-knot nematode (*Meloidogyne marylandi*) and osmotic adjustment in tall fescue. *Grass and Forage Science* 55, 166–172.

Eschweiler, J., Messelink, G., Kasal, T., Main, D., Grundler, F. and Schouten, A. (2014) Integration of fungal endophytes in a greenhouse environment to control insect pests. In: *59. Deutsche Pflanzenschutztagung*. Freiburg, Germany. Julius Kühn Archiv.

Escudero, N. and Lopez-Llorca, L.V. (2012) Effects on plant growth and root-knot nematode infection of an endophytic GFP transformant of the nematophagous fungus *Pochonia chlamydosporia*. *Symbiosis* 57, 33–42. doi:10.1007/s13199-012-0173-3

Foyer, C.H. and Noctor, G. (2009) Redox regulation in photosynthetic organisms: Signaling, acclimation, and practical implications. *Antioxidants & Redox Signaling* 11, 861–905. doi:10.1089/ars.2008.2177

Freeman, S. and Rodriguez, R.J. (1993) Genetic conversion of a fungal plant pathogen to a nonpathogenic, endophytic mutualist. *Science* 260, 75–78.

Frendo, P., Harrison, J., Norman, C., Jimenez, M. J.H., Van de Sype, G., Gilabert, A. and Puppo, A. (2005) Glutathione and homoglutathione play a critical role in the nodulation process of *Medicago truncatula*. *Molecular Plant-Microbe Interactions* 18, 254–259.

Frost, C.J., Mescher, M.C., Carlson, J.E. and De Moraes, C.M. (2008) Plant defense priming against herbivores: Getting ready for a different battle. *Plant Physiology* 146, 818–824.

Fuchs, J.-G., Moënne-Loccoz, Y. and Défago, G. (1997) Nonpathogenic *Fusarium oxysporum* strain Fo47 induces resistance to *Fusarium* wilt in tomato. *Plant Disease* 81, 492–496.

Gheysen, G. and Mitchum, M.G. (2011) How nematodes manipulate plant development pathways for infection. *Current Opinion in Plant Biology* 14, 415–421.

Giordano, D., Beta, T., Reyneri, A. and Blandino, M. (2017) Changes in the phenolic acid content and antioxidant activity during kernel development of corn (*Zea mays* L.) and relationship with mycotoxin contamination. *Cereal Chemistry* 94, 315–324. doi:10.1094/cchem-05-16-0155-r

Glazebrook, J. and Ausubel, F.M. (1994) Isolation of phytoalexin-deficient mutants of *Arabidopsis thaliana* and characterization of their interactions with bacterial pathogens. *Proceedings of the National Academy of Sciences of the United States of America* 91, 8955–8959. doi:10.1073/pnas.91.19.8955

Glynn, N.C., Hare, M.C., Parry, D.W. and Edwards, S.G. (2005) Phylogenetic analysis of EF-1 alpha gene sequences from isolates of *Microdochium nivale* leads to elevation of varieties *majus* and *nivale* to species status. *Mycological Research* 109, 872–880. doi:10.1017/s0953756205003370

Goswami, J., Pandey, R.K., Tewari, J.P. and Goswami, B.K. (2008) Management of root knot nematode on tomato through application of fungal antagonists, *Acremonium strictum and Trichoderma harzianum*. *Journal of Environmental Science and Health, Part B: Pesticides, Food Contaminants, and Agricultural Wastes* 43, 237–240. doi:10.1080/03601230701771164

Goverse, A. and Smant, G. (2014) The activation and suppression of plant innate immunity by parasitic nematodes. *Annual Review of Phytopathology* 52, 243–65. doi:10.1146/annurev-phyto-102313-050118

Griesbach, M. (1999) Occurrence of mutualistic fungal endophytes in bananas (*Musa* spp.) and their potential as biocontrol agents of the banana weevil *Cosmopolites sordidus* (Germar) (Coleoptera: Curculionidae) in Uganda. PhD. thesis, University of Bonn, Bonn, Germany.

Hallmann, J. and Sikora, R.A. (1994) Influence of *Fusarium oxysporum*, a mutualistic fungal endophyte, on *Meloidogyne incognita* infection of tomato. *Journal of Plant Diseases and Protection* 101, 475–481.

Hallmann, J. and Sikora, R.A. (1996) Toxicity of fungal endophyte secondary metabolites to plant parasitic nematodes and soil-borne plant pathogenic fungi. *European Journal of Plant Pathology* 102, 155–162.

Hallmann, J. and Sikora, R.A. (2011) Endophytic fungi. In: Davies, K.G. and Spiegel, Y. (eds) *Biological Control of Plant-Parasitic Nematodes: Building Coherence Between Microbial Ecology and Molecular Mechanisms*. Springer, Heidelberg, Germany.

Harimoto, Y., Hatta, R., Kodama, M., Yamamoto, M., Otani, H. and Tsuge, T. (2007) Expression profiles of genes encoded by the supernumerary chromosome controlling AM-toxin biosynthesis and pathogenicity in the apple pathotype of *Alternaria alternata*. Molecular Plant-Microbe Interactions 20, 1463–1476. doi:10.1094/mpmi-20-12-1463

Harrach, B.D., Baltruschat, H., Barna, B., Fodor, J. and Kogel, K.H. (2013) The mutualistic fungus *Piriformospora indica* protects barley roots from a loss of antioxidant capacity caused by the necrotrophic pathogen *Fusarium culmorum*. *Molecular Plant-Microbe Interactions* 26, 599–605.

Haverkort, A.J., Boerma, M., Velema, R. and Van De Waart, M. (1992) The influence of drought and cyst nematodes on potato growth. 4. Effects on crop growth under field conditions of four cultivars differing in tolerance. *Netherlands Journal of Plant Pathology* 98, 179–191.

Haverkort, A.J., De Ruijter, F.J., Boerma, M. and Van De Waart, M. (1996) Foliar calcium concentration of potato and its relation to genotype lateness and tolerance of cyst nematodes. *European Journal of Plant Pathology* 102, 317–324. doi:10.1007/Bf01878126

Heil, M. and Silva Bueno, J.C. (2007) Within-plant signaling by volatiles leads to induction and priming of an indirect plant defense in nature. *Proceedings of the National Academy of Sciences of the United States of America* 104, 5467–5472.

Heitefuss, R. (1997) Cell wall modifications in relation to resistance. In: Hartleb, H., Heitefuss, R. and Hoppe, H.H. (eds) *Resistance of Crop Plants Against Fungi*. Gustav Fischer Verlag, Stuttgart, Germany.

Hernández-Agramonte, I.M., Semmartin, M., Omacini, M., Durante, M., Gundel, P.E. and De Battista, J. (2018) A fungal endophyte of a palatable grass affects preference of large herbivores. *Austral Ecology* 43, 172–179. doi:10.1111/aec.12554

Hofmann, J., Hess, P.H., Szakasits, D., Blochl, A., Wieczorek, K., *et al.* (2009a) Diversity and activity of sugar transporters in nematode-induced root syncytia. *Journal of Experimental Botany* 60, 3085–3095. doi:10.1093/jxb/erp138

Hofmann, J., Kolev, P., Kolev, N., Daxböck-Horvath, S. and Grundler, F.M.W. (2009b) The *Arabidopsis thaliana* sucrose transporter gene *AtSUC4* is expressed in *Meloidogyne incognita*-induced root galls. *Journal of Phytopathology* 157, 256–261. doi:10.1111/j.1439-0434.2008.01468.x

Hölscher, D., Dhakshinamoorthy, S., Alexandrov, T., Becker, M., Bretschneider, T., *et al.* (2014) Phenalenone-type phytoalexins mediate resistance of banana plants (*Musa* spp.) to the burrowing nematode *Radopholus similis*. *Proceedings of the National Academy of Sciences of the United States of America* 111, 105–110. doi:10.1073/pnas.1314168110

Houterman, P.M., Speijer, D., Dekker, H.L., De Koster, C.G., Cornelissen, B.J.C. and Rep, M. (2007) The mixed xylem sap proteome of *Fusarium oxysporum*-infected tomato plants. *Molecular Plant Pathology* 8, 215–221.

Houterman, P.M., Ma, L., van Ooijen, G., de Vroomen, M.J., Cornelissen, B.J.C., Takken, F.L.W. and Rep, M. (2009) The effector protein Avr2 of the xylem-colonizing fungus *Fusarium oxysporum* activates the tomato resistance protein I-2 intracellularly. *The Plant Journal* 58, 970–978. doi:10.1111/j.1365-313X.2009.03838.x

Hoveland, C.S. (1993) Importance and economic significance of the *Acremonium* endophytes to performance of animals and grass plant. *Agriculture, Ecosystems & Environment* 44, 3–12. doi:10.1016/0167-8809(93)90036-O

Jaber, L.R. and Alananbeh, K.M. (2018) Fungal entomopathogens as endophytes reduce several species of *Fusarium* causing crown and root rot in sweet pepper (*Capsicum annuum* L.). *Biological Control* 126, 117–126. doi:10.1016/j.biocontrol.2018.08.007

Jaber, L.R. and Araj, S.E. (2018) Interactions among endophytic fungal entomopathogens (Ascomycota: Hypocreales), the green peach aphid *Myzus persicae* Sulzer (Homoptera: Aphididae), and the aphid endoparasitoid *Aphidius colemani* Viereck (Hymenoptera: Braconidae). *Biological Control* 116, 53–61. doi:10.1016/j.biocontrol.2017.04.005

Jaber, L.R. and Enkerli, J. (2017) Fungal entomopathogens as endophytes: Can they promote plant growth? *Biocontrol Science and Technology* 27, 28–41. doi:10.1080/09583157.2016.1243227

Jaber, L.R. and Ownley, B.H. (2018) Can we use entomopathogenic fungi as endophytes for dual biological control of insect pests and plant pathogens? *Biological Control* 116, 36–45. doi:10.1016/j.biocontrol.2017.01.018

Jaber, L.R. and Vidal, S. (2009) Interactions between an endophytic fungus, aphids and extrafloral nectaries: Do endophytes induce extrafloral-mediated defences in *Vicia faba*? *Functional Ecology* 23, 707–714. doi:10.1111/j.1365-2435.2009.01554.x

Jaber, L.R. and Vidal, S. (2010) Fungal endophyte negative effects on herbivory are enhanced on intact plants and maintained in a subsequent generation. *Ecological Entomology* 35, 25–36. doi:10.1111/j.1365-2311.2009.01152.x

Jacobs, S., Zechmann, B., Molitor, A., Trujillo, M., Petutschnig, E., *et al.* (2011) Broad-spectrum suppression of innate immunity is required for colonization of *Arabidopsis* roots by the fungus *Piriformospora indica*. *Plant Physiology* 156, 726–40. doi:10.1104/pp.111.176446

Jallow, M.F.A., Dugassa-Gobena, D. and Vidal, S. (2004) Indirect interaction between an unspecialized endophytic fungus and a polyphagous moth. *Basic and Applied Ecology* 5, 183–191. doi:10.1078/1439-1791-00224

Jumpponen, A. and Trappe, J.M. (1998) Dark septate endophytes: A review of facultative biotrophic root-colonizing fungi. *New Phytologist* 140, 295–310.doi:10.1046/j.1469-8137.1998.00265.x

Junker, C., Draeger, S. and Schulz, B. (2012) A fine line – Endophytes or pathogens in *Arabidopsis thaliana*. *Fungal Ecology* 5, 657–662. doi:10.1016/j.funeco.2012.05.002

Kauppinen, M., Saikkonen, K., Helander, M., Pirttilä, A.M. and Wäli, P.R. (2016) *Epichloë* grass endophytes in sustainable agriculture. *Nature Plants* 2.15224. doi:10.1038/nplants.2015.224

Kistler, H.C., Meinhardt, L.W. and Benny, U. (1996) Mutants of *Nectria haematococca* created by a site-directed chromosome breakage are greatly reduced in virulence toward pea. *Molecular Plant-Microbe Interactions* 9, 804–809.

Kuldau, G.A. and Yates, I.E. (2000) Evidence for *Fusarium* endophytes in cultivated and wild plants. In: Bacon, C.W. and White, J.F. Jr (eds) *Microbial Endophytes*. Marcel Dekker, Inc., New York.

Kunkel, B.A. and Grewal, P.S. (2003) Endophyte infection in perennial ryegrass reduces the susceptibility of black cutworm to an entomopathogenic nematode. *Entomologia Experimentalis et Applicata* 107, 95–104. doi:10.1046/j.1570-7458.2003.00048.x

Kunkel, B.A., Grewal, P.S. and Quigley, M.F. (2004) A mechanism of acquired resistance against an entomopathogenic nematode by *Agrotis ipsilon* feeding on perennial ryegrass harboring a fungal endophyte. *Biological Control* 29, 100–108. doi:10.1016/S1049-9644(03)00119-1

Kurtz, A., Schouten, A. and Sikora, R.A. (2008) Phylogenetic analysis of endophytic *Fusarium oxysporum* strains inducing systemic resistance against the burrowing nematode *Radopholus similis* in banana. In: Steinberg, C., Edel-Hermann, V., Friberg, H., Alabouvette, C. and Tronsmo, A. (eds) *Multitrophic Interactions in Soil: Influence of Abiotic and Biotic Factors on Biocontrol Agents*. *IOBC/WPRS Bulletin*, vol. 42.

Kyndt, T., Fernandez, D. and Gheysen, G. (2014) Plant-parasitic nematode infections in rice: Molecular and cellular insights. *Annual Review of Phytopathology* 52, 135–153. doi:10.1146/annurev-phyto-102313-050111

Kyndt, T., Nahar, K., Haeck, A., Verbeek, R., Demeestere, K. and Gheysen, G. (2017) Interplay between carotenoids, abscisic acid and jasmonate guides the compatible rice-*Meloidogyne graminicola* interaction. *Frontiers in Plant Science* 8, 951. doi:10.3389/fpls.2017.00951

Lane, G.A., Christensen, M.J. and Miles, C.O. (2000) Coevolution of fungal endophytes with grasses: The significance of secondary metabolites. In: Bacon, C.W. and White, J.F. Jr (eds) *Microbial Endophytes*. Marcel Dekker, Inc., New York.

Larriba, E., Jaime, M.D., Nislow, C., Martin-Nieto, J. and Lopez-Llorca, L.V. (2015) Endophytic colonization of barley (*Hordeum vulgare*) roots by the nematophagous fungus *Pochonia chlamydosporia* reveals plant growth promotion and a general defense and stress transcriptomic response. *Journal of Plant Research* 128, 665–678. doi:10.1007/s10265-015-0731-x

Le, H.T.T., Padgham, J.L., Hagemann, M.H., Sikora, R. and Schouten, A. (2016) Developmental and behavioral effects of the endophytic *Fusarium moniliforme* Fe14 toward *Meloidogyne graminicola* in rice. *Annals of Applied Biology* 169, 134–143.

Lee, K., Pan, J.J. and May, G. (2009) Endophytic *Fusarium verticillioides* reduces disease severity caused by *Ustilago maydis* on maize. *FEMS Microbiology Letters* 229, 31–37. doi:10.1111/j.1574-6968.2009.01719.x

Lievens, B., Rep, M. and Thomma, B.P.H.J. (2008) Recent developments in the molecular discrimination of *formae speciales* of *Fusarium oxysporum*. *Pest Management Science* 64, 781–788. doi:10.1002/ps.1564

Lin, S., Wu, X., Cao, J.Z. and Wang, F.L. (2013) Biocontrol potential of chitinase-producing nematophagous fungus *Acremonioum implicatum* against *Meloidogyne incognita*. *Acta Phytopathologica Sinica* 43, 509–517.

Lofgren, L.A., Leblanc, N.R., Certano, A.K., Nachtigall, J., Labine, K.M., *et al.* (2018) *Fusarium graminearum*: Pathogen or endophyte of North American grasses? *New Phytologist* 217, 1203–1212. doi:10.1111/nph.14894

Ludwig-Müller, J. (2015) Plants and endophytes: Equal partners in secondary metabolite production? *Biotechnology Letters* 37, 1325–1334. doi:10.1007/s10529-015-1814-4

Luis, J.G., Fletcher, W.Q., Echeverri, F. and Grillo, T.A. (1994) Phenalenone-type phytoalexins from *Musa-Acuminata* synthesis of 4-phenylphenalenones. *Tetrahedron* 50, 10963–10970. doi:10.1016/S0040-4020(01)85707-0

Ma, L.J, Freitag, M., Grabherr, M., Henrissat, B., Houterman, P.M. *et al.* (2010) Comparative genomics reveals mobile pathogenicity chromosomes in *Fusarium*. Nature 464, 367–373. doi:10.1038/nature08850

Maciá-Vicente, J.G., Jansson, H.B., Abdullah, S.K., Descals, E., Salinas, J. and Lopez-Llorca, L.V. (2008) Fungal root endophytes from natural vegetation in Mediterranean environments with special reference to *Fusarium* spp. *FEMS Microbiology Ecology* 64, 90–105. doi:10.1111/j.1574-6941.2007.00443.x

Malinowski, D.P. and Belesky, D.P. (2000) Adaptations of endophyte-infected cool-season grasses to environmental stresses: Mechanisms of drought and mineral stress tolerance. *Crop Science* 40, 923–940. doi:10.2135/cropsci2000.404923x

Martin, C., Schnee, S., Lintz, J., Vogelgsang, S., Gindro, K., Mauch-Mani, B. and Mascher, F. (2017) Role of ferulic acid in *Fusarium* head blight resistance of wheat spikes. Future IPM 3.0 Towards a Sustainable Agriculture, 15–20 October 2017. IOBC-WPRS, Riva del Garda, Italy.

Martín, J.A., Witzell, J., Blumenstein, K., Rozpedowska, E., Helander, M., Sleber, T.N. and Gil, L. (2013) Resistance to dutch elm disease reduces presence of xylem endophytic fungi in elms (*Ulmus* spp.). *PLOS ONE* 8, e56987. doi:10.1371/journal.pone.0056987

Martín, J.A., Macaya-Sanz, D., Witzell, J., Blumenstein, K. and Gil, L. (2015) Strong *in vitro* antagonism by elm xylem endophytes is not accompanied by temporally stable in planta protection against a vascular pathogen under field conditions. *European Journal of Plant Pathology* 142, 185–196. doi:10.1007/s10658-015-0602-2

Martinuz, A., Schouten, A. and Sikora, R.A. (2012a) Systemically induced resistance and microbial competitive exclusion: Implications on biological control. *Phytopathology* 102, 260–266. doi:10.1094/PHYTO-04-11-0120

Martinuz, A., Schouten, A. and Sikora, R.A. (2012b) Post-infection development of *Meloidogyne incognita* on tomato treated with the endophytes *Fusarium oxysporum* strain Fo162 and *Rhizobium etli* strain G12. *BioControl* 58, 95–104. doi:10.1007/s10526-012-9471-1

Martinuz, A., Zewdu, G., Ludwig, N., Grundler, F., Sikora, R.A. and Schouten, A. (2015) The application of *Arabidopsis thaliana* in studying tripartite interactions among plants, beneficial fungal endophytes and biotrophic plant-parasitic nematodes. *Planta* 241, 1015–1025. doi:10.1007/s00425-014-2237-5

Mastouri, F., Björkman, T. and Harman, G.E. (2010) Seed treatment with *Trichoderma harzianum* alleviates biotic, abiotic, and physiological stresses in germinating seeds and seedlings. *Phytopathology* 100, 1213–1221. doi:10.1094/phyto-03-10-0091

McKinnon, A.C., Saari, S., Moran-Diez, M.E., Meyling, N.V., Raad, M. and Glare, T.R. (2017) *Beauveria bassiana* as an endophyte: A critical review on associated methodology and biocontrol potential. *BioControl* 62, 1–17. doi:10.1007/s10526-016-9769-5

Menjivar, R.D., Hagemann, M.H., Kranz, J., Cabrera, J.A., Dababat, A.A. and Sikora, R.A. (2011) Biological control of *Meloidogyne incognita* on cucurbitaceous crops by the non-pathogenic endophytic fungus *Fusarium oxysporum* strain 162. *International Journal of Pest Management* 57, 249–253. doi:10.1080/09670874.2011.590239

Miao, V.P., Covert, S.F. and VanEtten, H.D. (1991) A fungal gene for antibiotic-resistance on a dispensable (B) chromosome. *Science* 254, 1773–1776. doi:10.1126/science.1763326

Miles, L.A., Lopera, C.A., Gonzalez, S., De Garcia, M.C.C., Franco, A.E. and Restrepo, S. (2012) Exploring the biocontrol potential of fungal endophytes from an Andean Colombian Paramo ecosystem. *BioControl* 57, 697–710. doi:10.1007/s10526-012-9442-6

Mostert, L., Crous, P.W. and Petrini, O. (2000) Endophytic fungi associated with shoots and leaves of *Vitis vinifera*, with specific reference to the *Phomopsis viticola* complex. *Sydowia* 52, 46–58.

Mukherjee, P.K., Buensanteai, N., Moran-Diez, M.E., Druzhinina, I.S. and Kenerley, C.M. (2012) Functional analysis of non-ribosomal peptide synthetases (NRPSs) in *Trichoderma virens* reveals a polyketide synthase (PKS)/NRPS hybrid enzyme involved in the induced systemic resistance response in maize. *Microbiology SGM* 158, 155–165. doi:10.1099/mic.0.052159-0

Müller, C.B. and Krauss, J. (2005) Symbiosis between grasses and asexual fungal endophytes. *Current Opinion in Plant Biology* 8, 450–456.

Naik, B.S., Shashikala, J. and Krishnamurthy, Y.L. (2009) Study on the diversity of endophytic communities from rice (*Oryza sativa* L.) and their antagonistic activities *in vitro*. *Microbiological Research* 164, 290–296. doi:10.1016/j.micres.2006.12.003

Olivain, C., Trouvelot, S., Binet, M.N., Cordier, C., Pugin, A. and Alabouvette, C. (2003) Colonization of flax roots and early physiological responses of flax cells inoculated with pathogenic and nonpathogenic strains of *Fusarium oxysporum*. *Applied and Environmental Microbiology* 69, 5453–5462.

Omacini, M., Chaneton, E.J., Ghersa, C.M. and Muller, C.B. (2001) Symbiotic fungal endophytes control insect host-parasite interaction webs. *Nature* 409, 78–81. doi:10.1038/35051070

Omacini, M., Semmartin, M., Perez, L.I. and Gundel, P.E. (2012) Grass-endophyte symbiosis: A neglected aboveground interaction with multiple belowground consequences. *Applied Soil Ecology* 61, 273–279. doi:10.1016/j.apsoil.2011.10.012

Ownley, B.H., Griffin, M.R., Klingeman, W.E., Gwinn, K.D., Moulton, J.K. and Pereira, R.M. (2008) *Beauveria bassiana*: Endophytic colonization and plant disease control. *Journal of Invertebrate Pathology* 98, 267–270. doi:10.1016/j.jip.2008.01.010

Ownley, B.H., Gwinn, K.D. and Vega, F.E. (2010) Endophytic fungal entomopathogens with activity against plant pathogens: Ecology and evolution. *BioControl* 55, 113–128. doi:10.1007/s10526-009-9241-x

Panaccione, D.G., Johnson, R., Morton, J., Kotcon, J. and Schardl, C. (2006) Ergot alkaloids are not essential for endophytic fungus-associated population suppression of the lesion nematode, *Pratylenchus scribneri*, on perennial ryegrass. *Nematology* 8, 583–590. doi:10.1163/156854106778614074

Papadopoulou, J. and Triantaphyllou, A.C. (1982) Sex-differentiation in *Meloidogyne incognita* and anatomical evidence of sex reversal. *Journal of Nematology* 14, 549–566.

Parisy, V., Poinssot, B., Owsianowski, L., Buchala, A., Glazebrook, J. and Mauch, F. (2007) Identification of *PAD2* as a γ-glutamylcysteine synthetase highlights the importance of glutathione in disease resistance of *Arabidopsis*. *The Plant Journal* 49, 159–172. doi:10.1111/j.1365-313X.2006.02938.x

Pérez-Martinez, J., Ploetz, R.C. and Konkol, J.L. (2018) Significant *in vitro* antagonism of the laurel wilt pathogen by endophytic fungi from the xylem of avocado does not predict their ability to control the disease. *Plant Pathology* 67, 1768–1776. doi:10.1111/ppa.12878

Petrini, O. (1986) Taxonomy of endophytic fungi of areal plant parts. In: Fokkema, N.J.J. and Van den Heuvel, J. (eds) *Microbiology of the Phyllosphere*. Cambridge Universtiy Press, Cambridge, UK, pp. 175–187.

Petrini, O. (1991) Fungal endophytes of tree leaves. In: Andrews, J.H. and Hirano, S.S. (eds) *Microbial Ecology of Leaves*. Springer, New York, pp. 179–197.

Pineda, A., Zheng, S.J., Van Loon, J.J.A., Pieterse, C.M.J. and Dicke, M. (2010) Helping plants to deal with insects: The role of beneficial soil-borne microbes. *Trends Plant Science* 15, 507–514.

Proite, K., Carneiro, R., Falcão, R., Gomes, A., Leal-Bertioli, S., Guimarães, P. and Bertioli, D. (2008) Post-infection development and histopathology of *Meloidogyne arenaria* race 1 on *Arachis* spp. *Plant Pathology* 57, 974–980. doi:10.1111/j.1365-3059.2008.01861.x

Qiang, X., Weiss, M., Kogel, K.H. and Schafer, P. (2012) *Piriformospora indica* – A mutualistic basidiomycete with an exceptionally large plant host range. *Molecular Plant Pathology* 13, 508–518. doi:10.1111/j.1364-3703.2011.00764.x

Reay, S.D., Brownbridge, M., Gicquel, B., Cummings, N.J. and Nelson, T.L. (2010) Isolation and characterization of endophytic *Beauveria* spp. (Ascomycota: Hypocreales) from *Pinus radiata* in New Zealand forests. *Biological Control* 54, 52–60. doi:10.1016/j.biocontrol.2010.03.002

Redman, R.S., Freeman, S., Clifton, D.R., Morrel, J., Brown, G. and Rodriguez, R.J. (1999) Biochemical analysis of plant protection afforded by a nonpathogenic endophytic mutant of *Colletotrichum magna*. *Plant Physiology* 119, 795–804.

Redman, R.S., Dunigan, D.D. and Rodriguez, R.J. (2001) Fungal symbiosis from mutualism to parasitism: Who controls the outcome, host or invader? *New Phytologist* 151, 705–716.

Redman, R.S., Sheehan, K.B., Stout, R.G., Rodriguez, R.J. and Henson, J.M. (2002) Thermotolerance generated by plant/fungal symbiosis. *Science* 298, 1581–1581.

Redman, R.S., Kim, Y.O., Woodward, C.J.D.A., Greer, C., Espino, L., Doty, S.L. and Rodriguez, R.J. (2011) Increased fitness of rice plants to abiotic stress via habitat adapted symbiosis: A strategy for mitigating impacts of climate change. *PLOS ONE* 6, e14823. doi:10.1371/journal.pone.0014823

Rep, M., van der Does, H.C., Meijer, M., van Wijk, R., Houterman, P.M., *et al.* (2004) A small, cysteine-rich protein secreted by *Fusarium oxysporum* during colonization of xylem vessels is required for I-3-mediated resistance in tomato. *Molecular Microbiology* 53, 1373–1383.

Richmond, D.S., Kunkel, B.A., Somasekhar, N. and Grewal, P.S. (2004) Top-down and bottom-up regulation of herbivores: *Spodoptera frugiperda* turns tables on endophyte-mediated plant defence and virulence of an entomopathogenic nematode. *Ecological Entomology* 29, 353–360. doi:10.1111/j.1365-2311.2004.00598.x

Rodriguez, R.J., Henson, J., Van Volkenburgh, E., Hoy, M., Wright, L., *et al*. (2008) Stress tolerance in plants via habitat-adapted symbiosis. *ISME Journal* 2, 404–416.

Rodriguez, R.J., White, J.F., Arnold, A.E. and Redman, R.S. (2009) Fungal endophytes: Diversity and functional roles. *New Phytologist* 182, 314–330.doi:10.1111/j.1469-8137.2009.02773.x

Roetschi, A., Si-Ammour, A., Belbahri, L., Mauch, F. and Mauch-Mani, B. (2001) Characterization of an *Arabidopsis-Phytophthora* pathosystem: resistance requires a functional PAD2 gene and is independent of salicylic acid, ethylene and jasmonic acid signalling. *The Plant Journal* 28, 293–305. doi:10.1046/j.1365-313X.2001.01148.x

Rouhier, N., Lemaire, S.D. and Jacquot, J.P. (2008) The role of glutathione in photosynthetic organisms: Emerging functions for glutaredoxins and glutathionylation. *Annual Review of Plant Biology* 59, 143–166.

Rumbos, C.I. and Kiewnick, S. (2006) Effect of plant species on persistence of *Paecilomyces lilacinus* strain 251 in soil and on root colonization by the fungus. *Plant and Soil* 283, 25–31. doi:10.1007/s11104-005-6804-1

Saikkonen, K., Faeth, S.H., Helander, M. and Sullivan, T.J. (1998) Fungal endophytes: A continuum of interactions with host plants. *Annual Review of Ecology and Systematics* 29, 319–343. doi:10.1146/annurev.ecolsys.29.1.319

Sánchez Márquez, S., Bills, G.F., Herrero, N. and Zabalgogeazcoa, I. (2012) Non-systemic fungal endophytes of grasses. *Fungal Ecology* 5, 289–297. doi:10.1016/j.funeco.2010.12.001

Schäfer, P., Pfiffi, S., Voll, L.M., Zajic, D., Chandler, P.M., *et al*. (2009) Manipulation of plant innate immunity and gibberellin as factor of compatibility in the mutualistic association of barley roots with *Piriformospora indica*. *The Plant Journal* 59, 461–474.doi:10.1111/j.1365-313X.2009.03887.x

Schardl, C.L., Leuchtmann, A. and Spiering, M.J. (2004) Symbioses of grasses with seedborne fungal endophytes. *Annual Review of Plant Biology* 55, 315–340.

Schlaeppi, K., Bodenhausen, N., Buchala, A., Mauch, F. and Reymond, P. (2008) The glutathione-deficient mutant *pad2-1* accumulates lower amounts of glucosinolates and is more susceptible to the insect herbivore *Spodoptera littoralis*. *The Plant Journal* 55, 774–786. doi:10.1111/j.1365-313X.2008.03545.x

Schouten, A. (2016) Mechanisms involved in nematode control by endophytic fungi. *Annual Review of Phytopathology* 54, 121–142.

Schulz, B. and Boyle, C. (2005) The endophytic continuum. *Mycological Research* 109, 661–686.

Schulz, B., Wanke, U., Draeger, S. and Aust, H.J. (1993) Endophytes from herbaceous plants and shrubs: Effectiveness of surface sterilization methods. *Mycological Research* 97, 1447–1450. doi:10.1016/s0953-7562(09)80215-3

Schulz, B., Sucker, J., Aust, H.J., Krohn, K., Ludewig, K., Jones, P.G. and Döring, D. (1995) Biologically-active secondary metabolites of endophytic *Pezicula* species. *Mycological Research* 99, 1007–1015. doi:10.1016/s0953-7562(09)80766-1

Schulz, B., Guske, S., Dammann, U. and Boyle, C. (1998) Endophyte-host interactions. II. Defining symbiosis of the endophyte-host interaction. *Symbiosis* 25, 213–227.

Schulz, B., Haas, S., Junker, C., Andree, N. and Schobert, M. (2015) Fungal endophytes are involved in multiple balanced antagonisms. *Current Science* 109, 39–45.

Selim, M. (2010) Biological, chemical and molecular studies on the systemic induced resistance in tomato against *Meloidogyne incognita* caused by the endophytic *Fusarium oxysporum*, Fo162. PhD. thesis, University of Bonn, Bonn, Germany

Shaw, M.W., Emmanuel, C.J., Emilda, D., Terhem, R.B., Shafia, A., *et al*. (2016) Analysis of cryptic, systemic *Botrytis* infections in symptomless hosts. *Frontiers in Plant Science* 7, 625. doi:10.3389/fpls.2016.00625

Shittu, H.O., Castroverde, D.C.M., Nazar, R.N. and Robb, J. (2009) Plant-endophyte interplay protects tomato against a virulent *Verticillium*. *Planta* 229, 415–426. doi:10.1007/s00425-008-0840-z

Shoresh, M. and Harman, G.E. (2008) The relationship between increased growth and resistance induced in plants by root colonizing microbes. *Plant Signaling & Behavior* 3, 737–739.

Shoresh, M., Yedidia, I. and Chet, I. (2005) Involvement of jasmonic acid/ethylene signaling pathway in the systemic resistance induced in cucumber by *Trichoderma asperellum* T203. *Phytopathology* 95, 76–84. doi:10.1094/PHYTO-95-0076

Shoresh, M., Harman, G.E. and Mastouri, F. (2010) Induced systemic resistance and plant responses to fungal biocontrol agents. *Annual Review of Phytopathology* 48, 21–43. doi:10.1146/annurev-phyto-073009-114450

Siddique, S., Radakovic, Z.S., De La Torre, C.M., Chronis, D., Novak, O., *et al.* (2015) A parasitic nematode releases cytokinin that controls cell division and orchestrates feeding site formation in host plants. *Proceedings of the National Academy of Sciences of the United States of America* 112, 12669–12674. doi:10.1073/pnas.1503657112

Siddiqui, I.A. and Shaukat, S.S. (2004) Systemic resistance in tomato induced by biocontrol bacteria against the root-knot nematode, *Meloidogyne javanica* is independent of salicylic acid production. *Journal of Phytopathology* 152, 48–54.

Sieber, T., Riesen, T., Muller, E. and Fried, P. (1988) Endophytic fungi in 4 winter-wheat cultivars (*Triticum aestivum* L.) differing in resistance against *Stagonospora nodorum* (Berk.) Cast. and Germ. = *Septoria nodorum* (Berk.) Berk. *Journal of Phytopathology* 122, 289–306.

Sikora, R.A. (1992) Management of the antagonistic potential in agricultural ecosystems for the biological control of plant parasitic nematodes. *Annual Review of Phytopathology* 30, 245–270. doi:10.1146/annurev.py.30.090192.001333

Singh, S. and Mathur, N. (2010) *In vitro* studies of antagonistic fungi against the root-knot nematode, *Meloidogyne incognita*. *Biocontrol Science and Technology* 20, 275–282. doi:10.1080/09583150903484318

Singh, U.B., Sahu, A., Sahu, N., Singh, B.P., Singh, R.K., *et al.* (2013) Can endophytic *Arthrobotrys oligospora* modulate accumulation of defence related biomolecules and induced systemic resistance in tomato (*Lycopersicon esculentum* Mill.) against root knot disease caused by *Meloidogyne incognita*. *Applied Soil Ecology* 63, 45–56. doi:10.1016/j.apsoil.2012.08.007

Stone, J.K., Bacon, C.W. and White, J.F. Jr (2000) An overview of endophytic microbes: Endopytism defined. In: Bacon, C.W. and White, J.F. Jr (eds) *Microbial Endophytes*. Marcel Dekker, Inc., New York/Basel, Switzerland.

Stone, J.K., Polishook, J.D. and White, J.F. (2004) Endophytic fungi. In: Mueller, G.M., Bills, G.F. and Foster, M.S. (eds) *Biodiversity of Fungi*. Academic Press, Burlington, Massachusetts, pp. 241–270. doi:10.1016/B978-012509551-8/50015-5

Tan, R.X. and Zou, W.X. (2001) Endophytes: A rich source of functional metabolites. *Natural Product Reports* 18, 448–459. doi:10.1039/b100918o

Tian, X., Yao, Y., Chen, G., Mao, Z., Wang, X. and Xie, B. (2014) Suppression of *Meloidogyne incognita* by the endophytic fungus *Acremonium implicatum* from tomato root galls. *International Journal of Pest Management* 60, 239–245. doi:10.1080/09670874.2014.958604

Tian, X.L., Cao, L.X., Tan, H.M., Zeng, Q.G., Jia, Y.Y., Han, W.Q. and Zhou, S.N. (2004) Study on the communities of endophytic fungi and endophytic actinomycetes from rice and their antipathogenic activities *in vitro*. *World Journal of Microbiology & Biotechnology* 20, 303–309.

Trillas, M. and Segarra, G. (2009) Interactions between nonpathogenic fungi and plants. In: Van Loon, L.C. (ed.) *Advances in Botanical Research*. Elsevier Ltd., Amsterdam/New York. doi:10.1016/S0065-2296(09) 51008–7

van der Does, H.C., Duyvesteijn, R.G.E., Goltstein, P.M., van Schie, C.C.N., Manders, E.M.M., Cornelissen, B.J.C. and Rep, M. (2008a) Expression of effector gene SIX1 of *Fusarium oxysporum* requires living plant cells. *Fungal Genetics and Biology* 45, 1257–1264. doi:10.1016/j.fgb.2008.06.002

van der Does, H.C., Lievens, B., Claes, L., Houterman, P.M., Cornelissen, B.J.C. and Rep, M. (2008b) The presence of a virulence locus discriminates *Fusarium oxysporum* isolates causing tomato wilt from other isolates. *Environmental Microbiology* 10, 1475–1485. doi:10.1111/j.1462-2920.2007.01561.x

van Loon, L.C., Bakker, P.A.H.M. and Pieterse, C.M.J.. (1998) Systemic resistance induced by rhizosphere bacteria. *Annual Review of Phytopathology* 36, 453–83.

VanEtten, H., Jorgensen, S., Enkerli, J. and Covert, S.F. (1998) Inducing the loss of conditionally dispensable chromosomes in *Nectria haematococca* during vegetative growth. *Current Genetics* 33, 299–303. doi:10.1007/s002940050340

Vargas, W.A., Djonović, S., Sukno, S.A. and Kenerley, C.M. (2008) Dimerization controls the activity of fungal elicitors that trigger systemic resistance in plants. *Journal of Biological Chemistry* 283, 19804–19815. doi:10.1074/jbc.M802724200

Viterbo, A., Harel, M., Horwitz, B.A., Chet, I. and Mukherjee, P.K. (2005) Trichoderma mitogen-activated protein kinase signaling is involved in induction of plant systemic resistance. *Applied and Environmental Microbiology* 71, 6241–6246. doi:10.1128/aem.71.10.6241-6246.2005

Vos, I.A., Moritz, L., Pieterse, C.M. and Van Wees, S.C. (2015) Impact of hormonal crosstalk on plant resistance and fitness under multi-attacker conditions. *Frontiers in Plant Science* 6, 639. doi:10.3389/fpls.2015.00639

Vu, T., Hauschild, R. and Sikora, R.A. (2006) *Fusarium oxysporum* endophytes induced systemic resistance against *Radopholus similis* on banana. *Nematology* 8, 847–852.

Wani, Z.A., Ashraf, N., Mohiuddin, T. and Riyaz-Ul-Hassan, S. (2015) Plant-endophyte symbiosis, an ecological perspective. *Applied Microbiology and Biotechnology* 99, 2955–2965. doi:10.1007/s00253-015-6487-3

Wasmann, C.C. and VanEtten, H.D. (1996) Transformation-mediated chromosome loss and disruption of a gene for pisatin demethylase decrease the virulence of *Nectria haematococca* on pea. *Molecular Plant-Microbe Interactions* 9, 793–803. doi:10.1094/mpmi-9-0793

Weyens, N., van der Lelie, D., Taghavi, S. and Vangronsveld, J. (2009) Phytoremediation: Plant-endophyte partnerships take the challenge. *Current Opinion in Biotechnology* 20, 248–254. doi:10.1016/j.copbio.2009.02.012

White, J.F. Jr, Belanger, F., Meyer, W., Sullivan, R.F., Bischoff, J.F. and Lewis, E.A. (2002) Clavicipitalean fungal epibionts and endophytes – Development of symbiotic interactions with plants. *Symbiosis* 33, 201–213.

Widinugraheni, S., Niño-Sánchez, J.. van der Does, H.C., van Dam, P., Garcia-Bastidas, F.A., *et al.* (2018) A SIX1 homolog in *Fusarium oxysporum* f.sp *cubense* tropical race 4 contributes to virulence towards Cavendish banana. *PLOS ONE* 13, e0205896. doi:10.1371/journal.pone.0205896

Wilson, D. (1995) Endophyte – The evolution of a term, and clarification of its use and definition. *Oikos* 73, 274–276. doi:10.2307/3545919

Win, P.M., Matsumura, E. and Fukuda, K. (2018) Diversity of tea endophytic fungi: Cultivar- and tissue preferences. *Applied Ecology and Environmental Research* 16, 677–695. doi:10.15666/aeer/1601_677695

Wu, H.-S., Liu, D.-Y., Ling, N., Bao, W., Ying, R.-R. and Shen, Q.-R. (2009) Influence of root exudates of watermelon on *Fusarium oxysporum* f. sp. *niveum*. *Soil Science Society of America Journal* 73, 1150–1156. doi:10.2136/sssaj2008.0266

Wu, L.S., Han, T., Li, W.C., Jia, M., Xue, L.M., Rahman, K. and Qin, L.P. (2013) Geographic and tissue influences on endophytic fungal communities of *Taxus chinensis* var. *mairei* in China. *Current Microbiology* 66, 40–48. doi:10.1007/s00284-012-0235-z

Wuyts, N., Swennen, R. and De Waele, D. (2006) Effects of plant phenylpropanoid pathway products and selected terpenoids and alkaloids on the behaviour of the plant-parasitic nematodes *Radopholus similis, Pratylenchus penetrans* and *Meloidogyne incognita*. *Nematology* 8, 89–101. doi:10.1163/156854106776179953

Xiang, C.B., Werner, B.L., Christensen, E.M. and Oliver, D.J. (2001) The biological functions of glutathione revisited in *Arabidopsis* transgenic plants with altered glutathione levels. *Plant Physiology* 126, 564–574.

Yan, X.N., Sikora, R.A. and Zheng, J.W. (2011) Potential use of cucumber (*Cucumis sativus* L.) endophytic fungi as seed treatment agents against root-knot nematode *Meloidogyne incognita*. *Journal of Zhejiang University, Science, B* 12, 219–225. doi:10.1631/jzus.B1000165

Yao, Y.R., Tian, X.L., Shen, B.M., Mao, Z.C., Chen, G.H. and Xie, B.Y. (2015) Transformation of the endophytic fungus *Acremonium implicatum* with GFP and evaluation of its biocontrol effect against *Meloidogyne incognita*. *World Journal of Microbiology & Biotechnology* 31, 549–556. doi:10.1007/s11274-014-1781-2

Yedidia, I., Benhamou, N. and Chet, I. (1999) Induction of defense responses in cucumber plants (*Cucumis sativus* L.) by the biocontrol agent *Trichoderma harzianum*. *Applied and Environmental Microbiology* 65, 1061–1070.

Yedidia, I., Benhamou, N., Kapulnik, Y. and Chet, I. (2000) Induction and accumulation of PR proteins activity during early stages of root colonization by the mycoparasite *Trichoderma harzianum* strain T-203. *Plant Physiology and Biochemistry* 38, 863–873. doi:10.1016/s0981-9428(00)01198-0

Yin, J., Chin, C., Ye, J., Zhao, W. and Li, G. (2012) An effective asparagus stem blight management program. *Acta Horticulturea* 950, 293–298. doi:10.17660/ActaHortic.2012.950.34

Zamioudis, C. and Pieterse, C.M.J. (2012) Modulation of host immunity by beneficial microbes. *Molecular Plant-Microbe Interactions* 25, 139–150.

Zhang, H.W., Song, Y.C. and Tan, R.X. (2006) Biology and chemistry of endophytes. *Natural Product Reports* 23, 753–771. doi:10.1039/b609472b

Zhang, Q., Zhang, J., Yang, L., Zhang, L., Jiang, D., Chen, W. and Li, G. (2014) Diversity and biocontrol potential of endophytic fungi in *Brassica napus*. *Biological Control* 72, 98–108. doi:10.1016/j.biocontrol.2014.02.018

3 Sources, Niches and Routes of Colonization by Beneficial Bacterial Endophytes

Stéphane Compant[1]* and Corinne Vacher[2]

[1]Bioresources Unit, Center for Health & Bioresources, AIT Austrian Institute of Technology GmbH, Tulln, Austria; [2]BIOGECO, INRA, University of Bordeaux, Pessac, France

Abstract

The plant individual is a holobiont as it hosts diverse microbial assemblages in and on vegetative, reproductive or disseminative organs. All plant compartments – roots, stems, leaves, flowers, fruits and seeds – have been shown to host microorganisms that can influence positively or negatively the plants performance. Some of these microorganisms thrive as endophytes inside plant tissues. Identifying the environmental sources of these microorganisms and the route they take to colonize plant tissues, visualizing their niches within their hosts and understanding how they make intimate associations with plants are of crucial importance in developing biocontrol and biofertilization approaches, both in organic and integrated protection systems. This chapter considers that the plant individual is part of a complex network of biotic interactions influenced by the environment in the phytobiome and provides a comprehensive review on the development of the interactions between plants and beneficial bacterial endophytes.

3.1 Introduction

Since the 19th century and the discovery of endophytes, i.e. microbes living inside plants, efforts have been made to identify which microbial taxa and assemblages inhabit plant tissues, depending on plant species and environmental conditions (Hardoim *et al.*, 2015). How the tissues are colonized, how the microorganisms are acquired from the environment and how they are transmitted from one generation to the next, and what influence they have on their hosts have been further revealed during the last decades (Compant *et al.*, 2016; Brader *et al.*, 2017; Kandel *et al.*, 2017). While some colonizers are known as pathogenic, others have been acknowledged as mutualists (Lemanceau *et al.*, 2017). The latter are of special interest for agriculture as they can improve agroecosystem health and productivity by alleviating abiotic stresses, reducing pathogen attacks and stimulating plant growth. A thorough understanding of the sources, niches and colonization routes of beneficial bacterial endophytes is required, however, for their successful application on crops and plantation forests (Turner *et al.*, 2013).

Most research performed so far on beneficial bacterial endophytes has focused on bacteria deriving from the rhizosphere and colonizing root tissues. However, other microenvironments on the plant surfaces, such as the anthosphere, carposphere, spermosphere, phyllosphere, calusphere, caulosphere or laimosphere, can also host beneficial microorganisms and constitute entry points toward internal plant tissues (Compant *et al.*, 2011,

* Corresponding author e-mail: stephane.compant@ait.ac.at

2016; Vacher *et al.*, 2016; Lemanceau *et al.*, 2017; Nelson *et al.*, 2018). As plants interact both with their abiotic environment and complex communities of organisms (forming the phytobiome; Beans, 2017; Leach *et al.*, 2017), beneficial endophytes can also come in other ways, such as neighbouring plants, and be transmitted by wind or animal vectors (Vacher *et al.*, 2016). In this chapter, we first review the knowledge on below-ground sources of colonization. We describe how the soil bacteria can reach the internal root tissues, and then eventually the above-ground plant organs. We then review the knowledge on the above-ground sources of colonization, which have been less studied so far. The colonization through natural openings in above-ground vegetative organs and the role played by insect vectors are discussed. Finally, a focus on the colonization of reproductive and disseminative organs (flowers, fruits and seeds) and the possibility of transmission of beneficial bacteria to the plant offspring is provided.

3.2 Below-ground Colonization Routes

3.2.1 From soil to inside roots

The soil is the main reservoir for bacterial endophytes. Many bacterial endophytes originate from the rhizosphere, the microenvironment surrounding roots, which is influenced by the presence of root exudates, rhizodeposits and microorganisms (Mendes *et al.*, 2013). Up to 40% of the photosynthates produced by the plant can be released in the rhizosphere, thereby attracting many microorganisms from the surrounding bulk soil (Lugtenberg and Kamilova, 2009). Some bacteria can further colonize specific zones of the root surfaces (i.e. the rhizoplane) as single or several cells (Fig. 3.1A–C) or by forming biofilms (Benizri *et al.*, 2001). Advanced visualization techniques revealed that strains can form lines along the grooves between root cells, multiply on them and then enter plant root tissues between two rhizodermal cells (Compant *et al.*, 2010; Fig. 3.1A–C). It has further been shown that *Pseudomonas fluorescens* 95rkG5 can colonize tomato roots in the root elongation zone (as single or dividing cells), in the root hair zone (as single or clustered cells) and in the collar zone, but not in the root tip zone (Gamalero *et al.*, 2005). Other bacteria are able to colonize root tips before establishing subcommunities inside root tissues (Brader *et al.*, 2017), as recently demonstrated for a strain isolated from grapevine plants and re-inoculated on *in vitro* plantlets (Lòpez-Fernàndez *et al.*, 2016).

Bacterial endophytes can actively or passively penetrate root tissues, depending on the strain and the root zone being colonized (James *et al.*, 2001; Mercado-Blanco, 2015). Several bacterial traits can favour endophytism, such as pili, flagella, fimbriae, nod factors, quorum sensing, cell-wall-degrading enzymes, twitching motility, lipopolysaccharide, in addition to several traits required for rhizosphere competence (Compant *et al.*, 2010). Interestingly, some (non-nodulating) bacteria have been visualized colonizing the interior of root hairs. For instance, *P. fluorescens* PICF7 and *Pseudomonas putida* PICP2 were observed inside root hairs of olive trees (Prieto *et al.*, 2011). Bacteria were detected as either single cells or clusters (colonies) attached to inner membrane structures of a few root hairs, before they reached the cortical cell layers (Prieto *et al.*, 2011). Once inside the rhizodermis, some bacteria colonize intercellular spaces while a few can be intracellular inside the cortical cell layers. James *et al.* (2002) showed, for instance, that a strain of *Herbaspirillum seropedicae* entered the roots through cracks at the point of lateral root emergence of rice plantlets. This strain subsequently colonized the root intercellular spaces, aerenchyma and cortical cells, and a few cells penetrated the stele and entered into the xylem vessels.

3.2.2 From roots to above-ground organs

Some bacteria can further progress within root tissues. Some of them remain in the root cortex, while others can reach the endodermis barrier and pass to the central cylinder up to the xylem vessels (Compant *et al.*, 2010; Fig. 3.1D). Once inside xylem vessels, some bacteria can pass from one element to the other (Fig. 3.1D) through the perforated

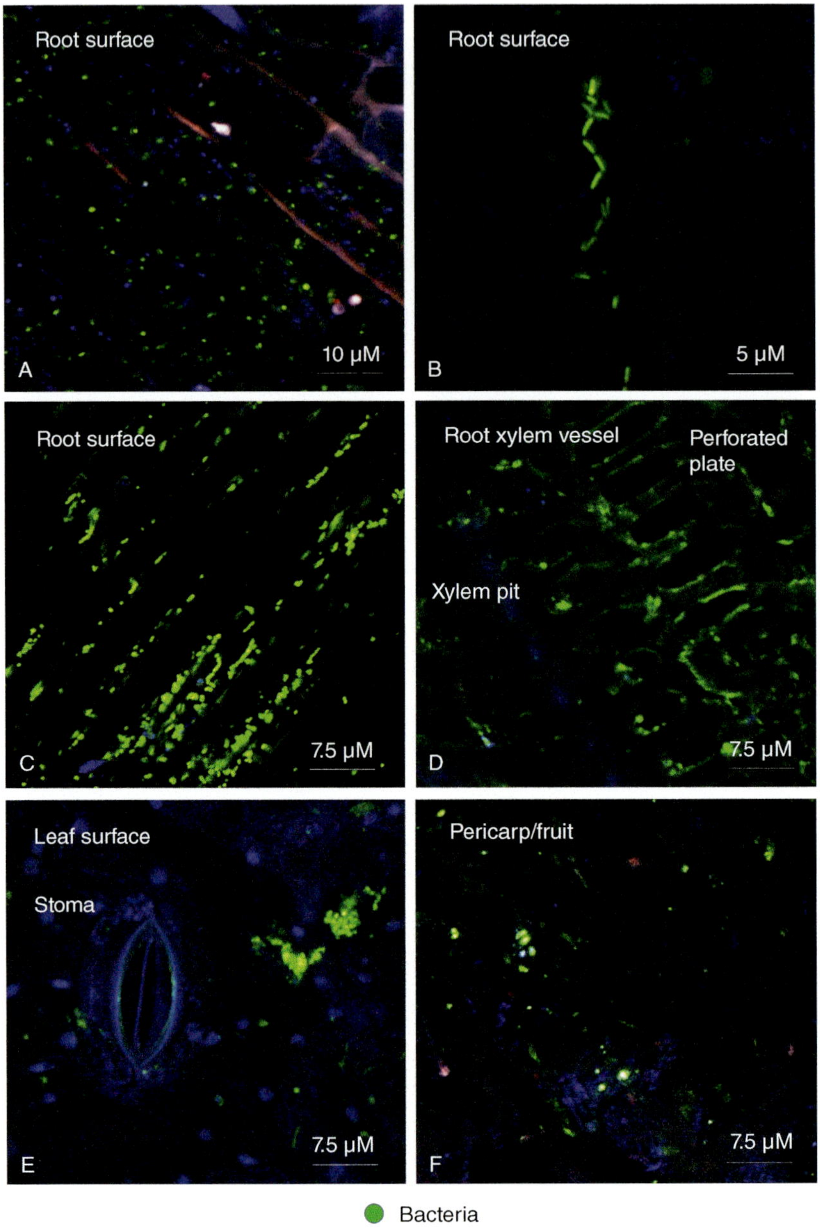

Fig. 3.1. Confocal microscopy photographs of bacteria (green fluorescent stained with Syto9®) colonizing root surfaces of a plant (A–B), or being inside root tissues (D), on leaf surfaces (E) and inner fruit tissues (F) of grapevine natural microbial communities (A–B, D–F) or salad inoculated with a beneficial strain (C).

plates of the xylem vessels, thus colonizing the plant systemically (Compant *et al.*, 2005, 2008). For instance, *Paraburkholderia phytofirmans* strain PsJN was visualized from roots up to the infructescence tissues of grapevine plants, after being inoculated into the soil (Compant *et al.*, 2008). *Azorhizobium caulinodans* ORS571 has been observed colonizing rice from the roots up to some leaves (Chi *et al.*, 2005). In poplar, *Pseudomonas* sp. PopHV6 was found in cuttings 10 weeks after cutting inoculation

(Germaine *et al.*, 2004). However, not all bacteria are able to migrate from below-ground to above-ground plant organs. A strain of *Pantoea agglomerans* 33.1 tagged with the *gfp* was not recovered from inside leaf tissues after seedlings inoculation but was found inside root and stem tissues of *Eucalyptus grandis* and the hybrid *E. grandis* × *E. globulus* (Ferreira *et al.*, 2008).

It is not yet clear why some bacteria are attracted by xylem vessels. However, an interesting study by Malfanova *et al.* (2013) demonstrated that some sugar compounds, especially l-arabinose, were present as traces inside xylem vessels of cucumber plants. These sugars could enable survival inside xylem of some endophytes such as *Pseudomonas* spp. strains. The time for bacterial colonization of above-ground plant organs, e.g. weeks or months (James *et al.*, 2002; Compant *et al.*, 2005), suggests that the perforated plates slow down the spread of the endophytes. Nevertheless, some bacteria finally reach sub-stomatal chambers in leaves, as they are close to xylem vessels inside leaf tissues (Compant *et al.*, 2010). For instance, *P. phytofirmans* strain PsJN has been observed inside substomatal chambers of grapevine leaves after dissemination inside the plant (Compant *et al.*, 2005). The cells even exited from stomata to attack the fungal pathogen *Botrytis cinerea* growing on leaves (Miotto-Vilanova *et al.*, 2016). These observations suggest that the plant can select endophytes to improve its resistance to pathogen attacks (Berg, 2009). Other bacteria have been observed in sub-stomatal chambers, such as the strain Z67 of *H. seropedicae* (James *et al.*, 2001, 2002). However, in this case, the possibility of a bacterial colonization from the phylloplane was not excluded (James *et al.*, 2001, 2002).

3.3 Above-ground Colonization Routes

3.3.1 Colonization through stomata, other natural openings and wounds

Diversified bacterial assemblages thrive on the surface of the plant aerial organs. A multitude of bacteria colonize the stem external environment and its surface, the caulosphere, the one from the bud, the calusphere, and the leaf one, the phyllosphere (Vorholt, 2012; Vacher *et al.*, 2016). These above-ground bacteria are less studied than those of the rhizosphere and have long been neglected in plant ecology, despite substantial evidence of the link between phyllosphere microbial communities and crucial functional traits such as photosynthetic strategy, hydraulics, reproduction or defence (Rosado *et al.*, 2018). For example, bacteria on the leaf surface can alter cuticular permeability and thus plant water loss through transpiration. They can also protect the plant against pathogens by competing with them or by priming the plant immune system (Remus-Emsermann and Schlechter, 2018). Phyllosphere microorganisms can be deposited on the leaf surface by wind, rainwater and irrigation water, or insects or they can colonize the flushing leaf after overwintering on twigs or in buds (Vacher *et al.*, 2016). Those that survive the selection exerted by the leaf microclimate and foliar traits can multiply and enter the tissues via stomata, hydathodes or wounds (Vorholt, 2012; Vacher *et al.*, 2016; Fig. 3.1E). Caulosphere bacteria can colonize stem internal tissues through lenticels, or stomata present on photosynthetic stems (Hardoim *et al.*, 2015; Brader *et al.*, 2017). Recently, a FISH analysis showed the aggregation of *Methylobacterium* PA1 cells in the substomatal chambers of *Arabidopsis thaliana*, after a phase of colonization of the phyllosphere (Peredo and Simmons, 2018). Another example of beneficial bacteria able to colonize the plant internal tissues through stomata is the nodulating bacteria belonging to the genus *Paraburkholderia*. These bacteria induce the formation of leaf nodules in about 450 dicotyledonous plant species, mostly growing in the tropics. They do not fix nitrogen but produce secondary metabolites that protect plants from herbivory. These endophytic bacteria are obligate symbionts and are maintained in a mucilage layer in buds and colonize young leaves through stomata. They are then transmitted to the plant offspring by colonizing inflorescences and then seeds (Pinto-Carbó *et al.*, 2018).

The ability to colonize the leaf internal tissues from the leaf surface is restricted to a few bacteria. Not all strains of non-pathogenic bacterial species can thrive as endophytes inside leaf internal tissues (Wilson *et al.*, 1999; Sabaratnam and Beattie, 2003). Moreover, the penetration of bacteria into leaf internal tissues through stomata is a process regulated by both the plant and the environment. Stomata can close when pathogenic bacteria are recognized by the plant, but this mechanism of plant defence is less effective when the relative humidity is high (Panchal and Melotto, 2018). Water on the leaf surface is indeed a key factor of the colonization process, since it influences both leaf physiology and the development of phyllosphere microorganisms. Water promotes the diffusion of nutrients to the leaf surface, through the cuticle, and can form films that create a connection between the leaf surface and the substomatal chambers (Vacher *et al.*, 2016; Dawson and Goldsmith, 2018). Heat can also influence the efficiency of colonization. It has been shown to reduce the number of leaf nodules in tropical plant species (Pinto-Carbó *et al.*, 2018). Symbiosis establishment should be considered therefore as a tripartite interaction between the plant, the endophytic bacterium and the abiotic environment.

3.3.2 Introduction by animal vectors

Endophytic bacteria can also be introduced into the internal tissues of above-ground plant organs by insect herbivores. For example, the leafhopper *Scaphoideus titanus*, which feeds on phloem sap in vines, is known to transmit the phytoplasma responsible for flavescence dorée. It has recently been shown that it can also transmit endophytic bacteria from one plant to the other by feeding on stems. The endophytic bacteria are then able to spread from the stems to the roots (Lòpez-Fernàndez *et al.*, 2017). Similar results were obtained for the phloem-sucking insect *Hyalesthes obsoletus*. This insect species carries a bacterial species capable of colonizing the phloem and reducing symptoms caused by phytoplasmas. Interestingly, this protective bacterium can be effectively introduced into vine plants by spraying the leaves with a bacterial culture, confirming that the leaves are a possible entry point for endophytic bacteria (Iasur-Kruh *et al.*, 2018). Interestingly, the bacterium *Propionibacterium acnes* type Zappa has been observed further colonizing the bark, the pith, and xylem vessels of several grapevine plants using FISH microscopy. Its role is unknown, but it has established a subpopulation in grapevine since the neolithic period and has diverged from human-associated populations of *P. acnes*, suggesting that not only insects but also other animals (including humans) can introduce endophytes inside plants (Campisano *et al.*, 2014).

3.3.3 Transmission from plants to plants

Plant individuals are usually not isolated. They live in association with other plant individuals, belonging or not to the same species. Neighbouring plants can be a reservoir of plant pathogens, but also a reservoir of beneficial endophytes. Samad *et al.* (2017) showed, for instance, that similar microbial taxa were associated with vines and weeds growing in the same vineyard and that some taxa had beneficial properties such as auxin, siderophore and HCN production and also some properties that would be of interest for biocontrol approaches. More information is currently needed, however, on the possibility for direct and indirect (through vectors) transmission of endophytic bacteria from one plant to its neighbour and on the influence of phylogenetic relatedness between plant species on the rate of transmission.

3.4 Colonization of Reproductive and Disseminative Organs and Vertical Transmission

3.4.1 Colonization of flowers

Flowers, fruits and seeds were considered as sterile up to the 2000s (Hallmann, 2001) due to very few isolates that could be isolated and cultivated, but there is increasing evidence that they are also colonized by beneficial bacterial endophytes. Flowers host diversified microbial assemblages, and endophytic bacteria have been visualized in their ovaries, epidermis and xylem tissues. For instance, bacteria belonging to the *Pseudomonas* and *Bacillus* genera have been observed in grapevine flowers using FISH microscopy (Compant *et al*., 2011). Most of the bacterial colonizers of flowers derive from the anthosphere, the external microenvironment of flowers (Compant *et al*., 2011). Bacteria have been observed colonizing inner flower tissues through stomata present on the surfaces of preflower buds (Compant *et al*., 2011). Bacteria can also enter inside flower tissues throughout stigma, by using pollen as a vector (Escobar-Rodríguez *et al*., 2018). Fürnkranz *et al*. (2012) showed, for instance, the presence of bacteria on pollen of pumpkin flowers. Ambika Manirajan *et al*. (2016) further visualized the presence of bacteria on pollen of birch, rye, rapes and autumn crocus. Pollen grains support diverse bacterial communities, the composition of which depends on the plant species and pollination type (Manirajan *et al*., 2018). The role of these bacteria in the reproductive process and their ability to be transmitted vertically are, however, poorly understood. To date, only a few studies suggest a possibility of transmission of endophytic bacteria from pollen to offspring (reviewed by Franck *et al*., 2017). It is highly possible that pollinator insects also transport bacteria from flower to flower, and some bacteria could then penetrate into plant tissues (Junker *et al*., 2011).

3.4.2 Colonization of fruits

Bacteria have been observed in the pericarp zones of fruits, corresponding to exo-, meso- and endocarp tissues (Fig. 3.1E). A differential colonization was revealed, depending on strains, taxa and preflower colonization sites (Glassner *et al*., 2015). These authors studied the colonization routes of native bacteria within fruits of several Cucurbitaceae and suggested that they derive from flowers. Some of these strains have biocontrol properties against melon pathogens (Glassner *et al*., 2015). Fruits can also have microwounds due to external factors such as insects or wind, allowing some carpospheric and carpoplane microorganisms to endophytically colonize fruits. Some bacteria can further derive from the soil as they have been visualized inside xylem vessels, albeit the soil being not the most important source of colonization of flowers and fruits (Compant *et al*., 2011).

3.4.3 Colonization of seeds and transmission of endophytic bacteria to the plant offspring

Bacteria have been visualized inside seeds as well, both inside the embryo (in the cotyledon and the root-hypocotyl tissues) and in the seed coat (Escobar-Rodríguez *et al*., 2018; Glassner *et al*., 2018). For instance, Compant *et al*. (2011) located bacteria by FISH analyses inside the tegument of seeds of grapevine. Non-culturable endophytic bacteria were also detected in seeds of cactus *Mammillaria fraileana* by scanning electron microscopy, and FISH enabled the location of bacteria inside the seed coat and embryo (Lopez *et al*., 2011). Glassner *et al*. (2018) further visualized bacteria inside the seed coat, cotyledon and root-hypocotyl tissues of melon seeds, and Escobar-Rodríguez *et al*. (2018) also described the presence of bacteria inside tomato seeds, especially on the root surface of the embryo, the cotyledon and the seed coat. Similar findings were obtained for kernels of wheat (Escobar-Rodríguez *et al*., 2018) and other plants such as *Anadenanthera colubrina* (Alibrandi et *al*., 2018).

These seed endophytes can derive from the tissues of either flowers or fruits, depending on their location. Bacteria present inside the seed coat might derive from the fruit and colonize the seed when it is still immature. Bacteria inside the cotyledon and root-hypocotyl embryo might derive from the colonized ovule at flowering. This could be the same for the perisperm/endosperm envelope surrounding the embryo. This thin layer can be massively colonized by bacteria as it is rich in nutritive compounds that can be used by bacteria (Glassner *et al*., 2018), and we can expect that among all bacteria some could be beneficial for their hosts.

Interestingly, the routes of colonization from flowers to seeds have been further revealed by using a beneficial strain of *P. phytofirmans*. This strain, PsJN, has been found in seeds (including the embryo) of cereals and other plants after flower inoculation (Mitter *et al.*, 2017) and can promote plant growth of the offspring. However, the strain was not recovered in the seeds of the offspring. This is not surprising, however, as different sources of colonization exist, and due to various environmental conditions and plant status, different routes of colonization can lead to different bacterial assemblages inside seeds (Escobar-Rodríguez *et al.*, 2018). Seeds harbour, however, a core microbiome with some bacteria being transmitted from one generation to the next with some of them helping the new plant generation (Nelson, 2018).

Once seeds reach the soil and germinate, some bacteria present from the soil can colonize the spermosphere, the microenvironment surrounding the seed once the seed has germinated. Similar to the rhizosphere, this zone contains exudates and other compounds attracting microbes (Nelson, 2018). After colonizing the spermoplane, the microbial colonizers then can enter inside plant tissues at the root, stem and crown levels using the same routes described before from the soil to inner root tissues. Bacteria have been also visualized as entering breaches due to stem and root development from the seeds. However, some bacteria can also exit from the seed when it germinates and colonize the soil surrounding the plant. Yang *et al.* (2017) showed that the seed microbiota has an early impact on the soil microbiota composition, while later on other colonizers replace the original population. They also demonstrated that some seed endophytes, belonging to the *Enterobacteriaceae* and *Paenibacillaceae* families, were abundant in plant roots in axenic systems but became less abundant when plants were grown in natural soil. Recently, Rahman *et al.* (2018) showed, with a FISH approach on young roots of barley plants grown under sterile conditions and from surface-sterilized seeds, a dense bacterial colonization from the root tip to the root hair zone after germination, suggesting the vertical transmission of some bacteria.

3.5 Conclusions

Microorganisms isolated from the internal tissues of plants (that are endophytes) have long been considered as a contaminant. However, several decades of research demonstrated that these endophytes readily interact with the plant host and may or may not improve its performance, depending on the physiological and genetic status of the plant, the microorganism, soil characteristics and the abiotic environmental conditions. Albeit some of them could be beneficial for the plant, many bacterial endophytes do not provide the expected beneficial effects once applied in the field, due to screening bias or ineffective colonization caused by non-appropriate environmental conditions and microhabitats (Compant *et al.*, 2010). A current challenge is to place the beneficial bacteria at the right place and time to improve agroecosystem health and productivity. To do so, knowledge on the environmental sources of beneficial bacteria, their routes of colonization and their niches within the plant is required. Here we showed that beneficial bacterial endophytes can derive from the soil, the surfaces of flowers, fruits and seeds and can also be transmitted by insects and other animals, other plants and humans. Interestingly, some bacterial endophytes originate from the soil, enter into the internal tissues of plant roots and then move up to the aerial plant organs through xylem vessels. These bacteria constitute a functional linkage between the below-ground and above-ground subsystems that should be included in future network analyses aimed at predicting the response of phytobiomes to global change (Ramirez *et al.*, 2018). Overall, future research on beneficial bacterial endophytes should take into account the various components of phytobiomes. It should go beyond the understanding of the molecular interactions enabling the plant to recognize the bacterial endophytes, and of the mechanisms enabling the bacteria to benefit the plant. Research at the community and ecosystem levels is needed. For instance, the environmental reservoirs of beneficial bacterial endophytes should be better explored. The impact of the abiotic environment on the success of colonization and on the output of the interaction should be further assessed.

Acknowledgements

Corinne Vacher received funding from the INRA metaprogramme 'Meta-Omics of Microbial Ecosystems' (MEM), the LABEX COTE (ANR-10-LABX-45) and the LABEX CEBA (ANR-10-LABX-25-01) for her research on the phyllosphere microbiota.

References

Alibrandi, P., Cardinale, M., Rahman, M.M., Strati, F., Ciná, P. *et al.* (2018) The seed endosphere of *Anadenanthera colubrina* is inhabited by a complex microbiota, including *Methylobacterium* spp. and *Staphylococcus* spp. with potential plant-growth promoting activities. *Plant and Soil* 422(1–2),81–99.doi:10.1007/s11104-017-3182-4

Ambika Manirajan, B., Ratering, S., Rusch, V., Schwiertz, A., Geissler-Plaum, R. *et al.* (2016) Bacterial microbiota associated with flower pollen is influenced by pollination type, and shows a high degree of diversity and species-specificity. *Environmental Microbiology* 18, 5161–5174. doi:10.1111/1462-2920.13524

Beans, C. (2017) Core concept: Probing the phytobiome to advance agriculture. *Proceedings of the National Academy of Sciences of the United States of America* 22, 114. doi:10.1073/pnas.1710176114

Benizri, E., Baudoin, E. and Guckert, A. (2001) Root colonization by inoculated plant growth rhizobacteria. *Biocontrol Science and Technology* 11, 557–574.

Berg, G. (2009) Plant-microbe interactions promoting plant growth and health: Perspectives for controlled use of microorganisms in agriculture. *Applied Microbiology and Biotechnology* 84, 11–18. doi:10.1007/s00253-009-2092-7

Brader, G., Compant, S., Vescio, K., Mitter, B., Trognitz, F. *et al.* (2017) Ecology and genomic insights of plant-pathogenic and -nonpathogenic endophytes. *Annual Review of Phytopathology* 55, 61–83. doi:10.1146/annurev-phyto-080516-035641

Campisano, A., Ometto, L., Compant, S., Pancher, M., Antonielli, L. *et al.* (2014) Interkingdom transfer of the acne causing agent, *Propionibacterium acnes*, from human to grapevine. *Molecular Biology and Evolution* 31, 1059–1065. doi:10.1093/molbev/msu075

Chi, F., Shen, S.H., Cheng, H.P., Jing, Y.X., Yanni, Y.G. *et al.* (2005) Ascending migration of endophytic rhizobia, from roots to leaves, inside rice plants and assessment of benefits to rice growth physiology. *Applied and Environmental Microbiology* 71(11), 7271–7278.

Compant, S., Reiter, B., Sessitsch, A., Nowak, J., Clément, C. *et al.* (2005) Endophytic colonization of *Vitis vinifera* L. by plant growth-promoting bacterium *Burkholderia* sp. strain PsJN. *Applied and Environmental Microbiology* 71, 1685–1693.

Compant, S., Kaplan, H., Sessitsch, A., Nowak, J., Ait Barka, E. *et al.* (2008) Endophytic colonization of *Vitis vinifera* L. by *Burkholderia phytofirmans* strain PsJN: From the rhizosphere to inflorescence tissues. *FEMS Microbiology Ecology* 63, 84–93.

Compant, S., Clément, C. and Sessitsch, A. (2010) Plant growth-promoting bacteria in the rhizo- and endosphere of plants. Their role, colonization, mechanisms involved and prospects for utilization. *Soil Biology and Biochemistry* 42, 669–678.

Compant, S., Mitter, B., Coli-Mull, J.G., Gangl, H. and Sessitsch, A. (2011) Endophytes of grapevine flowers, berries, and seeds: Identification of cultivable bacteria, comparison with other plant parts, and visualization of niches of colonization. *Microbial Ecology* 62, 188–197. doi:10.1007/s00248-011-9883-y

Compant, S., Saikkonen, K., Mitter, B., Campisano, A. and Mercado-Blanco, J. (2016) Editorial special issue: Soil, plants and endophytes. *Plant and Soil* 405(1), 1–11. doi:10.1007/s11104-016-2927-9

Dawson, T.E. and Goldsmith, G.R. (2018) The value of wet leaves. *New Phytologist* 219(4), 1156–1169. doi:10.1111/nph.15307

Escobar-Rodríguez, C., Mitter, B., Barret, M., Sessitsch, A. and Compant, S. (2018) Commentary: Seed bacterial inhabitants and their routes of colonization. *Plant and Soil* 422(1–2), 129–134. doi:10.1007/s11104-017-3368-9

Ferreira, A., Quecine, M., Lacava, P., Oda, S., Azevedo, J. *et al.* (2008) Diversity of endophytic bacteria from *Eucalyptus* species seeds and colonization of seedlings by *Pantoea agglomerans*. *FEMS Microbiology Letters* 287, 8–14. doi:10.1111/j.1574-6968.2008.01258.x

Franck, A.C., Saldierna Guzmán, J.P. and Shay, J.E. (2017) Transmission of bacterial endophytes. *Microorganisms* 5(4), 70. doi:10.3390/microorganisms5040070

Fürnkranz, M., Lukesch, B., Müller, H., Huss, H., Grube, M. *et al.* (2012) Microbial diversity inside pumpkins: Microhabitat-specific communities display a high antagonistic potential against phytopathogens. *Microbial Ecology* 63, 418–428. doi:10.1007/s00248-011-9942-4

Gamalero, E., Lingua, G., Tombolini, R., Avidano, L., Pivato, B. *et al.* (2005) Colonization of tomato root seedling by *Pseudomonas fluorescens* 92rkG5: Spatio-temporal dynamics localization, organization, viability and culturability. *Microbial Ecology* 50, 289–297.

Germaine, K., Keogh, E., Borremans, B., Lelie, D., Barac, T. *et al.* (2004) Colonisation of poplar trees by *gfp* expressing bacterial endophytes. *FEMS Microbiology Ecology* 48, 109–118. doi:10.1016/j.femsec.2003.12.009

Glassner, H., Zchori-Fein, E., Compant, S., Sessitsch, A., Katzir, N. *et al.* (2015) Characterization of endophytic bacteria from cucurbit fruits with potential benefits to agriculture in melons (*Cucumis melo* L.). *FEMS Microbiology Ecology* 91(7), pii: fiv074. doi:10.1093/femsec/fiv074

Glassner, H., Zchori-Fein, E., Yaron, S., Sessitsch, A., Sauer, U. *et al.* (2018) Bacterial niches inside seeds of *Cucumis melo* L. *Plant and Soil* 422(1–2), 101–113. doi:10.1007/s11104-017-3175-3

Hallmann, J. (2001) Plant interactions with endophytic bacteria. In Jegger, M.J. and Spence, N.J. (eds) *Biotic Interactions in Plant–Pathogen Associations*. CAB International, Wallingford, UK, pp. 87–119.

Hardoim, P.R., van Overbeek, L.S., Berg, G., Pirtillä, A.M., Compant, S. *et al.* (2015) The hidden world within plants: Ecological and evolutionary considerations for defining functioning of microbial endophytes. *Microbiology and Molecular Biology Reviews* 79(3), 293–320. doi:10.1128/MMBR.00050-14

Iasur-Kruh, L., Zahavi, T., Barkai, R., Freilich, S., Zchori-Fein, E. *et al.* (2018) *Dyella*-like bacterium isolated from an insect as a potential biocontrol agent against grapevine yellows. *Phytopathology* 108, 336–341. doi:10.1094/PHYTO-06-17-0199-R

James, E.K., Olivares, F.L., de Oliveira, A.L.M., dos Reis, F.B. Jr, da Silva, L.G. *et al.* (2001) Further observations on the interaction between sugar cane and *Gluconacetobacter diazotrophicus* under laboratory and greenhouse conditions. *Journal of Experimental Botany* 52, 747–760.

James, E.K., Gyaneshwar, P., Manthan, N., Barraquio, W.L., Reddy, P.M. *et al.* (2002) Infection and colonization of rice seedlings by the plant growth-promoting bacterium *Herbaspirillum seropedicae* Z67. *Molecular Plant-Microbe Interactions* 15, 894–906.

Junker, R.R., Loewel, C., Gross, R., Dötterl, S., Keller, A. *et al.* (2011) Composition of epiphytic bacterial communities differs on petals and leaves. *Plant Biology* 13(6), 918–924. doi:10.1111/j.1438-8677.2011.00454.x

Kandel, S.L., Joubert, P.M. and Doty, S.L. (2017) Bacterial endophyte colonization and distribution within plants. *Microorganisms* 5(4), pii: E77. doi:10.3390/microorganisms5040077

Leach, J.E., Triplett, L.R., Argueso, C.T. and Trivedi, P. (2017) Communication in the phytobiome. *Cell* 169(4), 587–596. doi:10.1016/j.cell.2017.04.025

Lemanceau, P., Barret, M., Mazurier, S., Mondy, S., Pivato, B. *et al.* (2017) Plant communication with associated microbiota in the spermosphere, rhizosphere and phyllosphere. *Advances in Botanical Research* 82, 101–133.

Lopez, B.R., Bashan, Y. and Bacilio, M. (2011) Endophytic bacteria of *Mammillaria fraileana*, an endemic rock-colonizing cactus of the southern Sonoran Desert. *Archives in Microbiology* 193(7), 527–541. doi:10.1007/s00203-011-0695-8

Lòpez-Fernàndez, S., Compant, S., Vrhosek, U., Bianchedi, P.L., Sessitsch, A. *et al.* (2016) Grapevine colonization by endophytic bacteria shifts secondary metabolism and suggests activation of defense pathways. *Plant and Soil* 405(1), 155–175. doi:10.1007/s11104-015-2631-1

Lòpez-Fernàndez, S., Mazzoni, V., Pedrazzoli, F., Pertot, I. and Campisano, A. (2017) A phloem-feeding insect transfers bacterial endophytic communities between grapevine plants. *Frontiers in Microbiology* 8, 834. doi:10.3389/fmicb.2017.00834

Lugtenberg, B. and Kamilova, F. (2009) Plant-growth-promoting rhizobacteria. *Annual Review of Microbiology* 63, 541–556.

Malfanova, N., Kamilova, F., Validov, S., Chebotar, V. and Lugtenberg, B. (2013) Is l-arabinose important for the endophytic lifestyle of *Pseudomonas* spp.? *Archives of Microbiology* 195(1), 9–17. doi:10.1007/s00203-012-0842-x

Manirajan, B.A., Maisinger, C., Ratering, S., Rusch, V., Schwiertz, A. *et al.* (2018) Diversity, specificity, co-occurrence and hub taxa of the bacterial-fungal pollen microbiome. *FEMS Microbiology Ecology* 94, 8. doi:10.1093/femsec/fiy112

Mendes, R., Garbeva, P. and Raaijmakers, J.M. (2013) The rhizosphere microbiome: Significance of plant beneficial, plant pathogenic, and human pathogenic microorganisms. *FEMS Microbiology Reviews* 37(5), 634–663. doi:10.1111/1574-6976.12028

Mercado-Blanco, J. (2015) Life of microbes inside the plant. In: Lugtenberg, B.J.J. (ed.) *Principles of Plant-Microbe Interactions*. Springer, Berlin, pp. 25–32. doi:10.1007/978-3-319-08575-3_5

Miotto-Vilanova, L., Jacquard, C., Courteaux, B., Wortham, L., Michel, J. *et al.* (2016) *Burkholderia phytofirmans* PsJN confers grapevine resistance against *Botrytis cinerea* via a direct antimicrobial effect combined with a better resource mobilization. *Frontiers in Plant Science* 7, 1236. doi:10.3389/fpls.2016.0123

Mitter, B., Pfaffenbichler, N., Flavell, R., Compant, S., Antonielli, L. *et al.* (2017) A new approach to modify plant microbiomes and traits by introducing beneficial bacteria at flowering into progeny seeds. *Frontiers in Microbiology* 8, 11. doi:10.3389/fmicb.2017.00011

Nelson, E.B. (2018) The seed microbiome: Origins, interactions, and impacts. *Plant and Soil* 422(1–2), 7–34. doi:10.1007/s11104-017-3289-7

Nelson, E.B., Simoneau, P., Barret, M., Mitter, B. and Compant, S. (2018) Editorial special issue: The soil, the seed, the microbes and the plant. *Plant and Soil* 422(1–2), 1–5. doi:10.1007/s11104-018-3576-y

Panchal, S. and Melotto, M. (2018) Stomate-based defense and environmental cues. *Plant Signaling & Behavior* 12(9), e1362517. doi:10.1080/15592324.2017.1362517

Peredo, E.L. and Simmons, S.L. (2018) Leaf-FISH: Microscale imaging of bacterial taxa on phyllosphere. *Frontiers in Microbiology* 8, 2669. doi:10.3389/fmicb.2017.02669

Pinto-Carbó, M., Gademann, K., Eberl, L. and Carlier, A. (2018) Leaf nodule symbiosis: Function and transmission of obligate bacterial endophytes. *Current Opinion in Plant Biology* 44, 23–31. doi:10.1016/j.pbi.2018.01.001

Prieto, P., Schiliro, E., Maldonado-Gonzalez, M.M., Valderrama, R., Barroso-Albarracin, J.B. *et al.* (2011) Root hairs play a key role in the endophytic colonization of olive roots by *Pseudomonas* spp. with biocontrol activity. *Microbial Ecology* 62, 435–445. doi:10.1007/s00248-011-9827-6

Rahman, M.M., Flory, E., Koyro, H.W., Abideen, Z., Schikora, A. *et al.* (2018) Consistent associations with beneficial bacteria in the seed endosphere of barley (*Hordeum vulgare* L.). *Systematic and Applied Microbiology* 41(4), 386–398. doi:10.1016/j.syapm.2018.02.003

Ramirez, K.S., Geisen, S., Morriën, E., Snoek, B.L. and van der Putten, W.H. (2018) Network analyses can advance above-belowground ecology. *Trends in Plant Science* 23(9), 759–768. doi:10.1016/j.tplants.2018.06.009

Remus-Emsermann, M.N.P. and Schlechter, R.O. (2018) Phyllosphere microbiology: At the interface between microbial individuals and the plant host. *New Phytologist* 218(4), 1327–1333. doi:10.1111/nph.15054

Rosado, B.H.P., Almeida, L.C., Alves, L.F., Lambais, M.R., Oliveira, R.S. (2018) The importance of phyllosphere on plant functional ecology: a phyllo trait manifesto. *New Phytologist* 219(4), 1145–1149. doi: 10.1111/nph.15235.

Sabaratnam, S. and Beattie, G.A. (2003) Differences between *Pseudomonas syringae* pv. syringae B728a and *Pantoea agglomerans* BRT98 in epiphytic and endophytic colonization of leaves. *Applied and Environmental Microbiology* 69, 1220–1228.

Samad, A., Trognitz, F., Compant, S., Antonielli, L. and Sessitsch, A. (2017) Shared and host-specific microbiome diversity and functioning of grapevine and accompanying weed plants. *Environmental Microbiology* 19(4), 1407–1424. doi:10.1111/1462-2920.13618

Turner, T.R., James, E.K. and Poole, P.S. (2013) The plant microbiome. *Genome Biology* 14(6), 209–219. doi:10.1186/gb-2013-14-6-209

Vacher, C., Hampe, A., Porte, A.J., Sauer, U., Compant, S. *et al.* (2016) The phyllosphere: Microbial jungle at the plant-climate interface. *Annual Review of Ecology, Evolution, and Systematics* 47, 1–24.

Vorholt, J.A. (2012) Microbial life in the phyllosphere. *Nature Reviews Microbiology* 10(12), 828–840. doi:10.1038/nrmicro2910

Wilson, M., Hirano, S.S. and Lindow, S.E. (1999) Location and survival of leaf-associated bacteria in relation to pathogenicity and potential for growth within the leaf. *Applied and Environmental Microbiology* 65(4), 1435–1443.

Yang, L., Danzberger, J., Schöler, A., Schröder, P., Schloter, M. *et al.* (2017) Dominant groups of potentially active bacteria shared by barley seeds become less abundant in root associated microbiome. *Frontiers in Plant Science* 8, 1005. doi:10.3389/fpls.2017.01005

4 Analysing Seed Endophytes for Biotechnology

Alessandro Bergna,[1] Tomislav Cernava,[2*] Christin Zachow[2] and Gabriele Berg[2]

[1]*Austrian Centre of Industrial Biotechnology, Graz, Austria;* [2]*Institute of Environmental Biotechnology, Graz University of Technology, Graz, Austria*

Abstract

Seed endophytes play a crucial role during the entire life cycle of plants due to their ability to promote germination and plant growth and provide defence against biotic and abiotic stress. The increasing interest related to these microorganisms for applications in sustainable agriculture requires the use of a wide spectrum of techniques to investigate their ecological role and to exploit their biotechnological potential. While the isolation of microorganisms is the most straightforward method to characterize and select microorganisms, molecular techniques represent an advantageous option for the discovery and tracking of uncultivable microbial species. This chapter shows that the concomitant employment of cultivation-dependent and cultivation-independent techniques represents the most sophisticated approach for the study of endophytic communities. In addition to a general assessment of developments in this field, the most frequently used tools are described in detail. Moreover, their possible integration as shown in various studies targeting seed endophytes is highlighted. We expect that novel products for biotechnology will become more feasible in the future due to the recent technological and methodological developments.

4.1 Introduction

Endophytes are defined as microorganisms that are able to asymptomatically reside within plant tissues for at least a part of their life cycle (Hardoim *et al.*, 2015). While these microorganisms are primarily known for their ability to enhance plant growth and defence, they also represent a significant source of natural metabolites and bioactive compounds of biotechnological interest (Tan and Zou, 2001; Gunatilaka, 2006; Aly *et al.*, 2013; Martinez-Klimova *et al.*, 2017; Gao *et al.*, 2018). Among endophytes, a subgroup of microorganisms able to reside in plant seeds and to be vertically transmitted to the successive plant generation represent a crucial starting inoculum of beneficial microbes for improved plant development and health (Puente *et al.*, 2009; Johnston-Monje and Raizada, 2011; Hardoim *et al.*, 2012). A better understanding of these microorganisms and of their vertical transmission will enhance the opportunities to exploit beneficial microbe–plant interactions in agriculture and horticulture (Berg and Raaijmakers, 2018).

The understanding of the plant microbiome, including endophytic communities, was revolutionized by the technological advancements in DNA sequencing and computational technologies of the last decade (Mendes *et al.*, 2011; Bulgarelli *et al.*, 2012; Rybakova *et al.*, 2016). However, the study of seed microbiomes can be considered to be still in its infancy as research efforts for their characterization are relatively recent (Barret *et al.*, 2015;

* Corresponding author e-mail: tomislav.cernava@tugraz.at

Klaedtke *et al.*, 2016). In the past, seed microbiology was focused on seed-borne pathogens and their detection and control (Munkvold, 2009). Due to the dormant phase of many microorganisms inside seeds, knowledge about seed endophytes obtained by cultivation was limited for a long time. Now it is possible to map the microbial community of interest with fast and cost-effective solutions by next-generation sequencing (NGS) technologies (Hugerth and Andersson, 2017), but it is important to emphasize that such technologies have limitations when studying endophytes. For this reason, several techniques are potentially employable to identify and characterize seed endophytes. All have advantages and disadvantages, but the combined results are indispensable for a holistic understanding of the ecological role of these microorganisms and their biotechnological applicability. In this chapter we provide a summary of the main methodologies that are employed today and that represent the fundamentals for the future of seed endophytes research. A schematic representation of the most commonly used approaches is provided in Fig. 4.1 and includes advantages and disadvantages connected with each strategy. In addition, an overview is provided and includes a selection of publications (Table 4.1) related to seed endophytes as well as methodologies that have been employed in these studies. The following sections include the most frequently applied approaches to study seed endophytes. Integrative approaches that combine different methodologies can more likely deliver deepening insights into the ecology and functioning of seed endophytes.

4.2 Isolation of Seed Endophytes

4.2.1 Seed surface sterilization

In order to study or extract seed endophytes from non-endophytic microorganisms, surface sterilization is required for removing microbes on the seed surface (Fig. 4.1). This initial and yet crucial step can be problematic, as the sterilization should be achieved without destructive effects on the host tissue. The procedure normally entails two washing steps: shaking in sodium hypochlorite and a final soaking in ethanol. Concentrations of sodium hypochlorite and ethanol and washing/shaking time vary based on the seed texture: we report valuable examples of seed surface sterilization methodologies for prominent crops.

- Maize (*Zea mays* L.) seeds can be subjected to two consecutive washing steps with 3% sodium hypochlorite for 10 min followed by 10 min washing in 95% ethanol for 10 min (Johnston-Monje and Raizada, 2011).
- Rice (*Oryza sativa* L.) seeds are surface-sterilized by washing for 25 min in a saline solution containing 0.12% sodium hypochlorite (NaClO) and 0.15% sodium hydroxide followed by a washing step with 2% sodium thiosulfate to remove surface-adhered NaClO (Hardoim *et al.*, 2012).
- Quinoa (*Chenopodium quinoa* Willd.) seeds are surface-sterilized by two consecutive incubations in 70% ethanol for 5 min followed by three washes in distilled water (Pitzschke, 2016).
- Cucurbitaceae seeds are surface-sterilized by washing them twice for 5 min in a sodium hypochlorite solution with a specific concentration (2.5–3.5%) based on seed coating texture for 5 min. After rinsing with autoclaved distilled water, the seeds were again washed with 95% ethanol for 5 min (Khalaf and Raizada, 2016).
- Bean (*Phaseolus vulgaris* L.) seeds are surface-sterilized by immersion in 95% ethanol for 3 min, followed by immersion in 20% NaClO for 20 min, and rinsed with sterile distilled water three times (Malinich and Bauer, 2018).

4.2.2 Seed activation and extraction of endophytes

The most straightforward approach to study the biotechnological capacities of endophytic microorganisms is the isolation and cultivation of the living strains (Fig. 4.1). This approach has the goal of extracting the

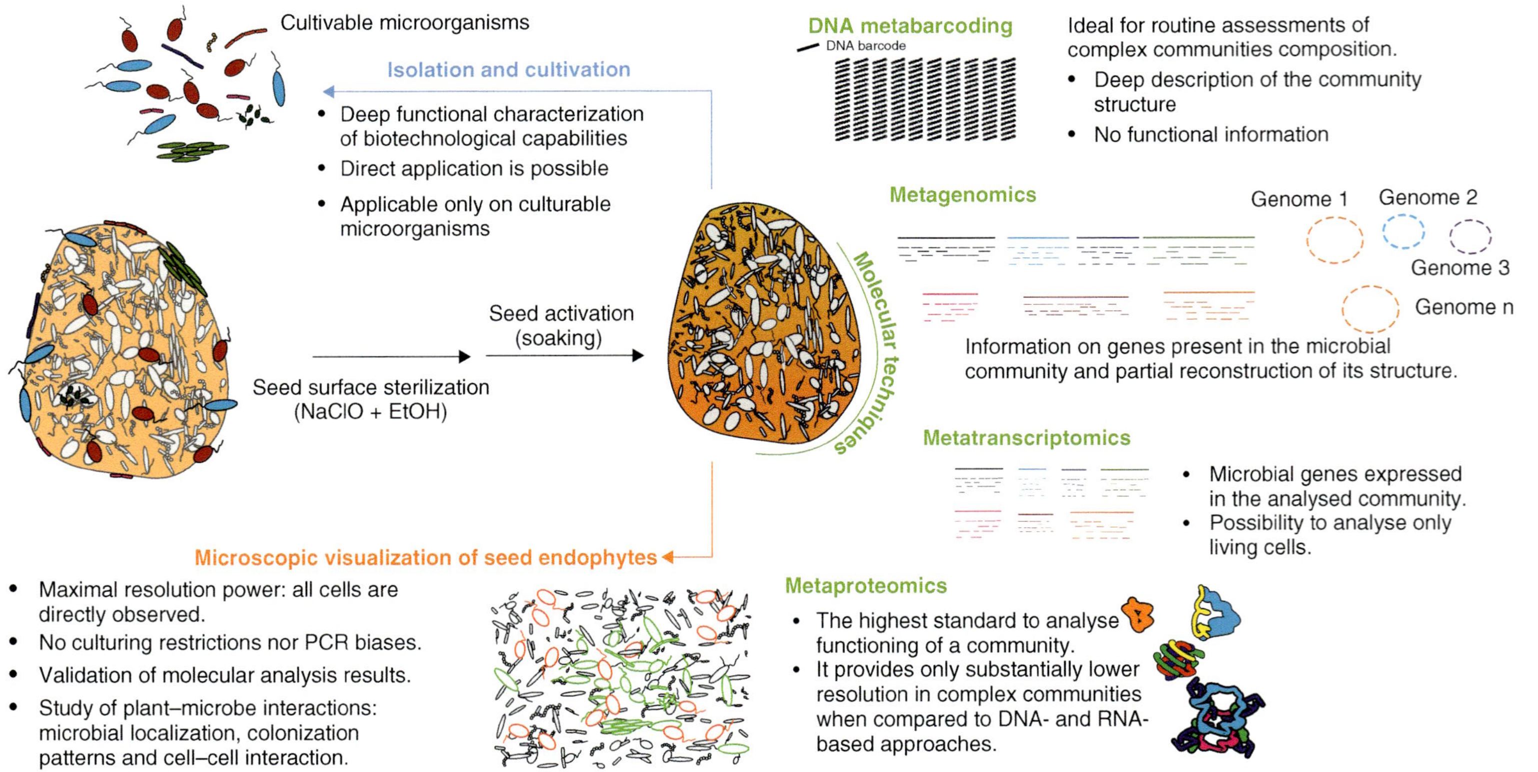

Fig. 4.1. Graphical representation of different methodologies employed in studies focusing on seed endophytes.

Table 4.1. Overview of current studies on the seed microbiome with the main endophytes-related methodologies employed.

Plant species/cultivar	Reference	Technologies employed
Maize (*Zea mays*)	Johnston-Monje and Raizada (2011)	Isolation and phenotyping, cloning, TRFLP
Rice (*Oryza sativa*)	Hardoim *et al.* (2012)	Isolation of seed endophytes, PCR-DGGE
Brassica and *Triticum* species	Links *et al.* (2014)	*cpn60* metabarcoding, qPCR quantification of microorganisms of interest, isolation of seed endophytes and characterization
Bean (*Phaseolus vulgaris*)	Klaedtke *et al.* (2016)	16S and ITS metabarcoding
Radish (*Raphanus sativus*)	Rezki *et al.* (2016)	Isolation of seed endophytes, qPCR, 16S, *gyr*B and ITS1 metabarcoding
Pumpkin (*Cucurbita pepo*)	Adam *et al.* (2016)	16S metabarcoding
Maize (*Z. mays*)	Johnston-Monje *et al.* (2016)	TRFLP and 16S metabarcoding
Sueda salsa	Qin *et al.* (2016)	ITS1 and ITS2 metabarcoding, fungal isolation and *in planta* characterization
Quinoa	Pitzschke (2016)	Microscopy, isolation of seed endophytes
Cucurbitaceae species	Khalaf and Raizada (2016)	Isolation of seed endophytes and 16S rRNA gene fingerprinting
Pepper (*Capsicum annuum*), soybean (*Glycine max*), *Triticum aestivum*	Mitter *et al.* (2017)	qPCR, 16S metabarcoding
Malvaceae species	Irizarry and White (2017)	Isolation of seed endophytes and *in planta* characterization
Oilseed rape (*Brassica napus*)	Rybakova *et al.* (2017)	16S metabarcoding, qPCR, isolation of seed endophytes and characterization, FISH-CLSM microscopy
Soybean (*G. max*)	Huang *et al.* (2018)	ITS1 and ITS2 metabarcoding, bacterial isolation and *in planta* characterization
Rice (*O. sativa*)	Walitang *et al.* (2017)	Isolation of seed endophytes and functional and genetic characterization of isolates
Radish (*R. sativus*)	Rezki *et al.* (2018)	*gyr*B and ITS metabarcoding
Muskmelon (*Cucumis melo*)	Glassner *et al.* (2018)	Scanning electron microscopy (SEM) and CLSM DOPE-FISH
Browntop millet (*Brachiaria* species)	Verma and White (2018)	Isolation of seed bacterial endophytes and *in vitro* characterization
Cucurbitaceae species	Khalaf and Raizada (2018)	Isolation of seed bacterial endophytes and *in vitro* characterization
Bean (*P. vulgaris*)	Malinich and Bauer (2018)	Cloning, isolation of seed bacterial endophytes, qPCR
Bean (*P. vulgaris*), Radish (*R. sativus*)	Torres-Cortés *et al.* (2018)	Shotgun metagenomics, bacterial isolation and *in planta* characterization
Barley (*Hordeum vulgare*)	Rahman *et al.* (2018)	Isolation of seed endophytes, 16S metabarcoding, FISH-CLSM
Crotalaria pumila	Sánchez-López *et al.* (2018)	16S metabarcoding
Phragmites australis	White *et al.* (2018)	Isolation of seed bacterial endophytes and *in planta* characterization
Salvia miltiorrhiza	Chen *et al.* (2018)	16S and ITS metabarcoding
Ground-ivy (*Glechoma hederacea*)	Vannier *et al.* (2018)	16S and 18S metabarcoding targeting bacteria/archaea and fungi

microorganisms alive and growing them under laboratory conditions to study their metabolic properties and capacities. From a technical point of view, the isolation of endophytes from seeds is more complicated than from other plant compartments due to the dormant state of this structure. For this reason, most of the current techniques for isolating microorganisms from seeds rely on an initial seed activation step under gnotobiotic conditions. To soften seeds and revive endophytic populations, washing and gently shaking the seeds is necessary; the liquid phase is usually distilled water or 0.85% NaCl. The soaking time, as for the surface sterilization, depends on the seed texture: 4 h is employed for rapeseeds (*Brassica napus*) (Rybakova *et al.*, 2017) while 48 h is necessary for maize seeds (Johnston-Monje and Raizada, 2011). With this process the seed switches from a dry and quiescent state to a hydrated and active state (Dekkers *et al.*, 2013) that enables endophytes to overcome dormancy and improves cultivation ratios and extraction by softening the seed. In order to extract the endophytic microorganisms from inner tissues of seeds, maceration is the preferred methodology. The softened activated seeds are grinded with autoclaved mortar and pestle and a liquid phase (buffer) is added (Berg *et al.*, 2013). This suspension is therefore plated in serial dilutions on the selected medium. Commonly employed media for isolating bacteria from plant tissues are tryptic soya agar (TSA), R2A and nutrient broth–yeast extract (Gardner *et al.*, 1982). Concerning the isolation of fungi, standard media include PDA (potato dextrose agar), malt extract–peptone–yeast extract and biomalt agar (Philipson and Blair, 1957; Schulz *et al.*, 1995; Hallmann *et al.*, 2006).

4.2.3 Assessment of yet uncultivable microorganisms

A second limitation for the isolation of seed endophytes is the high portion of 'so far uncultivable' microorganisms. The presence of dormant cells and insufficiently optimized cultivation media represents the main impediments for the isolation of endophytes from plant and seed tissues (Torsvik and Øvreås, 2002; Eevers *et al.*, 2015). Even if molecular and cultivation-independent techniques have undisputable higher screening power, in order to deeply understand the physiology of an endophyte, the cultivation of the microorganism in the laboratory is still required (Stewart, 2012). For this reason, even if time-consuming and expensive, attempts in the cultivation of 'so far uncultivable' microorganisms of biotechnological relevance are being carried out. Since the limitations for cultivation of these microorganisms are sometimes due to their reliance on the interaction with other beneficial microorganisms or with the biochemical surroundings, co-cultivation with helper strains and the recreation of the environment in the laboratory can sometimes result in their successful culturing (Ohno *et al.*, 2000; Nichols *et al.*, 2008).

4.2.4 Phenotyping

The main advantage of extracting living microorganisms from seeds lies in the possibility to directly test their properties in the laboratory. Several screenings have been developed in the last decade and can be adapted to the characteristics of the microorganism or of the plant. Shahzad *et al.* (2018) provided an almost complete summary of the functional attributes tested for both bacteria and fungi isolated from different seeds linking it to plant host and reference methodology. The assessment of microbial functioning can be tested *in vitro*, *in vivo* or *in situ* to assess the behaviour of the microorganism, respectively, without the host, inside the host tissues (also referred as *in planta*) or in a specific structure of the host. Specific assessments, e.g. the production of bioactive volatile compounds (Cernava *et al.*, 2015), can be integrated in deepening screenings based on the specific research question. These approaches represent the ground floor for the biotechnological employment of seed endophytes in several fields of agriculture. Endophytes are employable as:

- biofertilizers, for their ability to promote the acquisition of essential nutrients by, for example, solubilizing phosphorus, fixating nitrogen and producing siderophores that enhance iron uptake (Chhabra and Dowling, 2017);

- plant biostimultants, for the production of phytohormones and spermidine against abiotic stress (Berg, 2009); and
- biopesticides, for their ability to produce lytic enzymes, antibiotics, antimicrobial volatiles (Rybakova *et al.*, 2016).

From the biotechnological standpoint, it is evident that the cultivation of seed endophytes has great advantages, but also undeniable limitations connected with the impossibility to isolate the greatest portion of the seed endophytic community. For this reason, more investments are crucial for enhancing the cultivation efficiency from this promising plant compartment.

4.3 Molecular Techniques for the Analysis of Seed Endophytes

Since most endophytic microorganisms are uncultivable, studies of seed endophytes mostly rely on culture-independent techniques for their detection and identification (Liaqat and Eltem, 2016). Cultivation-independent techniques are primarily based on the extraction and analysis of target molecules (DNA, RNA, proteins, metabolites). The resulting sequences and information can be compared with microbial databanks to identify the microorganism's taxonomy or specific genes in its genome. In a similar way, the extraction and analysis of endophytic RNA and proteins can be used to identify active genes from living cells. While older techniques such as fingerprinting (e.g. terminal restriction fragment length polymorphism – TRFLP) and qPCR cannot provide deep and accurate insights at community level, NGS-based omics techniques are used to characterize highly complex microbial communities. Even if molecular techniques can represent powerful tools for the description of seed endophytes, they become powerless if used in poorly designed experiments. In fact, good experimental design, replication and appropriate methodology selection are essential for a precise interpretation of molecular data (Hallmann *et al.*, 1997; Prosser, 2010) regardless of the resolution power of the tool employed.

4.3.1 DNA extraction

In order to extract microbial genetic material from the seed endosphere, an initial surface sterilization is required. Similar to the procedure that allows the isolation of endophytes, according to the protocol described by Bragina *et al.* (2012), seeds can be homogenized with mortar and pestle and suspended in 0.85% NaCl. In contrast to isolation approaches, homogenization can also be done by using liquid nitrogen as no living cells are required. Pellets containing seed endophytes are collected by centrifugation and then used for total community DNA isolation, e.g. using the FastDNA® SPIN Kit for Soil and the FastPrep® Instrument (MP Biomedicals, Santa Ana, CA). This specific kit can be employed for the isolation of endophytic DNA (Compant *et al.*, 2011; Adam *et al.*, 2016; Rybakova *et al.*, 2017; Bergna *et al.*, 2018) as it contains a lysing matrix that allows treating complex tissues. It is important to specify that regardless of the intensity of pretreatments, plant genetic material will represent most of the DNA extracted with this and any other extraction procedures. Specific methodologies to target endophytic DNA are described in the following paragraphs.

4.3.2 Differentiation between living and dead cells

Since DNA can persist in the environment for relatively long periods after cell death (Josephson, 1993; Nocker *et al.*, 2007), surface sterilization alone would lead to the overestimation of the number of living cells. The use of propidium monoazide (PMA) after seed surface sterilization and homogenization leads to covalent binding of the photoreactive dye to accessible DNA from dead cells. This blocks the PCR amplification of genetic material belonging to dead microbial and damaged plant cells. Even if not yet broadly employed in the study of endophytes, this procedure enhances the probability to specifically amplify endophytic DNA (McKinnon, 2016).

4.3.3 DNA metabarcoding

DNA metabarcoding or amplicon sequencing is a rapid method for biodiversity assessment of highly diverse microbial communities (Fig. 4.1). Advances in sequencing technologies made this technique a fast and cost-effective solution that can now be considered as a routine assessment for endophytic microbial communities (Barret *et al.*, 2015; Berg *et al.*, 2015; Vandenkoornhuyse *et al.*, 2015; Adam *et al.*, 2016; Rybakova *et al.*, 2017). The approach relies on two main steps: (i) a mass PCR amplification of a single marker gene (DNA barcode) from environmental DNA followed by (ii) the sequencing of the amplicons with a high-throughput sequencing platform.

Primers and PCR reaction

The ribosomal operon is broadly accepted as the 'gold standard' for diversity assessment for its broad presence in all organisms and its favourable topology (Amann *et al.*, 1995). In fact, the alternating presence of conserved and highly variable regions is ideal for the construction of PCR primers, recognizing highly conserved regions for the amplification of the neighbouring variable regions (Sanschagrin and Yergeau, 2014). The genes employed for 16S, ITS, or 18S are to describe the composition of, respectively, bacterial, archaeal, fungal (Lindahl *et al.*, 2013), and micro-eukaryote communities (Lentendu *et al.*, 2014). The central role that this operon has been playing in microbial diversity assessments brought the construction of databases of unmatched size that are perfect for this methodology. Due to the difficulty of designing universal primers, primers for amplicon sequencing are in continuous evolution. The reference project is the Earth Microbiome Project (EMP, www.earthmicrobiome.org), providing primers designed for sequencing on Illumina platforms extensively used for the study of plant-associated microorganisms. Here we provide a short description of the key primers used for amplicon sequencing and of alternatives for the study of endophytes.

16S rRNA

This gene is extensively used for bacteria and archaea. The most commonly used primer pair in combination with metabarcoding approaches is provided by the EMP (www.earthmicrobiome.org/) and named 515F-806R (Caporaso *et al.*, 2012). It targets the V4 region of the 16S SSU rRNA gene. In addition to this primer pair, primer pairs targeting the V2 region or V4–V5 and V5–V7 are applicable for these types of studies (Beckers *et al.*, 2016). As a microbial genome can host multiple copies of the 16S rRNA gene with intragenomic variability, alternative target single-copy housekeeping genes have also been successfully employed on seed endophytic communities (Links *et al.*, 2014; Rezki *et al.*, 2016, 2018).

- *gyr*B is an example of an alternative bacterial marker developed by Barrett *et al.* in 2015 to overcome the low sequence divergence among related bacterial taxa of 16S rRNA gene (Větrovský and Baldrian, 2013) and to provide valuable insights into the taxonomic composition of the seed microbiota (Barret *et al.*, 2015). This DNA barcode is based on a portion of *gyr*B, a gene encoding the subunit of the DNA gyrase, frequently employed as a phylogenetic marker for many bacterial genera (Yamamoto and Harayama, 1995; Větrovský and Baldrian, 2013).
- In cases where resolution beyond the genus level and the confident identification of potentially novel taxa is desirable, cpn60 metabarcoding could represent a solution for *de novo* assembly of sequence data. This DNA barcode has the great advantage of simultaneously targeting bacteria and fungi with a unique primer providing a unified cross-domain view of the microbial community (Links *et al.*, 2012).

Although universal primers have been designed and tested for the cpn60 gene (Schellenberg *et al.*, 2011), the size of sequence databases of 16S rRNA gene greatly exceeds those of other bacterial genes. For this reason, the employment of 16S remains still the target of choice for studies in bacterial ecology and seed endophytes (Větrovský and Baldrian, 2013).

As already mentioned, when total community DNA (tcDNA) is extracted from seeds, microbial genetic material composes only a minor fraction of the obtained DNA. Since bacterial 16S rDNA primer pairs exhibit high affinity to plastid and mitochondrial DNA, a high proportion of NGS reads would include host-derived sequences. In order to facilitate PCR amplification of endophyte DNA, it is crucial to exclude non-target tcDNA from the PCR reaction. Two solutions are possible.

1. The employment of primer pairs (799F-1391R) targeting the V5–V7 hypervariable regions of the 16S rDNA. Primers targeting this region have been shown to produce very low amplification rate of non-target DNA across all plant compartments (Beckers *et al.*, 2016).
2. The employment of the peptide nucleic acid (PNA)-PCR clamping technique. This technique uses PNA oligomers with complementary sequences to mitochondria and plastid SSU rRNA genes. Their overlapping with the region in the 1492r primer-binding site suppresses the amplification of the two organelles (Sakai and Ikenaga, 2013).

Internal transcribed spacer

The internal transcribed spacer (ITS) region is extensively employed to identify fungal lineages and is formally the DNA barcode for fungi due to the presence of a rich and up-to-date database (Schoch *et al.*, 2012; Bates *et al.*, 2013). While microbiome projects utilize and endorse ITS1 subregion as a target using ITS1F (Gardes and Bruns, 1993) and ITS2 as primers (De Filippis *et al.*, 2017), recent studies propose to test different target genes simultaneously, recommending the use of ITS2 or the whole ITS region for metabarcoding. Unlike the ability of 16S universal primer to cover the bacterial domain almost completely, primers constructed on the ITS cannot be considered as phylogenetically inclusive for fungal genomes. In fact, the employment of this gene for amplicon sequencing can be biased by the preferential amplification of specific taxa. This is due to the length variability of the ITS1-2 region among different fungal genera and species (Esteve-Zarzoso *et al.*, 1999). For this reason, in the use of ITS subregions for amplicon sequencing there is no consistency in the choice of primer pairs (Tedersoo *et al.*, 2016). Finally, ITS primers could also be selected based on the relative proportion of fungal DNA and the expected dominant groups (Tedersoo *et al.*, 2015) or by using 'mock communities' that allow the evaluation of the reliability of each primer pair (Tessler *et al.*, 2017).

High-throughput sequencing platforms

Current NGS platforms are not optimized for the production of reads long enough to cover a whole marker gene in combination with a low error rate and high sequencing depth. Since using a single marker gene for inferring whole genome differences, sequencing quality has become a crucial factor. For these reasons, the choice of the sequencing platform comes together with the choice of the primer pair and is made in order to sequence the target region with the higher quality and depth. Due to a preferred utilization of primers provided by the EMP, studies focusing on seed endophytes mainly use Illumina sequencing technologies (MiSeq or HiSeq platforms). For a good understanding of the sequencing platforms and their use, we recommend the clear summary table provided by Tessler *et al.* (2017).

Computational data analysis for microbial community reconstruction

The inference of endophytic diversity using sequencing data from a single marker gene involves the clustering of sequencing reads into operational taxonomic units (OTUs). In this computational step, similar marker gene sequences are clustered and considered as belonging to the same taxon (Edgar, 2013). The taxonomy assignment of the OTU is retrieved by the comparison of a representative sequence of the OTU with a database. The main databases employed for this step are RDP (Cole *et al.*, 2014), SILVA (Quast *et al.*, 2013) and Greengenes (McDonald *et al.*, 2012). Various pipelines for such analyses are employable; however, the most recognized and broadly used is QIIME (Caporaso *et al.*, 2010; https://qiime2.org).

As a conclusion, DNA metabarcoding is a technique that allows characterization of the complex structure of microbial communities. Its reliance on a single non-functional gene does not allow the study of the functional potentiality of microorganisms. In addition, it should be taken into account that the application of this methodology on endophytic communities can be challenging during DNA extraction, PCR and data analysis steps. Nevertheless, this technique has to be accounted as a powerful tool to routinely study microbial communities with the possibility to detect indicator species and community shifts.

4.3.4 Omics technologies

In order to answer important ecological questions on functional roles of seed endophytic microbiomes, metagenomic and metatranscriptomic approaches are often applied (Alibrandi *et al.*, 2018). It is important to highlight that although the sequencing depth that can be reached with NGS instruments is steadily increasing, complex microhabitats (e.g. seed endosphere) still cannot be completely assessed with these methods (Myrold *et al.*, 2014).

Metagenomics

Metagenomics comprise the study of the genomic content within complex microbial communities (Wooley *et al.*, 2010; Fig. 4.1). This technique relies on the random fragmentation and sequencing of genomic DNA isolated (shotgun metagenomics) and allows reconstruction of the gene set of the microbial communities residing in a specific environment with no cultivability restriction. In addition, the possibility of mapping discovered genes to known microbial genomes allows the simultaneous study of both the composition and the functional capabilities of the community in a single experiment (Kurokawa *et al.*, 2007; Arumugam *et al.*, 2011). An example is the study conducted by Torres-Cortés *et al.* (2018) that, having both ecological and functional insights into the community of germinating seeds (bean and radish), was able to determine the functional traits connected with the modification and selection of the microbiome.

Metatranscriptomics

The same principle of metagenomics has been successfully applied to the study of mRNA. In fact, the mRNA is converted to cDNA and sequenced on an NGS platform (Fig. 4.1). The possibility to map activated microbial genes inside plant tissues is essential to understand the endophytic phenomenon (Kaul *et al.*, 2016) and the role of seed-associated microbes in plant growth and development. Even if several studies showed how beneficial seed-borne endophytes could defend the plant from stress (Truyens *et al.*, 2014; Khamchatra *et al.*, 2016; Shahzad *et al.*, 2016, 2017) or produce compounds that inhibit pathogen growth or strengthen plant resistance (Bonos *et al.*, 2005; Tayung *et al.*, 2012; Shahzad *et al.*, 2017), the employment of metatranscriptomics on seed endophytes has not yet been accomplished.

In order to select the most suitable NGS-based approach to study seed endophytes, it is crucial to evaluate the methodology in a broader context. While the ability of metagenomics to analyse the genomic content in a complex mixture of microorganisms avoiding PCR biases is of undoubtable importance (Wooley *et al.*, 2010), its ability to assess biodiversity and community ecology analyses is highly dependent on additional factors. Difficulties of this approach can be represented by the choice of sequencing depth and length. Sequencing depth is an important factor for shotgun metagenomics, as it determines its ability to discover new genes. Similarly, longer reads are more likely to cover full protein domains, and therefore allow to distinguish between closely related genes from different organisms. In fact, short reads are frequently misaligned leading to an inflation of both species count and diversity estimates (Caporaso *et al.*, 2012; Schulze-Schweifing *et al.*, 2014; Clooney *et al.*, 2016). From another point of view, the analysis of shotgun metagenomics data can be performed with different strategies with a variety of tools that can be employed for every computational step (Breitwieser *et al.*, 2017). In addition, the absence of recognized and unified pipelines, as QIIME (Caporaso *et al.*, 2010) is for DNA metabarcoding studies, makes the analysis more complicated and requires specialized training. However, well-curated databases, as those of major projects (as in the human microbiome), can lead shotgun metagenomics to have even more precise detection of species and diversity compared to DNA metabarcoding. Conversely, for environmental samples, shotgun metagenomics-based assessment of diversity allows identification of only half of the phyla and only 30% of the families when compared to DNA metabarcoding (Tessler *et al.*, 2017). This is mainly due to the lack of specific databases.

Metaproteomics

While proteomics is defined as the study of the different proteins expressed by an organism (Wilkins *et al.*, 1996), metaproteomics involves identification of the functional expression and metabolic activities within a microbial community (Siggins *et al.*, 2012; Fig. 4.1). From a technical point of view, proteomics is based on the employment of high-performance mass spectrometry (MS) to characterize the complete assemblage of proteins expressed by a microbial community. Similar to DNA- and RNA-based NGS approaches, the obtained data sets must be processed by bioinformatics, and peptide sequences must be aligned with specific database entries.

Metabolomics

This methodology relies on the assessment of all metabolites found in a specified sample. The technique is based on the ability of mass spectrometry to identify a large number of molecules by their specific masses and high accuracy. Based on the sensitivity and selectivity of both metabolite recovery and identification, a metabolomics approach can scan the whole set of metabolites present in the given environment (untargeted metabolomics) or focus on specific classes of metabolites (targeted metabolomics). Commonly, these two approaches are coupled. At first, untargeted metabolomics is used to scan the whole spectrum of metabolites and, after studying which functions are present, targeted studies allows one to quantify specific pathways and functions (Johnson *et al.*, 2016). Even if the technical complexity and the difficult interpretation of data limit the accessibility to this technique, the possibility to directly identify metabolites in complex samples makes this technique one of the most promising for the study of symbiotic relationship as it happens within endophytic communities (Kaul *et al.*, 2016).

Multi-omics approaches

Different omics techniques can be combined with each other in order to increase their informative value. For example, proteome-based studies are often incomplete without genomic information and the accuracy of assignments can be substantially increased when these data are added. For this reason, the ideal concept is the parallel bioinformatics assessment of several omics strategies with a multi-omics approach allowing microbial communities to be analysed from different points of view. Since these technologies are in continuous evolution, advantages and disadvantages of omics technologies are difficult to evaluate. As of today, even if the continuous improvement and expansion of databases will gradually resolve most of the problems connected with these technologies, the barrier to their application remains the cost, which relegates their employment only to big projects in advanced stages.

4.4 Microscopic Visualization of Seed Endophytes

Since omics methodologies are based on the extraction of nucleic acids or proteins, they cannot provide useful information on the microorganism localization at microscale level. This is the reason why microscopy is still valuable for complementing molecular microbiology tools as means for the visualization of the microbe-host systems (Cardinale, 2014). We here report the explanation of the main microscopy technique used for the study of seed endophytes.

4.4.1 Confocal laser scanning microscopy

The most frequently used microscopy approach to study plant and seed endophytes is confocal laser scanning microscopy (CLSM) (Cardinale, 2014; Pawley, 2006; Fig. 4.1). This is a widely applicable optical imaging technique for an accurate study for plant-microbe interactions. CLSM captures multiple two-dimensional images at different depths in a sample, allowing the *in situ* observation of host-associated microorganisms with an unprecedented accuracy.

In order to detect specific microorganisms, it is possible to employ the fluorescent *in situ* hybridization (FISH): a molecular cytogenetic technique that allows the identification and localization of cells in their microenvironment (Moter and Göbel, 2000). It uses the hybridization of designable DNA-probes labelled with fluorochromes able to bind with the complementary target sequence of choice. This technique is most frequently used for visualization of microbial colonization patterns (Moter and Göbel, 2000; Rudolf Amann *et al.*, 2001; Rybakova *et al.*, 2017), providing estimates of microbial abundance while avoiding quantification biases associated with cultivation or PCR (Bulgarelli *et al.*, 2012). The possibility to employ DNA-probes targeting specific taxonomic ranges allows a qualitative-quantitative study of microbial populations. While its resolution power in assessing diversity is not remotely comparable with that of molecular techniques, this approach represents an optimal tool for the validation of molecular analysis results while studying basic processes of plant–microbe interactions such as microbial localization, colonization pattern and cell–cell interaction (Cardinale, 2014). An example of the usage of this microscopy technique is provided in the study by Rybakova *et al.* (2017; Fig. 4.2). The authors used CLSM visualizations to investigate microbial colonization patterns in oilseed rape.

4.4.2 Scanning electron microscopy

Other microscopy-based techniques are less frequently applied for the study of seed endophytes; one example is scanning electron microscopy. This microscopy technique relies on the use of a focused beam of electrons scanning the surface of the sample.

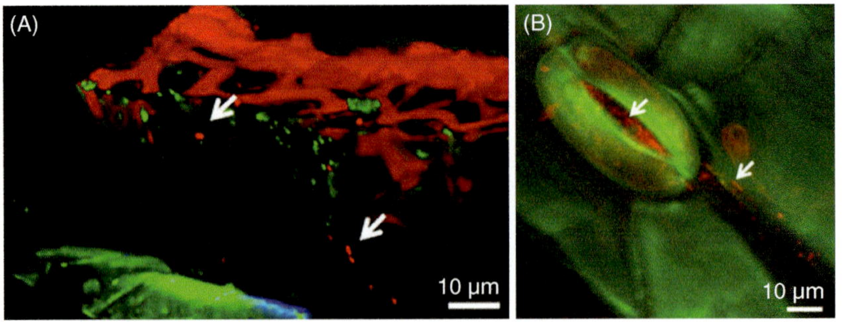

Fig. 4.2. Visualization of bacterial colonization patterns in oilseed rape (A) and seedlings (B). In the first visualization (A), differential BacLight LIVE/DEAD staining was used to visualize living (green) and dead (red) *Serratia plymuthica* cells. In the second micrograph (B), *Gammaproteobacteria* were localized in oilseed rape seedlings. White arrows highlight bacterial colonies. (From Rybakova *et al.*, 2017)

The remarkable magnification power of this microscope allows the visualization of the morphological features of cells by producing micrographs with unmatched three-dimensional quality on natural surfaces. Nevertheless, this technique provides less specific information than CLSM when studying sections of biological samples. Even though this technique provides clear pictures at high magnification levels, it is not possible to obtain information on the taxonomy simultaneously. Therefore, it is of limited use for the study of plant–microbe and cell–cell interactions, which reduces its employment in the study of seed endophytes (Golding *et al.*, 2016).

4.5 Concluding Remarks

Different methods can be employed for the study of seed endophytes. Integrative approaches currently provide the most suitable strategies to describe these microorganisms due to various limitations of single methodologies (Fig. 4.3). While microbial cell culturing seemingly represents the most suitable way to exploit the biotechnological potential of seed endophytes, the low applicability and time requirements for microbial isolation and cultivation are setting boundaries to this approach. Holistic descriptions of seed endophytic communities require the application of molecular techniques for the

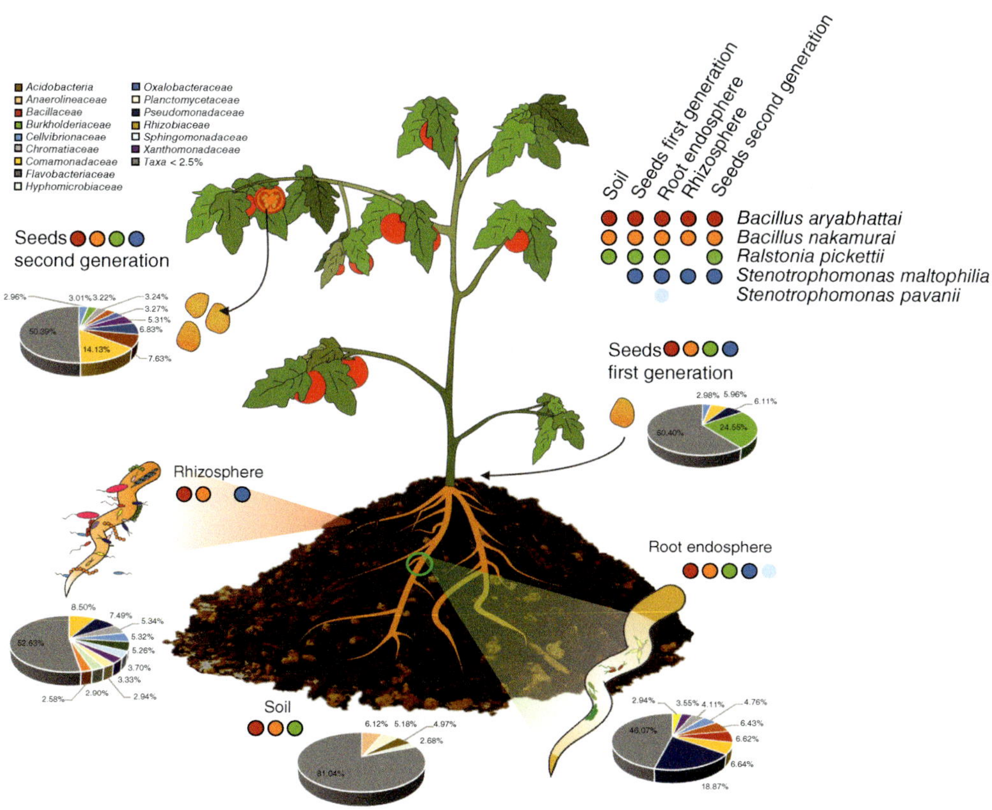

Fig. 4.3. Tracking of endophytes across a plant system. The integration of cultivation-dependent and cultivation-independent techniques provided insights into the allocation of beneficial bacteria across tomato plants. In a recent study (Bergna *et al.*, 2018), bacterial isolation was coupled with phenotyping and 16S metabarcoding to, respectively, detect plant beneficial bacteria and reconstruct the bacterial community of the tomato plant system. By merging these data, it was possible to reconstruct the association of beneficial bacteria to specific plant compartments. The identified key players are indicated for the specific plant compartments.

analysis of community structures and microbial functioning therein. Nevertheless, molecular techniques have also specific limitations. For example, the detection of genetic material from two microorganisms in the same niche could suggest specific interactions between them. However, the validation of their co-localization within the seed can be obtained only using microscopy techniques.

The investigation of seed-borne endophytes is in continuous evolution as are the methodologies. While the endophytic population in seeds has not yet been fully explored, the great applicability of these microorganisms for sustainable agriculture is attracting attention and funding. For this reason, it is feasible to believe that the characterization of this valuable plant compartment will progress very rapidly in the next decade.

As in a mosaic composed by precisely drawn tiles, integrating one methodology to the other shows that a combination of accurate methods is still the best solution to increase our understanding of the ecological role of microorganisms within these essential plant structures.

References

Adam, E., Bernhart, M., Müller, H., Winkler, J. and Berg, G. (2016) The *Cucurbita pepo* seed microbiome: Genotype-specific composition and implications for breeding. *Plant and Soil* 422, 35–49.

Alibrandi, P., Cardinale, M., Rahman, M.M., Strati, F., Ciná, P., *et al.* (2018) The seed endosphere of *Anadenanthera colubrina* is inhabited by a complex microbiota, including *Methylobacterium* spp. and *Staphylococcus* spp. with potential plant-growth promoting activities. *Plant and Soil* 422, 81–99.

Aly, A.H., Debbab, A. and Proksch, P. (2013) Fungal endophytes – Secret producers of bioactive plant metabolites. *Pharmazie* 68, 499–505.

Amann, R., Ludwig, W. and Schleifer, K. (1995) Phylogenetic identification and in situ detection of individual microbial cells without cultivation. *Microbiological Reviews* 59, 143–169. Available at: http://mmbr.asm.org/content/59/1/143.short.

Amann, R., Fuchs, B.M. and Behrens, S. (2001) The identification of microorganisms by fluorescence in situ hybridisation. *Current Opinion in Biotechnology* 12, 231–236.

Arumugam, M., Raes, J., Pelletier, E., Le Paslier, D., Yamada, T., *et al.* (2011) Enterotypes of the human gut microbiome. *Nature* 473, 174–180.

Barret, M., Briand, M., Bonneau, S., Préveaux, A., Valière, S., *et al.* (2015) Emergence shapes the structure of the seed microbiota. *Applied and Environmental Microbiology* 81, 1257–1266.

Bates, S.T., Clemente, J.C., Flores, G.E., Walters, W.A., Parfrey, L.W., *et al.* (2013) Global biogeography of highly diverse protistan communities in soil. *ISME Journal* 7, 652–659.

Beckers, B., Op De Beeck, M., Thijs, S., Truyens, S., Weyens, N., *et al.* (2016) Performance of 16s rDNA primer pairs in the study of rhizosphere and endosphere bacterial microbiomes in metabarcoding studies. *Frontiers in Microbiology* 7, 650.

Berg, G. (2009) Plant-microbe interactions promoting plant growth and health: Perspectives for controlled use of microorganisms in agriculture. *Applied Microbiology and Biotechnology* 84, 11–18.

Berg, G. and Raaijmakers, J.M. (2018) Saving seed microbiomes. *ISME Journal* 12, 1167–1170.

Berg, G., Zachow, C., Müller, H., Philipps, J. and Tilcher, R. (2013) Next-generation bio-products sowing the seeds of success for sustainable agriculture. *Agronomy* 3, 648–656.

Berg, G., Rybakova, D., Grube, M. and Ko berl, M. (2015) The plant microbiome explored: Implications for experimental botany. *Journal of Experimental Botany* 67, 995–1002.

Bergna, A., Cernava, T., Rändler, M., Grosch, R., Zachow, C. and Berg, G. (2018) Tomato seeds preferably transmit plant beneficial endophytes. *Phytobiomes* 2, 1–40.

Bonos, S.A., Wilson, M. M., Meyer, W. A. and Reed Funk, C. (2005) Suppression of red thread in fine fescues through endophyte-mediated resistance. *Applied Turfgrass Science* 2, 105–111.

Bragina, A., Berg, C., Cardinale, M., Shcherbakov, A., Chebotar, V. and Berg, G. (2012) *Sphagnum* mosses harbour highly specific bacterial diversity during their whole lifecycle. *ISME Journal* 6, 802–813.

Breitwieser, F.P., Lu, J. and Salzberg, S.L. (2017) A review of methods and databases for metagenomic classification and assembly. *Briefings in Bioinformatics* bbx120. doi:10.1093/bib/bbx120.

Bulgarelli, D., Rott, M., Schlaeppi, K., Ver Loren van Themaat, E., Ahmadinejad, N., *et al.* (2012) Revealing structure and assembly cues for *Arabidopsis* root-inhabiting bacterial microbiota. *Nature* 488, 91–95.

Caporaso, J.G., Kuczynski, J., Stombaugh, J., Bittinger, K., Bushman, F.D., *et al.* (2010) QIIME allows analysis of high-throughput community sequencing data. *Nature Methods* 7(5), 335–336.

Caporaso, J.G., Lauber, C.L., Walters, W.A., Berg-Lyons, D., Huntley, J., *et al.* (2012) Ultra-high-throughput microbial community analysis on the Illumina HiSeq and MiSeq platforms. *ISME Journal* 6, 1621–1624.

Cardinale, M. (2014) Scanning a microhabitat: Plant-microbe interactions revealed by confocal laser microscopy. *Frontiers in Microbiology* 5, 1–10.

Cernava, T., Aschenbrenner, I.A., Grube, M., Liebminger, S. and Berg, G. (2015) A novel assay for the detection of bioactive volatiles evaluated by screening of lichen-associated bacteria. *Frontiers in Microbiology* 6, 1–9.

Chen, H., Wu, H., Yan, B., Zhao, H., Liu, F., *et al.* (2018) Core microbiome of medicinal plant *Salvia miltiorrhiza* seed: A rich reservoir of beneficial microbes for secondary metabolism? *International Journal of Molecular Sciences* 19(3), 672.

Chhabra, S. and Dowling, D.N. (2017) Endophyte-promoted nutrient acquisition: Phosphorus and iron. In: Doty, S. (ed.) *Functional Importance of the Plant Microbiome*. Springer, Cham, Switzerland, pp. 21–41.

Clooney, A.G., Fouhy, F., Sleator, R.D., O'Driscoll, A., Stanton, C., Cotter, P.D. and Claesson, M.J. (2016) Comparing apples and oranges?: Next generation sequencing and its impact on microbiome analysis. *PLOS ONE* 11, 1–16.

Cole, J.R., Wang, Q., Fish, J.A., Chai, B., Mcgarrell, D.M., *et al.* (2014) Ribosomal Database Project: Data and tools for high throughput rRNA analysis. *Nucleic Acids Research* 42, 633–642.

Compant, S., Mitter, B., Colli-Mull, J.G., Gangl, H. and Sessitsch, A. (2011) Endophytes of grapevine flowers, berries, and seeds: Identification of cultivable bacteria, comparison with other plant parts, and visualization of niches of colonization. *Microbial Ecology* 62, 188–197.

De Filippis, F., Laiola, M., Blaiotta, G. and Ercolini D. (2017) Different amplicon targets for sequencing-based studies of fungal diversity. *Applied and Environmental Microbiology* 83, 1–9.

Dekkers, B.J.W., Pearce, S., Marshall, A., Widera, P., Gilbert, J., *et al.* (2013) Transcriptional dynamics of two seed compartments with opposing roles in *Arabidopsis* seed germination. *Plant Physiology* 163, 205–215.

Edgar, R.C. (2013) UPARSE: Highly accurate OTU sequences from microbial amplicon reads. *Nature Methods* 10, 996–998.

Eevers, N., Gielen, M., Sánchez-López, A., Jaspers, S., White, J.C., *et al.* (2015) Optimization of isolation and cultivation of bacterial endophytes through addition of plant extract to nutrient media. *Microbial Biotechnology* 8, 4.

Esteve-Zarzoso, B., Belloch, C., Uruburu, F. and Querol, A. (1999) Identification of yeasts by RFLP analysis of the 5.8S rRNA gene and the two ribosomal internal transcribed spacers. *International Journal of Systematic Bacteriology* 49, 329–337.

Gao, H., Li, G. and Lou, H.-X. (2018) Structural diversity and biological activities of novel secondary metabolites from endophytes. *Molecules* 23, 646.

Gardes, M. and Bruns, T.D. (1993) ITS primers with enhanced specificity for basidiomycetes – Application to the identification of mycorrhizae and rusts. *Molecular Ecology* 2, 113–118.

Gardner, J.M., Feldman, A.W. and Zablotowicz, R.M. (1982) Identity and behavior of xylem-residing bacteria in rough lemon roots of Florida citrus trees. *Applied and Environmental Microbiology* 43, 1335–1342.

Glassner, H., Zchori-Fein, E., Yaron, S., Sessitsch, A., Sauer, U., *et al.* (2018) Bacterial niches inside seeds of *Cucumis melo* L. *Plant and Soil* 422(1–2), 101–113.

Golding, C.G., Lamboo, L.L., Beniac, D.R. and Booth, T.F. (2016) The scanning electron microscope in microbiology and diagnosis of infectious disease. *Scientific Reports* 6, 1–8.

Gunatilaka, A.A.L. (2006) Natural products from plant-associated microorganisms: Distribution, structural diversity, bioactivity, and implications of their occurrence. *Journal of Natural Products* 69, 509–526.

Hallmann, J., Quadt-Hallmann, A., Mahaffee, W.F. and Kloepper, J.W. (1997) Bacterial endophytes in agricultural crops. *Canadian Journal of Microbiology* 43, 895–914.

Hallmann, J., Berg, G. and Schulz, B. (2006) Isolation procedures for endophytic microorganisms. *Microbial Root Endophytes* 9, 299–319.

Hardoim, P.R., Hardoim, C.C.P., van Overbeek, L.S. and van Elsas, J.D. (2012) Dynamics of seed-borne rice endophytes on early plant growth stages. *PLOS ONE* 7(2), e30438.

Hardoim, P.R., van Overbeek, L.S., Berg, G., Pirttilä, A.M., Compant, S., *et al.* (2015) The hidden world within plants: Ecological and evolutionary considerations for defining functioning of microbial endophytes. *Microbiology and Molecular Biology Reviews* 79, 293–320.

Huang, Y., Zhang, M., Deng, Z. and Cao, L. (2018) Evaluation of probiotic diversity from soybean (*Glycine max*) seeds and sprouts using Illumina-based sequencing method. *Probiotics and Antimicrobial Proteins* 10, 293–298.

Hugerth, L.W. and Andersson, A.F. (2017) Analysing microbial community composition through amplicon sequencing: From sampling to hypothesis testing. *Frontiers in Microbiology* 8, 1–22.

Irizarry, I. and White, J.F. (2017) Application of bacteria from noncultivated plants to promote growth, alter root architecture and alleviate salt stress of cotton. *Journal of Applied Microbiology* 122(4), 1110–1120.

Johnson, C.H., Ivanisevic, J. and Siuzdak, G. (2016) Metabolomics: Beyond biomarkers and towards mechanisms. *Nature Reviews Molecular Cell Biology* 17, 451–459.

Johnston-Monje, D. and Raizada, M.N. (2011) Conservation and diversity of seed associated endophytes in Zea across boundaries of evolution, ethnography and ecology. *PLOS ONE* 6, e20396.

Johnston-Monje, D., Lundberg, D.S., Lazarovits, G., Reis, V.M. and Raizada, M.N. (2016) Bacterial populations in juvenile maize rhizospheres originate from both seed and soil. *Plant and Soil* 405(1–2), 337–355.

Josephson K.L., Gerba, C.P. and Pepper, I.L. (1993) Polymerase chain reaction detection of nonviable bacterial pathogens. *Applied and Environmental Microbiology* 59, 3513–3515.

Kaul, S., Sharma, T. and Dhar, M.K. (2016) "Omics" tools for better understanding the plant–endophyte interactions. *Frontiers in Plant Science* 7, 1–9.

Khalaf, E.M.K. and Raizada, M.N. (2016) Taxonomic and functional diversity of cultured seed associated microbes of the cucurbit family. *BMC Microbiology* 16, 1–16.

Khalaf, E.M.K. and Raizada, M.N. (2018) Bacterial seed endophytes of domesticated cucurbits antagonize fungal and oomycete pathogens including powdery mildew. *Frontiers in Microbiology* 9, 42.

Khamchatra, N., Dixon, K.W., Tantiwiwat, S. and Piapukiew, J. (2016) Symbiotic seed germination of an endangered epiphytic slipper orchid, *Paphiopedilum villosum* (Lindl.) Stein. from Thailand. *South African Journal of Botany* 104, 76–81.

Klaedtke, S., Jacques, M.A., Raggi, L., Préveaux, A., Bonneau, S., *et al.* (2016) Terroir is a key driver of seed-associated microbial assemblages. *Environmental Microbiology* 18(6), 1792–1804.

Kurokawa, K., Itoh, T., Kuwahara, T., Oshima, K., Toh, H., *et al.* (2007) Comparative metagenomics revealed commonly enriched gene sets in human gut microbiomes. *DNA Research* 14, 169–181.

Lentendu, G., Wubet, T., Chatzinotas, A., Wilhelm, C., Buscot, F., *et al.* (2014) Effects of long-term differential fertilization on eukaryotic microbial communities in an arable soil: A multiple barcoding approach. *Molecular Ecology* 23, 3341–3355.

Liaqat, F. and Eltem, R. (2016) Identification and characterization of endophytic bacteria isolated from in vitro cultures of peach and pear rootstocks. *3 Biotech* 6, 1–8.

Lindahl, B.D., Nilsson, R.H., Tedersoo, L., Abarenkov, K., Carlsen, T., *et al.* (2013) Fungal community analysis by high-throughput sequencing of amplified markers – A user's guide. *New Phytologist* 199, 288–299.

Links, M.G., Dumonceaux, T.J., Hemmingsen, S.M. and Hill, J.E. (2012) The *Chaperonin-60* universal target is a barcode for bacteria that enables *de novo* assembly of metagenomic sequence data. *PLOS ONE* 7, 11.

Links, M.G., Demeke, T., Gräfenhan, T., Hill, J.E., Hemmingsen, S.M., *et al.* (2014) Simultaneous profiling of seed-associated bacteria and fungi reveals antagonistic interactions between microorganisms within a shared epiphytic microbiome on *Triticum* and *Brassica* seeds. *New Phytologist* 202, 542–553.

Malinich, E.A. and Bauer, C.E. (2018) The plant growth promoting bacterium *Azospirillum brasilense* is vertically transmitted in *Phaseolus vulgaris* (common bean). *Symbiosis* 76, 97–108.

Martinez-Klimova, E., Rodríguez-Peña, K. and Sánchez, S. (2017) Endophytes as sources of antibiotics. *Biochemical Pharmacology* 134, 1–17.

McDonald, D., Price, M.N., Goodrich, J., Nawrocki, E.P., DeSantis, T.Z., Hugenholtz, P., *et al.* (2012) An improved Greengenes taxonomy with explicit ranks for ecological and evolutionary analyses of bacteria and archaea. *ISME Journal* 6, 610–618.

McKinnon, A.C. (2016) Plant tissue preparation for the detection of an endophytic fungus in planta. In: Glare, T. and Moran-Diez, M. (eds) *Microbial-Based Biopesticides: Methods and Protocols. Methods in Molecular Biology*. Humana Press, New York, vol. 1477, pp. 167–173.

Mendes, R., Kruijt, M., De Bruijn, I., Dekkers, E., Van Der Voort, M., *et al.* (2011) Deciphering the rhizosphere microbiome for disease-suppressive bacteria. *Science* 332, 1097–1100.

Mitter, B., Pfaffenbichler, N., Flavell, R., Compant, S., Antonielli, L., *et al.* (2017) A new approach to modify plant microbiomes and traits by introducing beneficial bacteria at flowering into progeny seeds. *Frontiers in Microbiology* 8, 11.

Moter, A. and Göbel, U.B. (2000) Fluorescence *in situ* hybridization (FISH) for direct visualization of microorganisms. *Journal of Microbiological Methods* 41, 85–112.

Munkvold, G.P. (2009) Seed pathology progress in academia and industry. *Annual Review of Phytopathology* 47, 285–311.

Myrold, D.D., Zeglin, L.H. and Jansson, J.K. (2014) The potential of metagenomic approaches for understanding soil microbial processes. *Soil Science Society of America Journal* 78, 3.

Nichols, D., Lewis, K., Orjala, J., Mo, S., Ortenberg, R., *et al.* (2008) Short peptide induces an "uncultivable" microorganism to grow in vitro. *Applied and Environmental Microbiology* 74, 4889–4897.

Nocker, A., Sossa-fernandez, P., Burr, M.D. and Camper, A.K. (2007) Use of propidium monoazide for live/dead distinction in microbial ecology. *Applied and Environmental Microbiology* 73, 5111–5117.

Ohno, M., Shiratori, H., Park, M., Saitoh, Y., Kumon, *et al.* (2000) *Symbiobacterium thermophilum* gen. nov., sp. nov., a symbiotic thermophile that depends on co-culture with a *Bacillus* strain for growth. *International Journal of Systematic and Evolutionary Microbiology* 50, 1829–1832.

Pawley, J.B. (2006) Fundamental limits in confocal microscopy. In: Pawley, J.B. (ed.) *Handbook of Biological Confocal Microscopy*. Springer, Boston, Massachusetts, pp. 20–42.

Philipson, M.N. and Blair, I.D. (1957) Bacteria in clover root tissue. *Canadian Journal of Microbiology* 3, 125–129.

Pitzschke, A. (2016) Developmental peculiarities and seed-borne endophytes in Quinoa: Omnipresent, robust bacilli contribute to plant fitness. *Frontiers in Microbiology* 7, 1–15.

Prosser, J.I. (2010) Replicate or lie. *Environmental Microbiology* 12, 1806–1810.

Puente, M.E., Li, C.Y. and Bashan, Y. (2009) Endophytic bacteria in cacti seeds can improve the development of cactus seedlings. *Environmental and Experimental Botany* 66, 402–408.

Qin, Y., Pan, X. and Yuan, Z. (2016) Seed endophytic microbiota in a coastal plant and phytobeneficial properties of the fungus *Cladosporium cladosporioides*. *Fungal Ecology* 24, 53–60.

Quast, C., Pruesse, E., Yilmaz, P., Gerken, J., Schweer, T., *et al.* (2013) The SILVA ribosomal RNA gene database project: Improved data processing and web-based tools. *Nucleic Acids Research* 41, 590–596.

Rahman, M.M., Flory, E., Koyro, H.W., Abideen, Z., Schikora, A., *et al.* (2018) Consistent associations with beneficial bacteria in the seed endosphere of barley (*Hordeum vulgare* L.). *Systematic and Applied Microbiology* 41(4), 386–398.

Rezki, S., Campion, C., Iacomi-Vasilescu, B., Preveaux, A., Toualbia, Y., *et al.* (2016) Differences in stability of seed-associated microbial assemblages in response to invasion by phytopathogenic microorganisms. *PeerJ* 4, e1923.

Rezki, S., Campion, C., Simoneau, P., Jacques, M.A., Shade, A., *et al.* (2018) Assembly of seed-associated microbial communities within and across successive plant generations. *Plant and Soil* 422, 67–79.

Rybakova, D., Cernava, T., Köberl, M., Liebminger, S., Etemadi, M., *et al.* (2016) Endophytes-assisted biocontrol: Novel insights in ecology and the mode of action of *Paenibacillus*. *Plant and Soil* 405, 125–140.

Rybakova, D., Mancinelli, R., Wikström, M., Birch-Jensen, A.S., Postma, J., *et al.* (2017) The structure of the *Brassica napus* seed microbiome is cultivar-dependent and affects the interactions of symbionts and pathogens. *Microbiome* 5, 104.

Sakai, M. and Ikenaga, M. (2013) Application of peptide nucleic acid (PNA)-PCR clamping technique to investigate the community structures of rhizobacteria associated with plant roots. *Journal of Microbiological Methods* 92, 281–288.

Sánchez-López, A.S., Thijs, S., Beckers, B., González-Chávez, M.C., Weyens, N., *et al.* (2018). Community structure and diversity of endophytic bacteria in seeds of three consecutive generations of *Crotalaria pumila* growing on metal mine residues. *Plant and Soil* 422(1–2), 51–66.

Sanschagrin, S. and Yergeau, E. (2014) Next-generation sequencing of 16S ribosomal RNA gene amplicons. *Journal of Visualized Experiments* (90). doi:10.3791/51709

Schellenberg, J., Links, M.G., Hill, J.E., Hemmingsen, S.M., Peters, G.A., *et al.* (2011) Pyrosequencing of chaperonin-60 (*cpn60*) amplicons as a means of determining microbial community composition. In: Kwon, Y. and Ricke, S. (eds) *High-Throughput Next Generation Sequencing: Methods and Applications Methods in Molecular Biology (Methods and Protocols)*. Humana Press, Totowa, New Jersey, vol. 733, pp. 143–158.

Schoch, C.L., Seifert, K.A., Huhndorf, S., Robert, V., Spouge, J.L., *et al.* (2012) Nuclear ribosomal internal transcribed spacer (ITS) region as a universal DNA barcode marker for *Fungi*. *Proceedings of the National Academy of Sciences of the United States of America* 109, 6241–6246.

Schulz, B., Sucker, J., Aust, H.J., Krohn, K., Ludewig, K., *et al.* (1995) Biologically active secondary metabolites of endophytic *Pezicula* species. *Mycological Research* 99, 1007–1015.

Schulze-Schweifing, K., Banerjee, A. and Wade, W.G. (2014) Comparison of bacterial culture and 16S rRNA community profiling by clonal analysis and pyrosequencing for the characterization of the dentine caries-associated microbiome. *Frontiers in Cellular and Infection Microbiology* 4, 1–8.

Shahzad, R., Waqas, M., Khan, A.L., Asaf, S., Khan, M.A., *et al.* (2016) Seed-borne endophytic *Bacillus amyloliquefaciens* RWL-1 produces gibberellins and regulates endogenous phytohormones of *Oryza sativa*. *Plant Physiology and Biochemistry* 106, 236–243.

Shahzad, R., Khan, A.L., Bilal, S., Waqas, M., Kang, S.-M., *et al.* (2017) Inoculation of abscisic acid-producing endophytic bacteria enhances salinity stress tolerance in *Oryza sativa*. *Environmental and Experimental Botany* 136, 68–77.

Shahzad, R., Khan, A.L., Bilal, S., Asaf, S. and Lee, I. (2018) What is there in seeds? Vertically transmitted endophytic resources for sustainable improvement in plant growth. *Frontiers in Plant Science* 9, 1–10.

Siggins, A., Gunnigle, E. and Abram, F. (2012) Exploring mixed microbial community functioning: Recent advances in metaproteomics. *FEMS Microbiology Ecology* 80, 265–280.

Stewart, E.J. (2012) Growing unculturable bacteria. *Journal of Bacteriology* 194, 4151–4160.

Tan, R.X. and Zou, W.X. (2001) Endophytes: A rich source of functional metabolites. *Natural Product Reports* 18, 448–459.

Tayung, K., Sarkar, M. and Baruah, P. (2012) Endophytic fungi occurring in *Ipomoea carnea* tissues and their antimicrobial potentials. *Brazilian Archives of Biology and Technology* 55, 653–660.

Tedersoo, L., Anslan, S., Bahram, M., Põlme, S., Riit, T., *et al.* (2015) Shotgun metagenomes and multiple primer pair-barcode combinations of amplicons reveal biases in metabarcoding analyses of fungi. *MycoKeys* 10, 1–43.

Tedersoo, L., Bahram, M., Cajthaml, T., Põlme, S., Hiiesalu, I., *et al.* (2016) Tree diversity and species identity effects on soil fungi, protists and animals are context dependent. *ISME Journal* 10, 346–362.

Tessler, M., Neumann, J.S., Afshinnekoo, E., Pineda, M., Hersch, R., *et al.* (2017) Large-scale differences in microbial biodiversity discovery between 16S amplicon and shotgun sequencing. *Scientific Reports* 7, 1–14.

Torres-Cortés, G., Bonneau, S., Bouchez, O., Genthon, C., Briand, M., *et al.* (2018) Functional microbial features driving community assembly during seed germination and emergence. *Frontiers in Plant Science* 9, 1–16.

Torsvik, V. and Øvreås, L. (2002) Microbial diversity and function in soil: From genes to ecosystems. *Current Opinion in Microbiology* 5, 3.

Truyens, S., Weyens, N., Cuypers, A. and Vangronsveld, J. (2014) Bacterial seed endophytes: Genera, vertical transmission and interaction with plants. *Environmental Microbiology Reports* 7, 40–50.

Vandenkoornhuyse, P., Quaiser, A., Duhamel, M., Le Van, A. and Dufresne, A. (2015) The importance of the microbiome of the plant holobiont. *New Phytologist* 206, 1196–1206.

Vannier, N., Mony, C., Bittebiere, A.K., Michon-Coudouel, S., Biget, M., *et al.* (2018) A microorganisms' journey between plant generations. *Microbiome* 6(1), 79.

Verma, S.K. and White, J.F. (2018) Indigenous endophytic seed bacteria promote seedling development and defend against fungal disease in browntop millet (*Urochloa ramosa* L.). *Journal of Applied Microbiology* 124(3), 764–778.

Větrovský, T. and Baldrian, P. (2013) The variability of the 16S rRNA gene in bacterial genomes and its consequences for bacterial community analyses. *PLOS ONE* 8, 1–10.

Walitang, D.I., Kim, K., Madhaiyan, M., Kim, Y.K., Kang, Y., *et al.* (2017) Characterizing endophytic competence and plant growth promotion of bacterial endophytes inhabiting the seed endosphere of rice. *BMC Microbiology* 17(1), 209.

White, J.F., Kingsley, K.I., Kowalski, K.P., Irizarry, I., Micci, A., *et al.* (2018) Disease protection and allelopathic interactions of seed-transmitted endophytic pseudomonads of invasive reed grass (*Phragmites australis*). *Plant and Soil* 422(1–2), 195–208.

Wilkins, M.R., Sanchez, J.C., Gooley, A.A., Appel, R.D., Humphery-Smith, I., *et al.* (1996) Progress with proteome projects: Why all proteins expressed by a genome should be identified and how to do it. *Biotechnology and Genetic Engineering Reviews* 13, 19–50.

Wooley, J.C., Godzik, A. and Friedberg, I. (2010) A primer on metagenomics. *PLOS Computational Biology* 6, e1000667.

Yamamoto, S. and Harayama, S. (1995) PCR amplification and direct sequencing of *gyrB* genes with universal primers and their application to the detection and taxonomic analysis of *Pseudomonas putida* strains. *Applied and Environmental Microbiology* 61, 1104–1109.

5 Mitigating Climate Impacts on Crop Production via Symbiosis

Rusty Rodriguez,[1,2,3]* Alec Baird,[4] Sang Cho,[1] Zachery Gray,[1] Evan Groover,[5] Roman Harto,[1] Marian Hsieh,[1] Katie Malmberg,[1] Ryan Manglona,[1] Malia Mercer,[1] Natalie Nasman,[1] Tatiana Nicklason,[1] Melissa Rienstra,[1] Alex Van Inwegen,[1] Andy VanHooser[1] and Regina Redman[1,2]

[1]*Adaptive Symbiotic Technologies, Seattle, Washington, USA;* [2]*Symbiogenics, Seattle, Washington, USA;* [3]*University of Washington, Seattle, Washington, USA;* [4]*University of California, Los Angeles, California, USA;* [5]*University of California, Berkeley, California, USA*

Abstract

All plants in natural ecosystems are thought to be symbiotic with mutualistic fungal endophytes that can significantly improve plant fitness by enhancing root growth, fertility, nutrient acquisition and biotic or abiotic stress tolerance. Recently, Adaptive Symbiotic Technologies developed the product line BioEnsure® comprised of fungal endophytes that confer abiotic stress tolerance (drought, temperature, salinity) to food crops (www.adsymtech.com). These endophytes communicate similarly with monocots and eudicots to enhance crop production on marginal lands and mitigate the impacts of high daytime or nighttime temperatures on crop fertilization. Yield benefits in endophyte-colonized plants are remarkable and directly proportional to stress levels with average yield increases of 3–5% above control plants under low stress and >26% under high stress. The relationship between stress and yield enhancement was best exemplified in Rajasthan, India, where BioEnsure®-treated pearl millet and mung bean seeds were provided to 400 small landholding farmers. Under the hot, dry growing conditions that are typical in Rajasthan, the average yield increases were 29% for pearl millet and 64% for mung bean compared to untreated plants. This demonstrated the power of this technology to increase food security, animal fodder, carry-over seed and revenues. Interest in the USA is growing with BioEnsure®-treated seeds planted in 300,000 acres in 2017 and 900,000 acres in 2018, and >2,000,000 acres are projected for 2019.

5.1 Introduction

The greatest threats to agriculture in this century are abiotic stresses such as drought, temperature and salinity, all of which are increasing in frequency and severity as a result of climate change (https://climate.nasa.gov/effects/). In fact, the majority of crop production globally involves small landholders and dryland cultivation, and climate impacts on crop yields are already increasing poverty, famine, human migration and political instability (Brinkman and Hendrix, 2011; Deaton and Lipka, 2015; Skøt *et al.*, 2016; Simmons and Flowers, 2017). An obvious solution to this problem is to generate abiotic stress-tolerant

* Corresponding author e-mail: rjrodriguez@adsymtech.com

crops. Over the last several decades, a tremendous effort was undertaken to generate abiotic stress-tolerant crops either through breeding or through genetic modification, neither of which have been very fruitful. This is largely because the underlying strategies incorrectly assumed plants adapt themselves to stress in natural ecosystems (Gurian-Sherman, 2012; Komives and Kiraly, 2017; Nuccio *et al.*, 2018). Over the last 20 years, it has become clear that plants in high-stress habitats commonly adapt to stress via symbiotic associations with fungal endophytes (Lugtenberg *et al.*, 2016; Singh *et al.*, 2011; Lofgren *et al.*, 2018). In fact, without the fungal partners, native plants are no more adapted to abiotic stress than agricultural crops.

Based on how plants in nature adapt to stress, Adaptive Symbiotic Technologies (AST) developed BioEnsure®, a novel seed treatment containing mutualistic fungal endophytes that confer significant levels of abiotic stress tolerance to both monocot and eudicot crop plants (www.adsymtech.com; Redman and Rodriguez, 2017). Field results demonstrated that during periods of high drought, temperature and/or salinity stress, BioEnsure® increased crop yields on average from 26 to 60%, and under low to no stress, BioEnsure® increased yields by an average of 3–5%. This chapter will cover the science behind BioEnsure®, seven years of field-testing data and efforts to use this technology to enhance the food security of poor rural farmers in India in an effort to break the chain of poverty.

5.2 Fungal Endophytes and Abiotic Stress Tolerance of Plants

There are at least four classes of fungal endophytes associated with plants (Rodriguez *et al.*, 2009) that can be differentiated based on host range and colonization, mode of transmission, symbiotic function and genetic diversity. These endophytes can have profound impacts on plant physiology, health, fecundity, survival, adaptation and ecology. The fact that all plants in nature are thought to be symbiotic with fungal endophytes suggests that these symbiotic associations are a universal aspect of plant ecology. Interestingly, it appears that many endophytes confer some level of drought tolerance to plants, suggesting that this mutualistic benefit may be an evolutionarily conserved property of endophytes that may have played a role in the movement of plants onto land 450 million years ago (Redecker *et al.*, 2000; Krings *et al.*, 2007). In addition, many endophytes confer tolerance to other environmental stresses including temperature, salinity, heavy metals, toxic chemicals and low nutrients (Figs 5.1 and 5.2), traits which can develop in a habitat-specific manner (Rodriguez *et al.*, 2008).

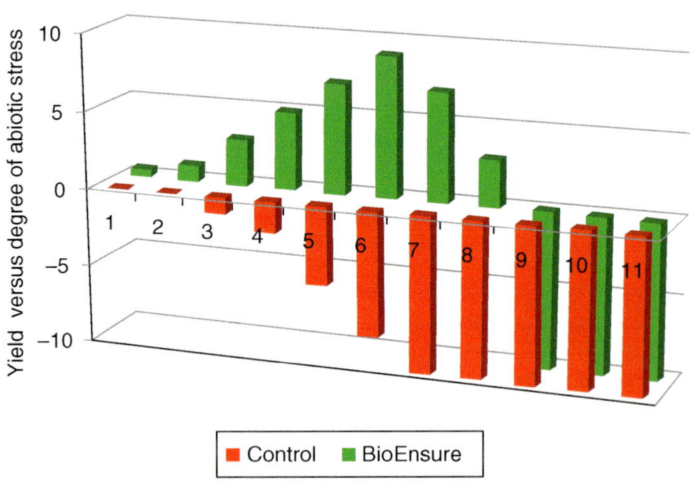

Fig. 5.1. Hypothetical relationship between stress during the growing season, stress tolerance conferred by endophytes and crop yield. The increasing numbers in the chart reflect increasing amounts of abiotic stress.

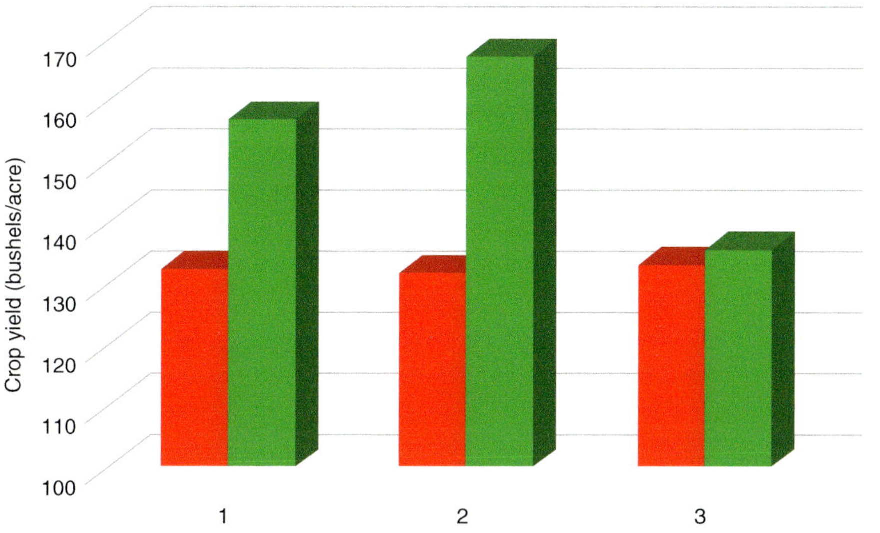

Fig. 5.2. Yield of untreated control (red) and BioEnsure-treated (green) corn plants versus irrigation level. Seeds of a commercial corn hybrid were treated with a rotary aspirating seed treater and plants grown with 100% (1), 75% (2), 50% (3) of recommended levels of water. Each bar represents an average of four replicate plots (4 rows × 30 ft), and plants were exposed to high heat stress throughout the growing season. Yield differences between untreated and BioEnsure-treated plants at 100% and 75% irrigation levels were statistically significant ($P < 0.1$) but not at the 50% irrigation level, where all plants were negatively impacted by the high level of stress. This was an independent study performed by RD4Ag in Yuma, Arizona.

The ecological significance of symbiotically conferred stress tolerance is highlighted by the fact that many plants thriving in high stress are not themselves stress-adapted even though they may have been there for extended periods of time. This was first observed with *Dichanthelium lanuginosum* (tropical panic grass) that thrives in geothermal soils that become very hot and dry during the summer months when temperatures can reach >60°C. Heat tolerance of this plant is based on a symbiotic communication with the fungal endophyte *Curvularia protuberata*. Remarkably, symbiotic communication is responsible for the fitness and survival of both partners as neither partner is heat tolerant when grown axenically (Redman *et al.*, 2002). To complicate things further, there is a double-stranded RNA virus in the fungus that is also required for heat tolerance of the plant. This three-way, cross domain of life, symbiosis is not obligate because the plant and fungus can establish symbioses with genetically unrelated partners (Márquez *et al.*, 2007). In fact, the vast majority of plant fungal symbioses do not involve obligate species–species interactions. The basis for this is unknown, but we hypothesize that it reflects an evolutionary need for adaptive flexibility that allows plants to colonize complex landscapes that vary in environmental stresses and for endophytes to identify optimal host physiology to optimize their fitness.

5.3 Alternating Symbiotic Lifestyles of Fungal Endophytes

The outcome of all symbiotic associations and the benefits realized is based on communication between the partners. Therefore, it is not surprising that either partner can alter the outcome of the association. In fact, individual fungal isolates can alter the outcome of symbiotic associations, from mutualism to parasitism, based on the genetic and/or ecological environment, a phenomenon defined as symbiotic lifestyle switching (Redman *et al.*, 2001; Alvarez-Loayza *et al.*, 2011; Lofgren *et al.*, 2018). The influence of host genetics on the outcome of symbioses has been reported for species of *Colletotrichum* and *Fusarium*. Several plant-pathogenic *Colletotrichum* species are able to colonize and express mutualistic lifestyles in asymptomatic host plants. For example, isolates of *Colletotrichum magna* that are virulent pathogens of cucurbit species can colonize tomato plants and express mutualistic lifestyles. In these symbiotic lifestyle switches, mutualism was confirmed by enhanced plant biomass, drought tolerance and disease protection (Redman *et al.*, 2001).

Fusarium graminearum is commonly found in the plant tissues and seeds of asymptomatic native prairie grasses in the Midwestern USA (Lofgren *et al.*, 2018). *F. graminearum* is known as a pathogen of agricultural grain crops such as wheat where it produces the toxin DON (vomitoxin) and causes head blight. In the native prairie grasses, there was no evidence for vomitoxin production even though *F. graminearum* was present. However, when *F. graminearum* isolates from native grasses were inoculated onto commercial wheat varieties, the plants were colonized and vomitoxin *was* produced. Although the symbiotic lifestyle of the isolates in native grasses has not been assessed, it is clear that a genetically induced lifestyle shift occurs in commercial wheat.

Environmental conditions can also result in a symbiotic lifestyle switch as reported for *a common symbiotic association* in tropical habitats between palm trees and the fungal endophyte *Diplodia mutila*. In this association, the fungus switches symbiotic lifestyle in response to light. When the plants are growing in shade, *D. mutila* expresses either a mutualistic or commensal lifestyle. However, when the plants are shifted into direct light, *D. mutila* expresses a pathogenic lifestyle (Alvarez-Loayza *et al.*, 2011).

Collectively, these studies highlight the importance of communication in the outcome of symbioses and the level of 'symbiotic plasticity' that can influence the ability of both partners to establish and thrive in natural habitats. More importantly, these observations indicate that there is still much to learn about symbiotic associations and their significance in the biology, ecology, adaptation and evolution of life on earth.

5.4 Endophyte Commercialization

Since the inception of agriculture, more than 10,000 years ago, abiotic stress has been a major concern of farmers. For thousands of years, agriculture expanded geographically without significant technological inputs. However, that changed during the industrial revolution with the introduction of steel plows and fuel-driven tractors. One of the unintended consequences of industrialization was an increase in atmospheric levels of CO_2 and other greenhouse gases that have caused increases in temperatures and altered rainfall patterns globally. This has resulted in an increase in the frequency and severity of severe drought and temperature events leading to a global concern over climate impacts on agriculture. The need for climate-resilient crops is at an all-time high and has been considered a technological barrier to food security. To overcome this agricultural barrier, AST began testing fungal endophytes for the ability to confer drought, temperature (high and low) and salt stress tolerance to food crops (www.adsymtech.com; Redman and Rodriguez, 2017). In 2012, AST began R&D efforts to overcome the many hurdles to commercialization including scale-up production, product shelf life, environmental vulnerabilities, regulatory approval and market penetration. At the same time, field testing began and has expanded each year to encompass 40 US states and many locations in South America, Africa, Europe, Australia and India. In 2018, experimental field testing was undertaken on more than 30,000 acres in the USA with both liquid and powder formulations used to treat seeds, applied in furrow during planting, or applied directly to plants via foliar spray or fertigation.

In 2017, AST commercialized its first product BioEnsure® as a microbial seed treatment to generate climate-resilient crops by virtue of symbiotic communication. While the amount of product sold in 2017 was used to treat 230,000 acres of various crops, the amount sold in 2018 was used on 900,000 acres. BioEnsure® crop yield enhancement in 2017 and 2018 exceeded expectations, and the projected sales for 2019 will cover more than 2,000,000 acres planted with both major and minor crop species.

5.5 BioEnsure® Field Performance

The potential benefits of any chemical or biological product on crop production can be visualized by monitoring seedling emergence, stand, seedling vigour, above- and below-ground biomass, chlorophyll levels, spectral reflectance patterns, fertilization, grain fill and time to maturity. Regardless of the metrics analysed, farmers are concerned about abiotic stress tolerance, yield and yield quality of crops. Laboratory and greenhouse studies between 2008 and 2012 identified a number of endophytes as candidates for commercialization to improve crop production. To determine if the lab and greenhouse results were meaningful, we began field testing with independent cooperators to test the efficacy of BioEnsure® in field trials with a diversity of crops (Table 5.1). As anticipated from controlled studies (Redman and Rodriguez, 2017), the BioEnsure® endophytes conferred abiotic stress tolerance to both monocot and eudicot crop plants. Moreover, field testing allowed us to develop a formulation of endophytes capable of generating climate-resilient crops with enhanced tolerance to multiple abiotic stresses (water, temperature, salinity, low nutrient).

Table 5.1. Benefits of BioEnsure treatment on crops in relation to plant biomass, stress tolerance and overall yield.

Crop	Stress tolerance	Yield
Alfalfa	+	+
Barley[a]	nyd	+
Blueberries	nyd	nyd
Canola	+	nyd
Corn[a]	+	+
Cotton	+	+
Cucumber	nyd	nyd
Dry beans[a]	nyd	nyd
Field peas	+	+
Guar	+	+
Leafy greens	nyd	nyd
Lentils[a]	+	+
Millet[a]	+	+
Mung bean	+	+
Okra	+	+
Onion	+	+
Pasture grass	+	+
Potato[a]	nyd	+
Rice[a]	+	+
Sesame	+	+
Sorghum	+	+
Soybean[a]	+	+
Sugar beets	+	+
Wheat[a]	+	+

[a]Some of the 20 major crops that sustain human life.
nyd = not yet determined.

Field test results have increased our understanding of the relationship between endophytes, abiotic stress and crop production (Fig. 5.1). The yield benefits conferred by BioEnsure® are small when stress levels are low. However, as stress levels increase, symbiotically enhanced yield differences increase until the stress levels reach a point where all plants are impacted. There appears to be three phases of the relationship between stress and yield. The first phase is when untreated control plant yields are not impacted by the stress but symbiotic crop yields increase significantly. The second phase is when control crop yields diminish significantly and the yields of symbiotic plants either are maintained or increase. The third phase is when the stress levels increase to a point where all crop yields are negatively impacted. For example, when BioEnsure® benefits were assessed in relation to recommended irrigation levels, corn yields of treated plants were higher at 75% of recommended irrigation levels compared to the 100% level (Fig. 5.2). The climatic conditions during the irrigation trial were very hot and dry, which resulted in significant symbiotic increases in yields even at 100% irrigation.

Remarkably, individual fungal endophytic isolates are able to confer similar levels of abiotic stress tolerance to monocots and eudicots (Rodriguez *et al*., 2008). These plant lineages diverged more than 145 million years ago (Wolfe *et al*., 1989; Yang *et al*., 1999; Chaw *et al*., 2004), suggesting that the symbiotic communication established was prior to that divergence. The ability of fungal endophytes to confer stress tolerance to these plant lineages is represented by field testing with corn and cotton in various locations in the USA and India where different levels of either water and/or temperature stress occurred during the growing season (Fig. 5.3).

In 2016, AST received funding from USAID to accelerate development of endophyte

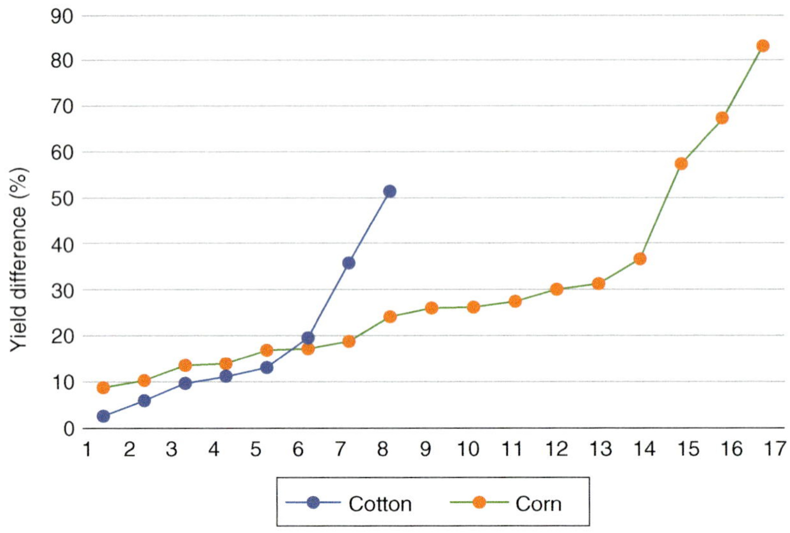

Fig. 5.3. BioEnsure yield increase differences compared with untreated controls. Increases observed in BioEnsure-treated versus untreated controls in monocot (corn) and eudicot (cotton). For each data set, abiotic stress increases from left to right and each dot represents a different field comparison. Field plots varied from 1 to 50 acres with 1–4 replicates for each treatment depending on plot sizes. Field tests were performed by farmers on their properties.

technology to enhance food security for small landholding farmers living below poverty levels around the world. Field testing was coordinated with 300 farmers in a remote region of Rajasthan, India, where dryland cultivation dominates and is dependent on annual monsoon rains. Due to climate change, monsoon rains have become less abundant and more inconsistent, resulting in drier and hotter growing seasons with temperatures commonly above 38°C. Over the last three years, AST staff travelled to Rajasthan each spring to treat seeds and work with farmers to field test BioEnsure® compared to untreated (check) seeds. Two staple crops (pearl millet and mung bean) were treated, allowing for comparison between a monocot and eudicot in the same soil types.

Based on discussions with the Rajasthani farmers (https://vimeo.com/192003746), we learned about several issues that limit crop production such as seed quality, low fertility soils and a perpetual lack of manure fertilizer. For example, most poor farmers in India cannot afford to purchase fresh market seeds, so they use 'carry-over' seeds form previous harvests. However, they lack proper facilities for seed storage, and carry-over seeds have lower germination rates and produce lower yields compared to fresh market seeds. Therefore, farmers assessed if BioEnsure® could improve the performance of carry-over seeds and compensate for low-fertility soils and a lack of fertilizer.

The yield results were greater than what farmers expected, and there were several important outcomes of BioEnsure®-conferred benefits to these plants:

1. Treated carry-over seeds yielded similarly or higher than fresh market seeds.
2. The biomass of pearl millet was increased by 25–50%.
3. Stress tolerance delayed plant senescence, allowing for secondary fertilization of mung bean.
4. The average yield increases were 29% for pearl millet and 56% for mung bean (Fig. 5.4).

5.6 Climate Mitigation and the Future of Poverty, Food Security and Political Stability

Food insecurity plays a critical role in human migration, poverty, famine and political unrest

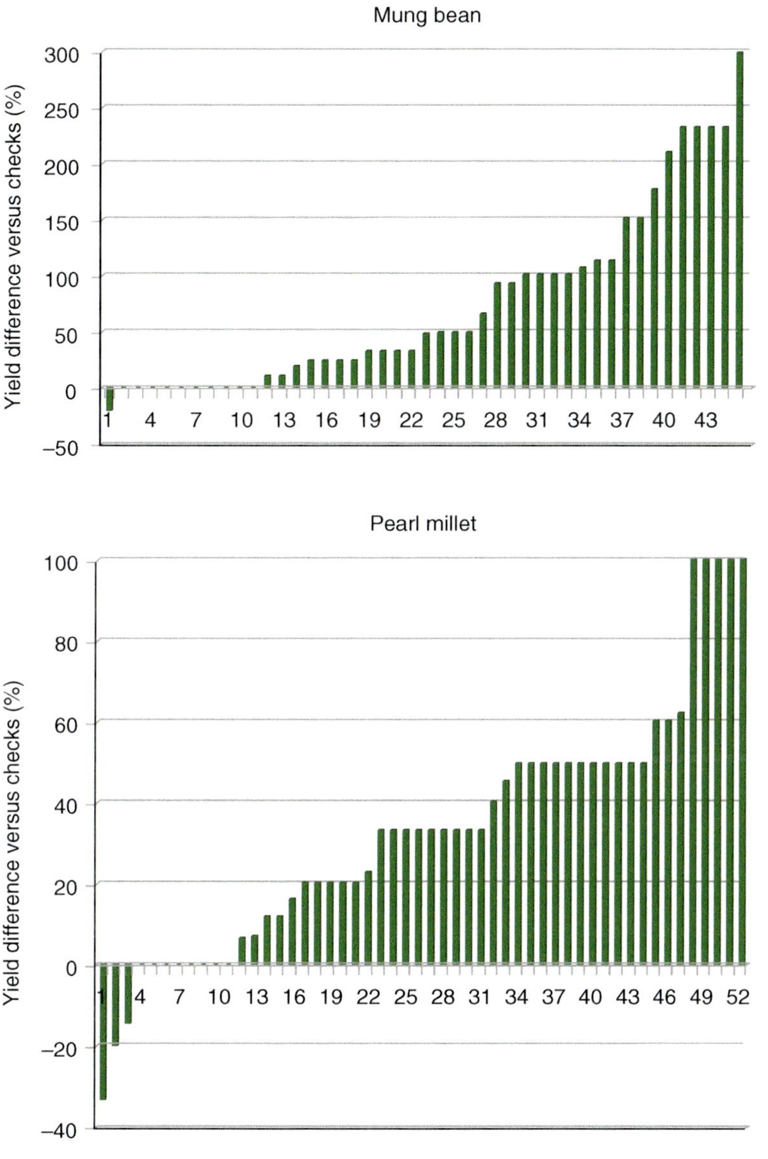

Fig. 5.4. In 2016, BioEnsure was tested on two staple crops in Rajasthan, India. Cultivation was dependent on rains from the annual monsoon and temperatures ranged from 35 to 45°C throughout the growing season. Each bar represents a field test with plots varying from 0.5 to 5 acres. Collectively, there were 98 farms involved in the testing. BioEnsure increased crop yields by an average 29% in pearl millet and 56% in mung bean. Yield differences compared to untreated checks were statistically significant (*T* test, *P* value < 0.001) for both crops. Seeds were treated in early June, planted by early July and harvested throughout October.

(Brinkman and Hendrix, 2011; Deaton and Lipka, 2015). Climate change has been destabilizing food security in many locations globally and will continue to worsen in the coming decades. Unfortunately, there is a strong relationship between poverty, food insecurity and political instability. This has been evident in Somalia and Syria where human migrations directly correlate to food insecurity. It is critical to point out that it is just a matter of time before food insecurity begins to impact more developed nations. For example, on 12 January 2018, the *Guardian News* reported that ongoing food shortages over a three-year span in Venezuela left citizens desperate and outbreaks of looting and mob violence ensued. The only realistic option for improving food security for poor farmers is to mitigate the impacts of climate change with abiotic stress-tolerant crops.

The discovery that fungal endophytes could adapt plants to abiotic stress was fundamental to developing climate mitigation biotechnology. Adaptive Symbiotic Technologies developed BioEnsure® with the intent of mitigating impacts of climate change long enough for governments and societies to implement measures necessary to reverse climate change. Independent field testing demonstrated the ability of BioEnsure® to mitigate climate impacts on crop production on both highly productive and marginal lands. Based on results in India, we conclude that the implications of endophyte technology are much greater than just increasing crop yields.

Improving the performance of carry-over seed translates to the following benefits for farmers:

- Decreased input costs
- Security of seed availability since there is not enough replacement seed produced in India
- Increased yields for sustenance and potential revenues
- Increased fodder for the animals and milk production, a primary protein source
- This technology is being used to develop a female empowerment programme to train females as seed treaters.

During the next two decades, climate change will lead to decreased food security and increased political instability, conflicts, human migrations and poverty. It is our hope that symbiotic technologies, such as BioEnsure®, can be used to mitigate impacts of climate stress on agriculture to break the chain of poverty and increase food security globally.

References

Alvarez-Loayza, P., White, J.F. Jr, Torres, M.S., Gil, N., Svenning, J.-C., Balslev, H. and Kristiansen, T. (2011) Light converts endosybiotic fungus to pathogen, influencing seedling survival and recruitment of host. *PLOS ONE* 6(1), e16386.

Brinkman, H.-J. and Hendrix, C.S. (2011) Food Insecurity and Violent Conflict: Causes, Consequences, and Addressing the Challenges. Occasional Paper No. 24. World Food Programme, Rome. Available at: https://ucanr.edu/blogs/food2025/blogfiles/14415.pdf (accessed 29 May 2019).

Chaw, S., Chang, C., Chen, H. and Li, W. (2004) Dating the monocot-dicot divergence and the origin of core eudicots using whole chloroplast genomes. *Journal of Molecular Evolution* 58, 424–441.

Deaton, B.J. and Lipka, B. (2015) Political instability and food security. *Journal of Food Security* 3, 29–33.

Gurian-Sherman, D. (2012) High and Dry: Why Genetic Engineering Is Not Solving Agriculture's Drought Problem in a Thirsty World. Union of Concerned Scientists, Cambridge, Massachusetts. Available at: https://www.ucsusa.org/food_and_agriculture/our-failing-food-system/genetic-engineering/high-and-dry.html (accessed 22 July 2019).

Komives, T. and Kiraly, Z. (2017) From golden rice to drought-tolerant maize and new techniques to control plant disease – Can we expect a breakthrough in crop production? *Ecocycles* 3, 1–5.

Krings, M., Taylor, T.N., Hass, H., Kerp, H., Dotzler, N. and Hermsen, E.J. (2007) Fungal endophytes in a 400 million-yr-old land plant: Infection pathways, spatial distribution and host responses. *New Phytologist* 174, 648–657.

Lofgren, L.A., LeBlanc, N.R., Certano, A.K., Nachtigall, J., LaBine, K.M. *et al.* (2018) *Fusarium graminearum*: Pathogen or endophyte of North American grasses? *New Phytologist* 217(3), 1203–1212.

Lugtenberg, B.J.J., Caradus, J.R. and Johnson, L.J. (2016) Fungal endophytes for sustainable crop production. *FEMS Microbiology Ecology* 92, 1–17.

Márquez, L.M., Redman, R.S., Rodriguez, R.J. and Roossinck, M.J. (2007) A virus in a fungus in a plant: Three-way symbiosis required for thermal tolerance. *Science* 315, 513–515.

Nuccio, M.L., Paul, M., Bate, N.J., Cohn, J. and Cutler, S.R. (2018) Where are the drought tolerant crops? An assessment of more than two decades of plant biotechnology effort in crop improvement. *Plant Science* 273, 110–119.

Redecker, D., Kodner, R. and Graham, L.E. (2000) Glomalean fungi from the Ordovician. *Science* 289, 1920–1921.

Redman, R.S. and Rodriguez, R.J. (2017) The symbiogenic Tango: Achieving climate resilient crops via mutualistic plant-fungal relationships. In: Doty, S.L. (ed.) *Functional Importance of the Plant Endophytic Microbiome: Implications for Agriculture, Forestry and Bioenergy*. Springer, pp. 71–88.

Redman, R.S., Dunigan, D.D. and Rodriguez, R.J. (2001) Fungal symbiosis: From mutualism to parasitism, who controls the outcome, host or invader? *New Phytologist* 151, 705–716.

Redman, R.S., Sheehan, K.B., Stout, R.G., Rodriguez, R.J. and Henson, J.M. (2002) Thermotolerance generated by plant/fungal symbiosis. *Science* 298, 1581.

Rodriguez, R.J., Henson, J., Van Volkenburgh, E., Hoy, M., Wright, L., *et al.* (2008) Stress tolerance in plants via habitat-adapted symbiosis. *ISME Journal* 2, 404–416.

Rodriguez, R.J., White, J.F.J., Arnold, A.E. and Redman, R.S. (2009) Fungal endophytes: Diversity and functional roles. *New Phytologist (Tansley Review)* 182, 314–330.

Simmons, E. and Flowers, K. (2017) Recurring Storms: Food Insecurity, Political Instability, and Conflict. CSIS, Washington DC. Available at: www.csis.org/analysis/recurring-storms-food-insecurity-political-instability-and-conflict (accessed 29 May 2019).

Singh, L.P., Gill, S.S. and Tuteja, N. (2011) Unraveling the role of fungal symbionts in plant abiotic stress tolerance. *Plant Signaling & Behavior* 6(2), 175–191.

Skøt, J., Lipper, L., Thomas, G., Agostini, A., Bertini, R. *et al.* (2016) The State of Food and Agriculture. Food and Agriculture Organization of the United Nations, Rome. Available at: http://www.fao.org/3/a-i6030e.pdf (accessed 29 May 2019).

Wolfe, K.H., Gouy, M., Yang, Y., Sharp, P.M. and Li, W. (1989) Date of the monocot-dicot divergence estimated from chloroplast DNA sequence data. *Proceedings of the National Academy of Sciences of the United States of America* 86, 6201–6205.

Yang, Y.W., Lai, K.N., Tai, P.Y. and Li, W.H. (1999) Rates of nucleotide substitution in angiosperm mitochondrial DNA sequences and dates of divergence between *Brassica* and other angiosperm lineages. *Journal of Molecular Evolution* 48, 597–604.

6 Endophytes as Novel Pest Control Agents: Myth or Reality?

Alan C. Gange,* Amanda F. Currie and Jessica-Rose Spong

School of Biological Sciences, Royal Holloway, University of London, Egham, Surrey, UK

Abstract

Endophytic fungi are ubiquitous in nature and their potential for pest control in grassland systems is well understood. However, their role as novel pest control agents in herbaceous crops is far less clear. These fungi can be broadly split into two groups: unspecialized species (including saprotrophs and latent pathogens) and entomopathogens. The literature on the interactions of these fungi with herbivorous insects is biased towards studies in a few plant families and with a few insect species. This fact notwithstanding, we suggest that infection of plants by these fungi elicits dramatic chemical changes within their hosts, which have the potential to reduce insect and pathogen attack. However, the effects of fungal infection on insects are context-specific, being influenced by the identity of the insect and plant, the existing community of fungi within a plant, the habitat in which it is growing and the plant age. Unspecialized endophytes can reduce the performance of sucking insects, but effects are only seen when seeds are inoculated. This result may reflect poor experimental technique but corresponds well to fungal biology, as these endophytes can be transmitted through seeds from one plant generation to the next. Endophytic entomopathogens show more consistent detrimental effects on insects and plant pathogens and can even provide growth benefits in the absence of antagonists. We conclude that a better understanding of the biochemical and molecular changes elicited by endophytes in plants is required, so that these can be harnessed in future pest control strategies. Endophytes will not replace conventional pesticides in the near future, but could be incorporated into future integrated control programmes, thereby reducing the reliance on synthetic chemicals.

6.1 Introduction

Every living plant ever examined seems to harbour endophytic fungi, species that reside within the tissues of plants for various periods of time and which cause no symptoms of disease. This definition is a broad one and comprises disparate groups of fungi with varying lifestyles. These include 'true endophytes', latent pathogens, pathogens of other hosts, general saprotrophs and entomopathogenic species (Currie *et al.*, 2014).

Clavicipitaceous fungi are found in the Poaceae and exhibit vertical transmission, in which the fungus grows systemically within the host tissues and ultimately into the seeds. The fungus is transmitted into the developing seedling, potentially never leaving its host. These fungi are often referred to as 'true endophytes' (Wani *et al.*, 2015). However, external sporulation does occur, probably because germination seems to represent a bottleneck for fungal growth, meaning that the transmission process is imperfect (Afkhami and Rudgers, 2008). Due to the close association between the fungus and host, this is often considered a mutualistic relationship with consequences for both invertebrate and vertebrate herbivores. These relationships are reviewed by Caradus and

* Corresponding author e-mail: a.gange@rhul.ac.uk

Johnson in Chapter 7 of this book and not considered further here.

In this chapter, we concentrate on the endophytes that occur in forbs and grasses, and which have a much wider host range than clavicipitaceous endophytes. Our focus is on whether these fungi can act as plant protection agents against foliar pests and diseases. Firstly, we describe some general patterns of fungal occurrence at various spatial scales, to examine how fungal diversity differs between and within hosts. We then consider the effects that these fungi have on insect herbivores, to understand the potential of these endophytes as novel pest control agents. Within this group, entomopathogenic (insect-killing) fungi stand apart from the others, both taxonomically and biologically. The discovery of the endophytic lifestyle of these fungi is relatively recent, and there are several comprehensive reviews of how these fungi exist within plants (Vega *et al.*, 2008, 2009; Vidal and Jaber, 2015; Vega, 2018). Within these studies, work with *Beauveria bassiana* is by far the most prevalent, with this fungus involved in 87% of experiments (Vega, 2018).

We also consider non-entomopathogenic fungi (which hereafter we refer to as 'unspecialized endophytes') and present some results of a meta-analysis we conducted to determine how both groups of fungi, acting as endophytes, may influence insect herbivores. We also describe how the fungi may affect plant pathogen resistance, and consider whether endophytes could be used to provide dual control of pests and pathogens. Throughout, we look at the current barriers to these fungi being novel pest control agents and how this control may change from a myth into reality.

6.2 The Nature of the Endophyte Literature

We conducted a Web of Science search in July 2018 using search terms such as endophyt* AND (plant family name), NOT bacteri*, to determine the amount of published research on fungal endophytes across plant families. The search was repeated with bacteria included to check that we were not excluding papers that reported on both fungal and bacterial endophytes. We also repeated it with various combinations to exclude the clavicipitaceous fungal literature, using terms such as 'NOT grass', 'NOT *Epichloe*', NOT *Neotyphodium* and NOT (major grass genera, including *Achnatherum, Balansia, Lolium, Festuca, Schedonorus, Stipa* etc.). We selected the 30 plant families with most references, all of which contain substantial numbers of non-woody species, major crop and medicinal plants, those of conservation interest, invasive species and those which have interest as sources of secondary metabolites. These were: Amaranthaceae, Amaryllidaceae, Apiaceae, Apocynaceae, Boraginaceae, Brassicaceae, Cactaceae, Campanulaceae, Caryophyllaceae, Compositae, Convolvulaceae, Crassulaceae, Cucurbitaceae, Euphorbiaceae, Gentianaceae, Geraniaceae, Lamiaceae, Leguminosae, Liliaceae, Orchidaceae, Orobanchaceae, Papaveraceae, Plantaginaceae, Polygonaceae, Primulaceae, Ranunculaceae, Rosaceae, Scrophulariaceae, Solanaceae and Violaceae. We then obtained the known number of species in each family from http://www.theplantlist.org/ (accessed 2 July 2018). We found a total of 1874 manuscripts published during the period 1950–2018 which dealt with some aspect of endophyte biology. These covered a wide range of topics, from simple surveys, through studies of endophyte effects on plant growth, pests and diseases, to examinations of metabolite production and gene expression. Our overall aim was to determine which plant families had received most attention, and why.

The number of studies per plant family is heavily skewed (Fig. 6.1A) and represents a classic 'hollow curve' seen in many sampling distributions. Leguminosae have received the most attention, with 26% of the literature being focused on this family. Indeed, 61% of the literature encompasses just five families; Leguminosae, Compositae, Orchidaceae, Rosaceae and Solanaceae. There is a positive relation between the number of endophyte publications per plant family and the number of species within that family ($F_{1,28} = 22.3$, $P < 0.001$, $R^2 = 44.3\%$; Fig. 6.1B). Like in many macroecological studies, endophyte researchers tend to work with organisms that are more abundant (Gaston and Blackburn, 2008). However,

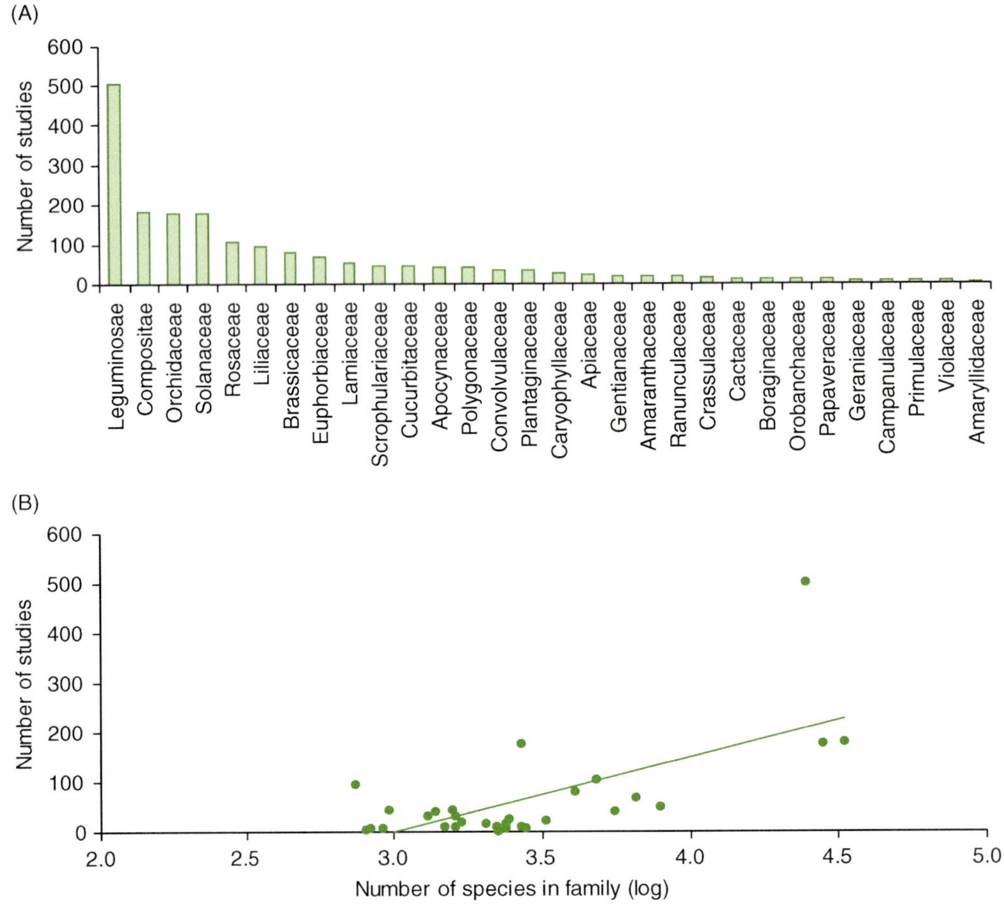

Fig. 6.1. Some of the bias that exists in the endophyte literature. (A) The majority of studies concentrate on just a few plant families, showing a classic 'hollow curve'. (B) The relation between the number of endophyte publications per plant family and the logarithm of species number in the family. There is a positive relation, indicating that researchers tend to work more with species in large plant families.

this graph is instructive as it provides a good indication of the driving force behind a lot of endophyte research. Firstly, plant families which contain important crops, such as Brassicaceae, Leguminosae and Solanaceae, have received much attention (i.e. their points lie well above the fitted line), wherein much research has been devoted to the use of these fungi as enhancers of resistance to pests and diseases. Other well-studied families include Lamiaceae, which are rich in essential oils, and Apocynaceae, Convolvulaceae, Liliaceae, Plantaginaceae and Scrophulariaceae, all of which contain medicinal plants and are potential sources of new drugs through secondary metabolite production. The common theme running through all of this literature is that endophytes have been studied because they can alter the chemical content of their hosts, often to the detriment of pests and diseases, through production or induction of secondary metabolites, a topic which has been reviewed recently (Nisa *et al.*, 2015; Gao *et al.*, 2018).

6.3 Endophyte Distributions within Plants

Many studies have produced lists of endophytes within plants, usually describing fungal abundance as 'infection frequency' (commonly defined as the number of isolates obtained through culturing, as a function of the total number of isolates) or in recent years as operational taxonomic units (OTUs). Both methods have their advantages and disadvantages, which were discussed by Dissanayake *et al.* (2018). Problems of culture dependency, lack of discrimination using internal transcribed spacer (ITS) and lack of fungal sequences notwithstanding, virtually all studies produce species abundance distributions that exhibit hollow curves, similar to the pattern seen in Fig. 6.1A. In other words, each endophyte community tends to be dominated by a few abundant species, with a long tail to the distribution, consisting of many rare species (e.g. Wearn *et al.*, 2012; Ek-Ramos *et al.*, 2013; Dissanayake *et al.*, 2018). In all cases, diversity is generally high and species counts of around 100 per plant species (not per plant) are not uncommon (Wearn *et al.*, 2012; Garcia *et al.*, 2013).

All of these surveys have confirmed the original assertion (Petrini, 1986) that many of these unspecialized endophytes exhibit a broad host range. Furthermore, as most of these fungi infect plants via airborne spores (Sánchez Márquez *et al.*, 2012), one might assume that there would be little difference in the endophyte communities within plant species that grow together in the same community, as the hosts are exposed to the same spore rain. Very few studies have tested this assumption, but when they have it was found not to be true, and even closely related plant species growing in close proximity to each other harbour different endophyte communities (Gange *et al.*, 2007). It is unclear if this is due to a failure of certain fungi to infect the plant or if the host can exert a degree of control over fungal growth. It is likely that both are true, as Redman *et al.* (2001) found that the ability of a fungus to infect a plant depended on previous infection events and that the overall lifestyle (e.g. pathogenic or mutualistic) displayed by the endophytes was controlled by the plant. These are important considerations to take into account in inoculation experiments, as they imply that not all fungal infection events will be successful.

It has long been known that when placed in mixed cultures, endophytes often display antagonism towards each other, with the production of many different metabolites (Chagas *et al.*, 2013). Given that virtually every host plant is infected by more than one endophyte at any time, interactions between the fungi themselves are a likely cause of much variation in plant secondary chemistry (Kusari *et al.*, 2012). There are remarkably few studies of the relations between the abundance of different fungal species within a plant, but when found, these are often negative (Gange *et al.*, 2007). This suggests that the effects of inoculating any plant with endophytes will depend upon the background community that is already present. Systemic chemical changes that can be observed when a plant is infected by an endophyte (e.g. Hartley *et al.*, 2015) are thus likely to be context-dependent, influenced by the resident endophytes and the environment (Yang *et al.*, 2014). Furthermore, the beneficial effects of any endophyte upon its host plant is habitat-dependent (Rodriguez *et al.*, 2008), indicating that if endophytes are to be exploited as pest control agents, then the combination of fungus, habitat and plant needs to be carefully evaluated (Suryanarayanan, 2013).

The age of a plant is a further factor that needs to be taken into account. Seasonal changes in leaf endophyte communities have been observed in a study of three forb species (*Cirsium arvense*, *Plantago lanceolata* and *Rumex acetosa*) (Wearn *et al.*, 2012), upland cotton (*Gossypium hirsutum*) (Ek-Ramos *et al.*, 2013), *Arabidopsis thaliana* (Garcia *et al.*, 2013) and sugar beet (*Beta vulgaris*) (Shi *et al.*, 2016). In all cases, endophyte abundance and diversity built up over the course of a season, with a microbial succession occurring within the plant tissues, as these age (Hodgson, 2010). Competition between the fungi (and associated metabolite production) is likely to be the main driver of these successional changes within leaves. However, as a season progresses, the density of airborne spores and thus the intensity of the endophyte spore rain to which a plant is exposed also increases (Rodríguez-Rajo *et al.*, 2005).

Not only is there a degree of host specificity in the community structure, there is also organ specificity. For example, in *C. arvense, P. lanceolata* and *R. acetosa*, Wearn *et al.* (2012) found that there was remarkably little overlap in the endophyte communities found in the roots and shoots of the plants. Indeed, in experiments investigating the effect of endophyte inoculation on insect herbivores, Jaber and Vidal (2010) inoculated the roots of broad bean (*Vicia faba*) with *Acremonium strictum*, but failed to recover the fungus from foliar tissues, even though good root colonization was achieved. An absence of systemic growth appears to be a feature of unspecialized endophytes, and the fact that a particular fungal species can be isolated from different leaves or different organs of a plant likely represents multiple infection events, rather than systemic growth (Yan *et al.*, 2015). However, entomopathogenic fungi appear to provide an exception to this rule. For example, *Beauveria bassiana* was successfully recovered from leaves of *Papaver somniferum* when it was originally applied to the roots (Quesada-Moraga *et al.*, 2006), and this fungus, as well as other entomopathogens, were recovered from leaves of *V. faba* following inoculation of seeds (Akello and Sikora, 2012). Vega (2018) lists other studies in which movement of an entomopathogen through the plant has been observed. It is important to note that these studies were performed in insect-free conditions. If an entomopathogen resulted in mycosis of a herbivore, then the spores released could provide a further source of inoculum for uninfected tissues on the host plant. Taken together, these results indicate that systemic, acropetal growth of entomopathogens must occur in a variety of host plants and this characteristic will be very useful in their exploitation as pest control agents.

The facts that endophyte infection may be determined by the resident fungi in a plant and that fungal presence induces systemic chemical changes mean that the starting points in a plant's life, i.e. the seed and seedling stages, may be critical for determining the success of fungal inoculation in any manipulative experiment. It has generally been assumed that the unspecialized endophyte community of foliar tissues is transmitted horizontally, by airborne spores (Sánchez Márquez *et al.*, 2012). However, it is now known that both entomopathogenic and non-entomopathogenic fungi can be transmitted vertically, through seeds (Hodgson *et al.*, 2014; Quesada-Moraga *et al.*, 2014). However, just as with clavicipitaceous fungi, germination presents a problem for fungal growth and the process is imperfect (Hodgson *et al.*, 2014). Seeds contain a wide variety of endophytic fungi, recently reviewed by Nelson (2018) and Shahzad *et al.* (2018) and in Chapter 4 of this volume. However, the use of seed endophytes has yet to be well explored in agriculture, due in part to incompatibility with fungicide seed treatments and a lack of knowledge of the transmission process (Le Cocq *et al.*, 2017). Such fungi could have novel uses in the protection of stored seed products against insects, but to date, this topic has been ignored, even though clavicipitaceous fungal presence in grass seeds reduces insect attack (Bamisile *et al.*, 2018).

The above characters of host range, diversity, location of infection, intra-plant growth and transmission method were used in the comprehensive review of Rodriguez *et al.* (2009) to ascribe endophytes to different classes (I–IV). Class I covered clavicipitaceous fungi, while Class IV described dark septate endophytes confined to roots. The main distinguishing factors for Class II and III endophytes were that Class II occur in all plant parts, while Class III are confined to foliar parts; that Class II show extensive growth in plants, while Class III form highly localized infections; and that Class II endophytes show horizontal and vertical transmission, while Class III fungi are only transmitted horizontally. Furthermore, Class II endophytes seem to occur in temperate herbaceous plants, while Class III fungi comprise the hyper-diverse communities found in tropical forests.

These are valuable distinctions, but since this review was published, it is clear that the research on different aspects of endophyte biology has meant that the dividing lines between categories have become distinctly blurred. In Table 6.1, we summarize the features of Class II and III endophytes provided by Rodriguez *et al.* (2009) and highlight the characteristics of the unspecialized

Table 6.1. Part of Table 6.1 from Rodriguez *et al.* (2009), redrawn to accommodate recent discoveries of endophyte lifestyles. The criteria for each attribute are those given in the original paper and characteristics highlighted in grey indicate those that can be applied to non-clavicipitaceous endophytes in herbaceous plants. Note: in Transmission category, 'both' means horizontal and vertical transmission.

	Non-entomopathogenic fungi		Entomopathogenic fungi	
	Class II	Class III	Class II	Class III
Host range	Broad	Broad	Broad	Broad
Diversity	Low	High	Low	High
Tissues	Shoot and root	Shoot	Shoot and root	Shoot
In-plant growth	Extensive	Limited	Extensive	Limited
Transmission	Both	Horizontal	Both	Horizontal

endophytes and entomopathogenic fungi that occur in herbaceous plants.

It is clear that entomopathogenic fungi satisfy all the criteria of Class II endophytes, but the other fungi that occur within herbaceous plants are not so easily classified; they exhibit features of both Classes II and III. Indeed for tissue location, the criterion for these species could be redefined as 'shoot or root', since growth between the two structures is extremely limited.

6.4 Endophytes and Insect Herbivores

We searched the Web of Science (ISI) electronic bibliographic database for the period 1950–2017 using a variety of search terms including endophyt* AND insect, endophyt* and herbivor* and all combinations of endophyt* with entomopathogen* and different insect orders and feeding modes (monophagous, oligophagous and polyphagous) (Gange *et al.*, 2019). We excluded all studies of clavicipitaceous fungi, focusing on the unspecialized endophytes and entomopathogens that occur in herbaceous plants. We also checked the major recent review papers (Rodriguez *et al.*, 2009; Currie *et al.*, 2014, Vidal and Jaber, 2015; Bamisile *et al.*, 2018; Fernandez-Conradi *et al.*, 2018; Vega, 2018) and citations thereof. All included studies had performed a manipulative experiment in which plants were inoculated with endophyte(s), insects subsequently reared upon the plant and some measure of performance (e.g. growth, survival, size, fecundity, abundance, etc.) taken.

We obtained a database comprising 527 experiments, spread over 55 manuscripts. Of these, 61% involved entomopathogenic endophytes and 39% non-entomopathogens. We conducted meta-analysis on this data set, calculating an effect size, Hedges *d* and its variance (Koricheva *et al.*, 2013). Meta-regression was used to test for the effects of different moderator variables, such as insect order, and the method of fungal inoculation. Full details of all methods and tests of bias are given in Gange *et al.* (2019). Briefly, effect sizes are considered significant if their 95% confidence intervals do not overlap zero, and two effect sizes are different if their confidence intervals do not overlap.

Firstly, it is useful to examine the bias in the literature and to compare the percentage of studies that have used different insect orders to previous similar analyses using mycorrhizas and true endophytes (Hartley and Gange, 2009). Overall, 83% of herbaceous endophyte studies have taken place in controlled or semi-controlled conditions, rather than the field, remarkably similar to the figures obtained for mycorrhizal fungi (80%) and clavicipitaceous endophytes (85%). Hartley and Gange (2009) found that the literature was dominated by studies involving Lepidoptera, Coleoptera and Hemiptera, with the majority of work with true endophytes using *Spodoptera frugiperda* (fall armyworm) and *Listronotus bonariensis* (Argentine stem weevil). The pattern here is not the same (Fig. 6.2), but there is also a significant association between

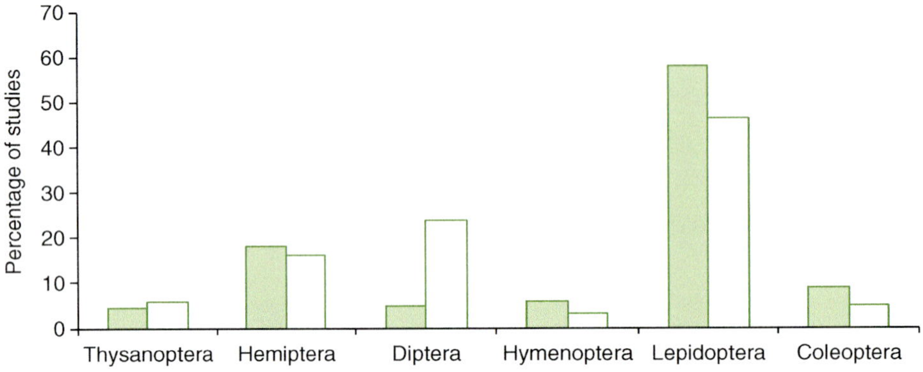

Fig. 6.2. Per cent of studies in the literature involving insects and either non-entomopathogenic (grey bars) or entomopathogenic endophytes (open bars).

fungal type and insect order ($\chi^2 = 23.8$, d.f. = 5, $P < 0.001$). Both data sets are dominated by studies on Lepidoptera, but more experiments involving Diptera have been performed with entomopathogens. The main difference between the current data set and that of Hartley and Gange (2009) is that relatively few studies have examined the effects of non-clavicipitaceous endophytes on Coleoptera.

As with mycorrhizas and true endophytes, certain insect species dominate the literature. For non-entomopathogenic fungi, a total of 24 insect species have been used, but 25% of all experiments have involved the polyphagous lepidopteran pest *Helicoverpa armigera*. The second most 'popular' species is *Eldana saccharina*, a stem-chewing pest of sugar cane (*Saccharum officinarum*), used in 14% of experiments. Meanwhile, entomopathogenic studies have involved 28 insect species, 16 of which are common with the non-entomopathogenic studies. Here, *H. armigera* is also the most studied insect (18% of experiments), followed by the lepidopteran pest *Sesamia calamistis* and the leaf-mining dipteran, *Liriomyza huidobrensis*. Curiously, only one study (Ramirez-Rodriguez and Sánchez-Peña, 2016) has involved *S. frugiperda* and endophytic *B. bassiana* and none have involved *L. bonariensis*. There is therefore very little overlap in this endophyte literature with that of true endophytes and mycorrhizal fungi (Hartley and Gange, 2009).

Notwithstanding the taxonomic bias that exists in the literature, there are some very interesting results in the meta-analysis. Perhaps not surprisingly, all orders of insects are detrimentally affected by entomopathogenic fungi, though to differing degrees (Fig. 6.3); Hymenoptera and Thysanoptera seem to show the greatest effects. Here, studies involving Hymenoptera include galling insects, and it may be that as the insect is unable to move around the plant, it becomes an 'easy target' for the pathogen. Non-entomopathogenic endophytes are found inside insect galls and the structure does not represent a barrier to fungal growth (Gange *et al.*, 2002).

Meanwhile, non-entomopathogenic fungi only have detrimental effects on the performance of sucking insects, in the orders Hemiptera and Thysanoptera. The former involves a variety of aphid species (particularly *Aphis fabae* (bean aphid) and *A. gossypii* (cotton aphid)), which are phloem feeders. The most likely reason for these effects is that a wide variety of secondary metabolites are transported in the phloem (Wink, 2010), which would be ingested by the insects. Phenolic compounds are just one example of secondary metabolites that are produced by endophytes (Negreiros de Carvalho *et al.*, 2016) and which are active against aphids, enhancing plant resistance (Kaur *et al.*, 2017). These results are interesting because it has been suggested that the relatively loose association between unspecialized fungi, determined mostly by the random infection of plants by airborne spores, should not lead to any mutualistic relationship between plant and fungus (Faeth, 2002). However, the fact that these endophytes can be vertically

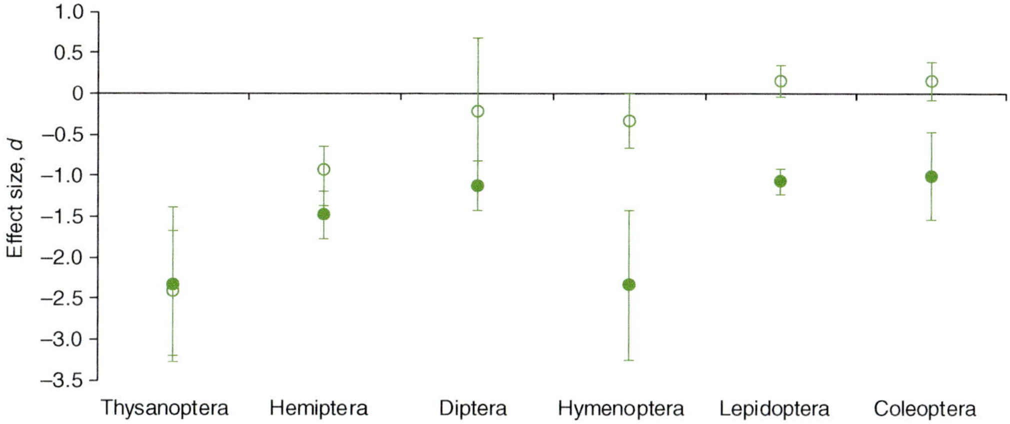

Fig. 6.3. Effects of endophytic fungi on insects from different orders. Values are mean effect size (Hedges *d*) and negative values means that the presence of the fungi is detrimental to insect performance. Vertical lines represent 95% confidence intervals, and the effect size is considered significant when these do not overlap zero. Non-overlap of confidence intervals (CIs) indicates that the effect sizes differ between the two groups of fungi. Entomopathogens (filled symbols) are detrimental to all orders of insects, but non-entomopathogens (open symbols) only reduce the growth and performance of sucking insects (Hemiptera and Thysanoptera).

transmitted through seeds (Hodgson *et al.*, 2014) lends support to the idea that such a mode of transmission is evidence for cooperation between plants and these fungi (Sachs *et al.*, 2004) and implies that those fungi that occur as seed inhabitants may exhibit the most potential as novel agents of plant protection. At this point, we should not lose sight of the fact that the term 'unspecialized endophyte' covers latent pathogens and fungi which are pathogenic in other hosts (Wani *et al.*, 2015). Symptomatic pathogen infection generally reduces insect performance (Fernandez-Conradi *et al.*, 2018) through crosstalk of defensive pathways. Thus, it would not be surprising if these fungi altered the chemical profile of their host when occurring as asymptomatic endophytes. Such a situation is known to occur in woody hosts, where a pathogen may have a long latent, endophytic period (Gange, 1996).

Perhaps the most surprising result from this analysis is that the method by which the fungi are inoculated onto plants has a dramatic effect on the outcome (Fig. 6.4). No matter what plant part is infected by an entomopathogen, there are detrimental effects on foliar-feeding insects. The systemic growth of the fungi and metabolite production means that applications to the soil or as seed treatments are likely to be effective in pest control (Quesada-Moraga *et al.*, 2009; Akello and Sikora, 2012). However, the result with non-entomopathogens is quite different and probably holds the key to understanding why the use of these fungi as novel pest control agents has yet to become a reality.

Intuitively, for fungi that are thought to infect plants mostly by airborne spores, one might assume that the appropriate way to perform an experiment would be to spray leaves with spores, contained in an aqueous suspension. However, neither this method nor application to roots seems to produce any effect on insect herbivores (Fig. 6.4). Instead, application to seeds has a significant, detrimental effect on subsequent insect performance. As already discussed, these fungi do not exhibit systemic growth through their hosts (Yan *et al.*, 2015), so the effect must be one of systemic metabolite translocation. Indeed, this would explain why some experiments with these fungi and insects showed poor recovery rates of inoculated species from new growth foliage, and yet effects on insects and metabolomic changes were observed in this foliage (Gange *et al.*, 2012; Hartley *et al.*, 2015). This likely explains the inconsistency reported from previous experiments

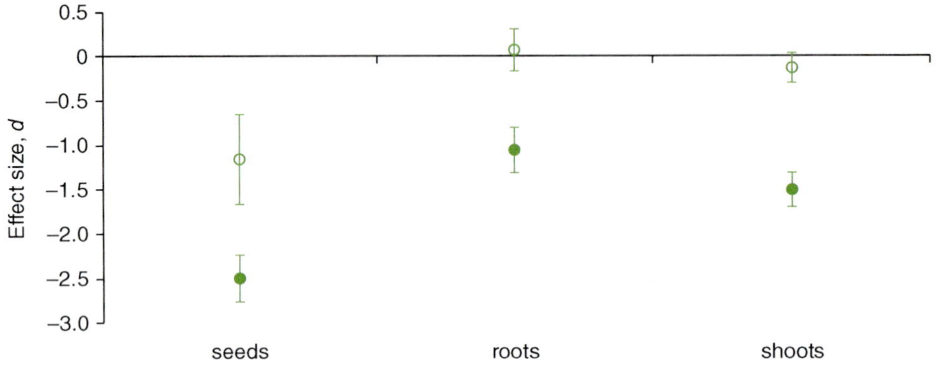

Fig. 6.4. The outcome of experiments is affected by the experimental design. Endophytic entomopathogens (filled symbols) reduce the growth of foliar-feeding insects no matter what part of the plant they are applied to. However, inoculation of seeds with unspecialized endophytes (open symbols) is the only method that results in reduced insect growth from these fungi.

(Suryanarayanan, 2013) and may go some way towards making the use of these fungi as pest control agents a reality. It also reinforces the fact that vertical transmission of these fungi may be very important (Hodgson *et al*., 2014) and suggests that those seedlings which possess endophytes will be chemically different and better protected than those without the fungi. However, it is also possible that in field situations, seedlings may acquire endophytes from leaf litter, a method that is known to provide pathogen resistance in tropical ecosystems (Christian *et al*., 2017).

One must question why leaf inoculation often seems to result in failure of endophytes to infect and a lack of effects on insects. It is possible that the resident fungal community prevented infection by the inoculated species. However, while this might cause variation in effects on insects, one would not expect it to result in failure across an array of plant species. Indeed, it could even amplify effects on insects, through metabolites produced by the fungal–fungal interactions (Pan and May, 2009). Instead, the most likely reason is that it is simply a failure of technique, and humidity at the leaf surface was too low. Plant pathologists use dew chambers to increase humidity and obtain successful infections, yet no endophyte-insect study has used such a chamber. Some studies used polythene bags to increase humidity, but many took no account of leaf wetness when applying endophyte inoculum. The use of polythene bags is fraught with difficulty, as it has long been known that they emit ethylene, which could have unintended effects on plant growth and development (Scott and Wills, 1972; Abeles *et al*., 1992). Many plant pathogens require particular periods of temperature and humidity to successfully infect; one of the most famous examples being that for late blight of potato, caused by *Phytophthora infestans* and known as the Smith Period (recently reformulated as the 'Hutton Criteria', McEwan, 2016). The distinction between an endophytic lifestyle and a pathogenic one may be just a mutation at a single locus (Freeman and Rodriguez, 1993; Eaton *et al*., 2010), and thus there is no reason why endophytes do not require the same levels of humidity for infection. Thus, we must conclude that foliar applications of non-entomopathogenic endophytes to plants in the field are unlikely to be of much use in pest control, being too susceptible to environmental conditions.

Bioencapsulation of fungi on seeds may offer an alternative infection method, but presents a number of challenges (Schoebitz *et al*., 2013), with success being determined by the biology of the microbe, production costs, storage and handling. However, it has recently been found that successful endophytic infection of tomato (*Lycopersicon esculentum*) plants can be achieved following seed encapsulation of the entomopathogenic fungus *Metarhizium brunneum* (Krell *et al.*, 2018a). Even so, application of endophytes to seeds presents many problems, particularly establishing the microbe in field conditions (O'Callaghan, 2016), and there is clearly much to be done if seed inoculation is to become a reliable method of endophyte use in agricultural and horticultural systems (Le Cocq *et al.*, 2017).

6.5 Endophytes and Higher Trophic Levels

One problem with conventional insecticides is that they are often broad spectrum and have unintended deleterious effects on higher trophic levels, including beneficial insects such as predators and parasitoids. Endophytes, being internal within plants, may therefore offer much promise in a combined approach of pest control involving higher trophic levels. To date, very few studies have examined the effects of endophyte presence on parasitoid insects, and conflicting results have been obtained. Exner and Vidal (1996) found that tomato plants infected by *Acremonium strictum* had no effect on parasitism rates of whiteflies (*Trialeurodes vaporariorum*) attacked by *Encarsia formosa*, but that searching behaviour was altered (and extended) on infected plants, suggesting some form of response to plant volatiles. Meanwhile, Kaur *et al.* (2015) found that infection of cauliflower (*Brassica oleracea*) by *Aspergillus flavus* and *A. niger* extended development time and reduced adult longevity and fecundity of *Bracon hebetor* attacking *Spodoptera litura*. Metabolite production by the fungi within the plant is likely to be the explanation, with possible sequestration by the larvae impacting on the parasitoid. Perhaps the most intriguing example is that of Contreras-Cornejo *et al.* (2018), in which roots of maize (*Zea mays*) were infected with *Trichoderma atroviride*. Parasitism rates of *Spodoptera frugiperda* by *Campoletis sonorensis* were higher when larvae were fed foliage of infected plants. Attraction of parasitoids was linked to the compound 6-Pentyl-2H-pyran-2-one (6-PP), released into the atmosphere by the subterranean fungus. However, this experiment was performed in small containers, and it remains to be seen whether these interactions can be detected in field conditions. If they can, then such an attraction offers great promise as a novel form of pest control and suggests that these fungi could be incorporated into 'push-pull' strategies of pest control, in which they have yet to appear (Eigenbrode *et al.*, 2016).

Entomopathogenic fungi also seem to be compatible with parasitoid insects, with both Akutse *et al.* (2014) and Gathage *et al.* (2016) finding that parasitism rates were mostly unaffected when prey were fed on plants infected by *B. bassiana*. Akutse *et al.* (2014) did find some evidence for a reduction of parasitism rates in one wasp species, and this may be similar to the finding of Bixby-Brosi and Potter (2012) who showed that the effects of a true endophyte (*Neotyphodium lolii*) on parasites of the polyphagous moth *Agrotis ipsilon* depended on the identity of the parasitoid species. The slow-developing communal hymenopteran species *Copidosoma bakeri* was detrimentally affected by endophyte presence, while the fast-developing solitary dipteran *Linnaemya comta* was unaffected. These results suggest that the use of endophytes in multitrophic pest control systems may often be context-specific, with outcomes determined by the identity of the fungi, the plant host, the herbivore and the parasitoid.

The complexity of these interactions may explain why, to date, there is just a single study that has manipulated endophyte presence and examined the effects on communities of insects and their parasitoids, in a crop situation. Zarate *et al.* (2015) applied a mixed inoculum, including endophytic *Trichoderma* spp. to the roots of a cabbage crop. Over two seasons, levels of predatory and parasitic insects were higher on the plants which contained the endophyte. Further such experiments are urgently required to understand how endophytic fungi could be used in field crops, and whether any control of insect pests is brought about by direct effects of metabolite production in plants or indirect effects through higher trophic levels, or both.

6.6 Endophytes and Plant Pathogens

A comprehensive review of the interactions between endophytes and plant pathogenic fungi was published by Busby *et al.* (2016). This covers wild plants (many of which are tree species), plants of agricultural interest and invasive plants. There is therefore a large degree of overlap with this review, and the reader is directed there for more information. Busby *et al.* (2016) did not perform a meta-analysis of the topic, and it would appear that this is a subject that would be highly suitable, particularly when one includes the papers that have been published since their review.

In many respects, the outcomes of the interactions between endophytes and pathogens are very similar to those described for insect herbivores described above. Agricultural plants dominate the literature, but the outcome of experiments is frequently context-specific. A wide variety of unspecialized endophytes can provide disease protection, though the interactions between them and pathogens can range from disease antagonism through to facilitation. When antagonism is reported, the most likely explanation is secondary metabolite production or induction by the endophytic fungi (Busby *et al.*, 2016). An excellent recent example is that of de Vries *et al.* (2018) who found that a range of unspecialized endophytes isolated from several herbaceous plants could provide resistance to the potato pathogen *P. infestans*. Most importantly, one isolate, *Phoma eupatorii* (isolated from *Eupatorium cannabinum*) exhibited broad-spectrum activity against the pathogen, offering the potential for the control of genetically diverse pathogen strains. A number of ubiquitous fungi are mentioned in the Busby *et al.* (2016) review that occur as endophytes, including species in the genera *Alternaria, Chaetomium, Cladosporium, Colletotrichum* and *Trichoderma*. Interestingly, all may provide disease suppression, but they have also been reported to reduce insect performance (Gange *et al.*, 2012; Thakur *et al.*, 2013; Zhou *et al.*, 2016). However, to date, no study has examined the simultaneous effects of unspecialized endophytes on an insect and pathogen attacking the same plant.

However, of particular interest is the recent discovery that entomopathogenic fungi can also provide resistance to plant diseases, offering the potential for multiple uses of these species in integrated pest management programmes (Ownley *et al.*, 2010; Jaber and Ownley, 2018). In one case, infection of tomato plants by *B. bassiana* provided simultaneous resistance against incidence of leaf curl virus and its vector, the whitefly *Bemisia tabaci* (El-Deeb *et al.*, 2012). Various mechanisms are thought to play a role in disease suppression, including competition for space and nutrients, secondary metabolite production and induction of systemic plant resistance. These entomopathogens produce an array of secondary metabolites in cultures and within plants (Gibson *et al.*, 2014), with *B. bassiana* having been the subject of much research in this respect. However, much remains to be discovered, since the levels of one important chemical, beauvericin, are frequently insufficient within plants to make disease suppression a consistent reality (Jaber and Ownley, 2018).

Of even greater interest is that entomopathogens seem to offer growth benefits to plants, even in the absence of pests or pathogens. Jaber and Enkerli (2017) found that inoculation of bean (*Vicia faba*) plants with *Beauveria brongniartii*, *B. bassiana* or *Metarhizium brunneum* produced increases in both shoot and root weight, though effects were inconsistent across experiments. Recently, Krell *et al.* (2018b) have suggested that the beneficial effect of *M. brunneum* on the growth of potato is due to the mitigation of nutrient stress. Indeed, amelioration of nutrient, drought or salt stress is a frequent mechanism by which non-entomopathogenic endophytes improve plant growth (Lata *et al.*, 2018). Perhaps one of the best known fungi in this respect is *Piriformospora indica*, which can confer resistance to a wide range of biotic and abiotic stresses (Oelmüller *et al.*, 2009). However, the type of stress tolerance provided by symbiotic fungi depends upon the habitat in which the plant is growing, a phenomenon, termed ‘habitat-adapted symbiosis’ (Rodriguez *et al.*, 2008). For example, endophytes isolated from plants growing in geothermal habitats confer heat, but not salt tolerance, while fungi isolated from plants in coastal habitats confer salt but not heat tolerance. Furthermore, the same fungi isolated from habitats without these stresses did not confer such tolerances (Rodriguez *et al.*, 2008). In addition, the success of fungal inoculation and effects on a plant will be also be governed by the identity of the fungal community already there (Pan and May, 2009). More integrated, multitrophic studies are urgently needed, to fully understand the context-specificity of endophyte–plant–antagonist interactions and more are needed for these fungi to become consistent and reliable plant protection agents.

One instance where it may be desirable for endophytes to increase pest or pathogen susceptibility in a plant is in the biological control of weeds. There are extremely few studies to date on this topic, with the best examples being by Kurose *et al.* (2012) and David *et al.* (2016). In both cases, some (but not all) endophytes increased pathogen incidence or insect attack. Perhaps the lack of published studies is due to the fact that crosstalk between defensive pathways, active against pathogens and insects, is a frequent phenomenon (Pineda *et al.*, 2013), which may negate any significant results. However, the fact that endophytes often elicit a defence reaction in the host that is also active against a pathogen and/or an insect means that most interactions between these organisms will be negative, and so the use of endophytes in biological control using other agents is likely to be limited.

6.7 Conclusions

Herbaceous plants host an array of endophytic fungi, many of which show great potential for use as pest control agents. However, our knowledge of such systems beyond grasses is poor but perhaps best developed with the entomopathogenic fungi. These species, particularly *B. bassiana*, can exist as endophytes, conferring plant growth promotion benefits in the absence of antagonists and having detrimental effects on foliar-feeding pest insects and pathogenic fungi. Effects often seem to be context-specific and a 'one-size-fits-all' strategy is unlikely to be developed in the near future. However, metabolite production, either by the fungi or induced in plants by fungal presence, offers the most likely mechanism that may be harnessed for pest and disease control. Successful infection of plants with endophytes is determined by a range of biotic and abiotic factors. Prominent amongst the former is the composition of the endophyte community already within the plant. Antagonistic endophyte–endophyte interactions are common, but biochemical characterization of the outcomes may enable us to turn this apparent barrier into a predictable way of altering plant biochemistry and increasing resistance to pests and diseases. Since crosstalk between defensive pathways is a common phenomenon, there is the possibility of using endophytes to provide simultaneous protection against pests and diseases if the 'right' combination of fungi for a particular plant can be found.

Non-entomopathogenic fungi are ubiquitous and have been greatly overlooked for their use in pest control. In order for these fungi to become a reality in agriculture, the method of inoculation (seed, rather than foliar or root) needs to be considered. By identifying particular naturally occurring consortia of fungi, it should be possible to use these alongside entomopathogens, in a combined strategy that produces reliable and consistent effects on pests. Characterization of seed-inhabiting fungi (i.e. those that show vertical transmission) may hold the key to their development. It is unlikely that any of these fungal endophytes will replace conventional pesticides in the near future, so careful consideration needs to be given to their compatibility with insecticides (Perez-Gonzalez and Sánchez-Peña, 2017), fungicides (Prior *et al.*, 2017) and other biological controls, such as predators and parasitoids (Gathage *et al.*, 2016). To date, there is extremely limited knowledge of such compatibility. Harnessing metabolite production, either by the plants or the endophytes themselves when in competition, is likely to be a useful method that can be adopted in a range of integrated pest control programmes.

It should be noted that 83% of the literature on endophyte interactions in herbaceous plants comes from controlled experiments in constant-environment rooms or glasshouses, and there is a clear need to take such experiments into the field. It is of course possible that many field experiments have been tried but 'failed' and were not published, a phenomenon often referred to as the 'file drawer problem'. However, meta-analytical techniques are available to deal with publication bias, which is why such analyses need to be conducted rather than simple literature reviews or vote-counting procedures (Gurevitch *et al.*, 2018).

We have tried to show that endophytic fungi offer great opportunities for novel forms of pest control research and its application in agricultural systems. A full understanding of the complex systems involved will only come about through collaborative efforts of ecologists, plant molecular biologists and biochemists. Thus, we believe that using endophytes as plant protection agents is certainly not a myth and has the potential to become reality in the near future.

Acknowledgements

We are grateful to the Natural Environment Research Council (grant number NE/N00244X/1) and the Nuffield Foundation for funding our work with endophytes, insects and plant pathogens.

References

Abeles, F.B., Morgan, P.W. and Saltveit, M.E. (1992) *Ethylene in Plant Biology*. Academic Press, San Diego, California. doi:10.1016/C2009-0-03226-7

Afkhami, M.E. and Rudgers, J.A. (2008) Symbiosis lost: Imperfect vertical transmission of fungal endophytes in grasses. *American Naturalist* 172, 405–416. doi:10.1086/589893

Akello, J. and Sikora, R. (2012) Systemic acropedal influence of endophyte seed treatment on *Acyrthosiphon pisum* and *Aphis fabae* offspring development and reproductive fitness. *Biological Control* 61, 215–221. doi:10.1016/j.biocontrol.2012.02.007

Akutse, K.S., Fiaboe, K.K.M., Van den Berg, J., Ekesi, S. and Maniania, N.K. (2014) Effects of endophyte colonization of *Vicia faba* (Fabaceae) plants on the life-history of leafminer parasitoids *Phaedrotoma scabriventris* (Hymenoptera: Braconidae) and *Diglyphus isaea* (Hymenoptera: Eulophidae). *PLOS ONE* 9, e109965: doi:109910.101371/journal.pone.0109965

Bamisile, B.S., Dash, C.K., Akutse, K.S., Keppanan, R. and Wang, L.D. (2018) Fungal endophytes: Beyond herbivore management. *Frontiers in Microbiology* 9, 544. doi:10.3389/fmicb.2018.00544

Bixby-Brosi, A.J. and Potter, D.A. (2012) Endophyte-mediated tritrophic interactions between a grass-feeding caterpillar and two parasitoid species with different life histories. *Arthropod-Plant Interactions* 6, 27–34. doi:10.1007/s11829-011-9163-2

Busby, P.E., Ridout, M. and Newcombe, G. (2016) Fungal endophytes: Modifiers of plant disease. *Plant Molecular Biology* 90, 645–655. doi:10.1007/s11103-015-0412-0

Chagas, F.O., Dias, L.G. and Pupo, M.T. (2013) A mixed culture of endophytic fungi increases production of antifungal polyketides. *Journal of Chemical Ecology* 39, 1335–1342. doi:10.1007/s10886-013-0351-7

Christian, N., Herre, E.A., Mejia, L.C. and Clay, K. (2017) Exposure to the leaf litter microbiome of healthy adults protects seedlings from pathogen damage. *Proceedings of the Royal Society B-Biological Sciences* 284, 20170641, doi:10.1098/rspb.2017.0641

Contreras-Cornejo, H.A., Del-Val, E., Macias-Rodriguez, L., Alarcon, A., Gonzalez-Esquivel, C.E., *et al.* (2018) *Trichoderma atroviride*, a maize root associated fungus, increases the parasitism rate of the fall armyworm *Spodoptera frugiperda* by its natural enemy *Campoletis sonorensis*. *Soil Biology & Biochemistry* 122, 196–202. doi:10.1016/j.soilbio.2018.04.013

Currie, A.F., Wearn, J.A., Hodgson, S., Wendt, H., Broughton, S.J., *et al.* (2014) Foliar fungal endophytes in herbaceous plants: A marriage of convenience? In: Verma, V.C. and Gange, A.C. (eds) *Advances in Endophytic Research*. Springer, New Delhi, pp. 61–81. doi:10.1007/978-81-322-1575-2_3

David, A.S., Quiram, G.L., Sirota, J.I. and Seabloom, E.W. (2016) Quantifying the associations between fungal endophytes and biocontrol-induced herbivory of invasive purple loosestrife (*Lythrum salicaria* L.). *Mycologia* 108, 625–637. doi:10.3852/15-207

de Vries, S., von Dahlen, J.K., Schnake, A., Ginschel, S., Schulz, B., *et al.* (2018) Broad-spectrum inhibition of *Phytophthora infestans* by fungal endophytes. *FEMS Microbiology Ecology* 94, fiy037. doi:10.1093/femsec/fiy037

Dissanayake, A.J., Purahong, W., Wubet, T., Hyde, K.D., Zhang, W., *et al.* (2018) Direct comparison of culture-dependent and culture-independent molecular approaches reveal the diversity of fungal endophytic communities in stems of grapevine (*Vitis vinifera*). *Fungal Diversity* 90, 85–107. doi:10.1007/s13225-018-0399-3

Eaton, C.J., Cox, M.P., Ambrose, B., Becker, M., Hesse, U., Schardl, C.L. and Scott, B. (2010) Disruption of signaling in a fungal-grass symbiosis leads to pathogenesis. *Plant Physiology* 153, 1780–1794. doi:10.1104/pp.110.158451

Eigenbrode, S.D., Birch, A.N.E., Lindzey, S., Meadow, R. and Snyder, W.E. (2016) A mechanistic framework to improve understanding and applications of push-pull systems in pest management. *Journal of Applied Ecology* 53, 202–212. doi:10.1111/1365-2664.12556

Ek-Ramos, M.J., Zhou, W.Q., Valencia, C.U., Antwi, J.B., Kalns, L.L., *et al.* (2013) Spatial and temporal variation in fungal endophyte communities isolated from cultivated cotton (*Gossypium hirsutum*). *PLOS ONE* 8, e66049, doi:10.1371/journal.pone.0066049

El-Deeb, H.M., Lashin, S.M. and Arab, Y.A.-S. (2012) Reaction of some tomato cultivars to tomato leaf curl virus and evaluation of the endophytic colonisation with *Beauveria bassiana* on the disease incidence and its vector, *Bemisia tabaci*. *Archives of Phytopathology and Plant Protection* 45, 1538–1545. doi:10.1080/03235408.2012.681246

Exner, A. and Vidal, S. (1996) Interactions between plant pathogens, endophytes, whiteflies and *Encarsia formosa*. *Mitteilungen Der Deutschen Gesellschaft Für Allgemeine Und Angewandte Entomologie* 11, 751–754.

Faeth, S.H. (2002) Are endophytic fungi defensive plant mutualists? *Oikos* 98, 25–36. doi:10.1034/j.1600-0706.2002.980103.x

Fernandez-Conradi, P., Jactel, H., Robin, C., Tack, A.J.M. and Castagneyrol, B. (2018) Fungi reduce preference and performance of insect herbivores on challenged plants. *Ecology* 99, 300–311. doi:10.1002/ecy.2044

Freeman, S. and Rodriguez, R.J. (1993) Genetic conversion of a fungal plant pathogen to a non-pathogenic, endophytic mutualist. *Science* 260, 75–78. doi:10.1126/science.260.5104.75

Gange, A.C. (1996) Positive effects of endophyte infection on sycamore aphids. *Oikos* 75, 500–510. doi:10.2307/3545892

Gange, A.C., Croft, R. and Wu, W. (2002) Gall insect and endophytic fungal co-occurrence in a xeric and mesic site. *Ecological Entomology* 27, 362–365. doi:10.1046/j.1365-2311.2002.00410.x

Gange, A.C., Dey, S., Currie, A.F. and Sutton, B.C. (2007) Site- and species-specific differences in endophyte occurrence in two herbaceous plants. *Journal of Ecology* 95, 614–622. doi:10.1111/j.1365-2745.2007.01245.x

Gange, A.C., Eschen, R., Wearn, J.A., Thawer, A. and Sutton, B.C. (2012) Differential effects of foliar endophytic fungi on insect herbivores attacking a herbaceous plant. *Oecologia* 168, 1023–1031. doi:10.1007/s00442-011-2151-5

Gange, A.C., Koricheva, J., Currie, A.F., Jaber, L. and Vidal, S. (2019). Meta-analysis of the role of entomopathogenic and unspecialised fungal endophytes as plant bodyguards. *New Phytologist* doi: 10.1111/nph.15859

Gao, H., Li, G. and Lou, H.X. (2018) Structural diversity and biological activities of novel secondary metabolites from endophytes. *Molecules* 23, 646. doi:10.3390/molecules23030646

Garcia, E., Alonso, A., Platas, G. and Sacristan, S. (2013) The endophytic mycobiota of *Arabidopsis thaliana*. *Fungal Diversity* 60, 71–89. doi:10.1007/s13225-012-0219-0

Gaston, K.J. and Blackburn, T.M. (2008) *Pattern and Process in Macroecology*. Wiley, Oxford, UK. doi:10.1002/9780470999592

Gathage, J.W., Lagat, Z.O., Fiaboe, K.K.M., Akutse, K.S., Ekesi, S., *et al.* (2016) Prospects of fungal endophytes in the control of *Liriomyza* leafminer flies in common bean *Phaseolus vulgaris* under field conditions. *BioControl* 61, 741–753. doi:10.1007/s10526-016-9761-0

Gibson, D.M., Donzelli, B.G.G., Krasnoff, S.B. and Keyhani, N.O. (2014) Discovering the secondary metabolite potential encoded within entomopathogenic fungi. *Natural Product Reports* 31, 1287–1305. doi:10.1039/c4np00054d

Gurevitch, J., Koricheva, J., Nakagawa, S. and Stewart, G. (2018) Meta-analysis and the science of research synthesis. *Nature* 555, 175–182. doi:10.1038/nature25753

Hartley, S.E. and Gange, A.C. (2009) Impacts of plant symbiotic fungi on insect herbivores: Mutualism in a multitrophic context. *Annual Review of Entomology* 54, 323–342. doi:10.1146/annurev.ento.54.110807.090614

Hartley, S.E., Eschen, R., Horwood, J.M., Gange, A.C. and Hill, E.M. (2015) Infection by a foliar endophyte elicits novel arabidopside-based plant defence reactions in its host, *Cirsium arvense*. *New Phytologist* 205, 816–827. doi:10.1111/nph.13067

Hodgson, S. (2010) Foliar fungal endophyte dynamics in herbaceous hosts. PhD. thesis, Royal Holloway, University of London.

Hodgson, S., de Cates, C., Hodgson, J., Morley, N.J., Sutton, B.C., *et al.* (2014) Vertical transmission

of fungal endophytes is widespread in forbs. *Ecology and Evolution* 4, 1199–1208. doi:10.1002/ece3.953
Jaber, L.R. and Enkerli, J. (2017) Fungal entomopathogens as endophytes: Can they promote plant growth? *Biocontrol Science and Technology* 27, 28–41. doi:10.1080/09583157.2016.1243227
Jaber, L.R. and Ownley, B.H. (2018) Can we use entomopathogenic fungi as endophytes for dual biological control of insect pests and plant pathogens? *Biological Control* 116, 36–45. doi:10.1016/j.biocontrol.2017.01.018
Jaber, L.R. and Vidal, S. (2010) Fungal endophyte negative effects on herbivory are enhanced on intact plants and maintained in a subsequent generation. *Ecological Entomology* 35, 25–36. doi:10.1111/j.1365-2311.2009.01152.x
Kaur, H., Salh, P.K. and Singh, B. (2017) Role of defense enzymes and phenolics in resistance of wheat crop (*Triticum aestivum* L.) towards aphid complex. *Journal of Plant Interactions* 12, 304–311. doi:10.1080/17429145.2017.1353653
Kaur, T., Singh, B., Kaur, A. and Kaur, S. (2015) Endophyte-mediated interactions between cauliflower, the herbivore *Spodoptera litura*, and the ectoparasitoid *Bracon hebetor*. *Oecologia* 179, 487–494. doi:10.1007/s00442-015-3358-7
Koricheva, J., Gurevitch, J. and Mengersen, K. (2013) *Handbook of Meta-Analysis in Ecology and Evolution*. Princeton University Press, Princeton, New Jersey. doi:10.1515/9781400846184
Krell, V., Jakobs-Schoenwandt, D., Vidal, S. and Patel, A.V. (2018a) Encapsulation of *Metarhizium brunneum* enhances endophytism in tomato plants. *Biological Control* 116, 62–73. 10.1016/j.biocontrol.2017.05.004.
Krell, V., Unger, S., Jakobs-Schoenwandt, D. and Patel, A.V. (2018b) Endophytic *Metarhizium brunneum* mitigates nutrient deficits in potato and improves plant productivity and vitality. *Fungal Ecology* 34, 43–49. doi:10.1016/j.funeco.2018.04.002
Kurose, D., Furuya, N., Tsuchiya, K., Tsushima, S. and Evans, H.C. (2012) Endophytic fungi associated with *Fallopia japonica* (Polygonaceae) in Japan and their interactions with *Puccinia polygoni-amphibii* var. *tovariae*, a candidate for classical biological control. *Fungal Biology* 116, 785–791. doi:10.1016/j.funbio.2012.04.011
Kusari, S., Hertweck, C. and Spitellert, M. (2012) Chemical ecology of endophytic fungi: Origins of secondary metabolites. *Chemistry & Biology* 19, 792–798. doi:10.1016/j.chembiol.2012.06.004
Lata, R., Chowdhury, S., Gond, S.K. and White, J.F. (2018) Induction of abiotic stress tolerance in plants by endophytic microbes. *Letters in Applied Microbiology* 66, 268–276. doi:10.1111/lam.12855
Le Cocq, K., Gurr, S.J., Hirsch, P.R. and Mauchline, T.H. (2017) Exploitation of endophytes for sustainable agricultural intensification. *Molecular Plant Pathology* 18, 469–473. doi:10.1111/mpp.12483
McEwan, G. (2016) New criteria to replace Smith periods in potato blight forecasting. *Horticulture Week*, 9 May.
Negreiros de Carvalho, P.L., Silva, E.D., Chagas-Paula, D.A., Luiz, J.H.H. and Ikegaki, M. (2016) Importance and implications of the production of phenolic secondary metabolites by endophytic fungi: A mini-review. *Mini-Reviews in Medicinal Chemistry* 16, 259–271. doi:10.2174/1389557515666151016123923
Nelson, E.B. (2018) The seed microbiome: Origins, interactions, and impacts. *Plant and Soil* 422, 7–34. doi:10.1007/s11104-017-3289-7
Nisa, H., Kamili, A.N., Nawchoo, I.A., Shafi, S., Shameem, N., *et al.* (2015) Fungal endophytes as prolific source of phytochemicals and other bioactive natural products: A review. *Microbial Pathogenesis* 82, 50–59. doi:10.1016/j.micpath.2015.04.001
O'Callaghan, M. (2016) Microbial inoculation of seed for improved crop performance: Issues and opportunities. *Applied Microbiology and Biotechnology* 100, 5729–5746. doi:10.1007/s00253-016-7590-9
Oelmüller, R., Sherameti, I., Tripathi, S. and Varma, A. (2009) *Piriformospora indica*, a cultivable root endophyte with multiple biotechnological applications. *Symbiosis* 49, 1–17. doi:10.1007/s13199-009-0009-y
Ownley, B.H., Gwinn, K.D. and Vega, F.E. (2010) Endophytic fungal entomopathogens with activity against plant pathogens: Ecology and evolution. *BioControl* 55, 113–128. doi:10.1007/s10526-009-9241-x
Pan, J.J. and May, G. (2009) Fungal-fungal associations affect the assembly of endophyte communities in maize (*Zea mays*). *Microbial Ecology* 58, 668–678. doi:10.1007/s00248-009-9543-7
Pérez-González, O. and Sánchez-Peña, S.R. (2017) Compatibility in vitro and in vivo of the entomopathogenic fungi *Beauveria bassiana* and *Hirsutella citriformis* with selected insecticides. *Southwestern Entomologist* 42, 707–718. doi:10.3958/059.042.0309
Petrini, O. (1986) Taxonomy of endophytic fungi of aerial plant tissues. In: Fokkema, N.J. and van den Heuvel, J. (eds) *Microbiology of the Phyllosphere*. Cambridge University Press, Cambridge, UK, pp. 175–187.
Pineda, A., Dicke, M., Pieterse, C.M.J. and Pozo, M.J. (2013) Beneficial microbes in a changing environment: Are they always helping plants

to deal with insects? *Functional Ecology* 27, 574–586. doi:10.1111/1365-2435.12050

Prior, R., Mittelbach, M. and Begerow, D. (2017) Impact of three different fungicides on fungal epi- and endophytic communities of common bean (*Phaseolus vulgaris*) and broad bean (*Vicia faba*). *Journal of Environmental Science and Health Part B-Pesticides Food Contaminants and Agricultural Wastes* 52, 376–386. doi:10.1080/03601234.2017.1292093

Quesada-Moraga, E., Landa, B.B., Muñoz-Ledesma, J., Jiménez-Diáz, R.M. and Santiago-Álvarez, C. (2006) Endophytic colonisation of opium poppy, *Papaver somniferum*, by an entomopathogenic *Beauveria bassiana* strain. *Mycopathologia* 161, 323–329. doi:10.1007/s11046-006-0014-0

Quesada-Moraga, E., Muñoz-Ledesma, F.J. and Santiago-Álvarez, C. (2009) Systemic protection of *Papaver somniferum* L. against *Iraella luteipes* (Hymenoptera: Cynipidae) by an endophytic strain of *Beauveria bassiana* (Ascomycota: Hypocreales). *Environmental Entomology* 38, 723–730. doi:10.1603/022.038.0324

Quesada-Moraga, E., López-Díaz, C. and Landa, B.B. (2014) The hidden habit of the entomopathogenic fungus *Beauveria bassiana*: First demonstration of vertical plant transmission. *PLOS ONE* 9, e89278. doi:10.81371/journal.pone.0089278

Ramirez-Rodriguez, D. and Sánchez-Peña, S.R. (2016) Endophytic *Beauveria bassiana* in *Zea mays*: Pathogenicity against larvae of fall armyworm, *Spodoptera frugiperda*. *Southwestern Entomologist* 41, 875–878. doi:10.3958/059.041.0330

Redman, R.S., Dunigan, D.D. and Rodriguez, R.J. (2001) Fungal symbiosis from mutualism to parasitism: Who controls the outcome, host or invader? *New Phytologist* 151, 705–716. doi:10.1046/j.0028-646x.2001.00210.x

Rodriguez, R.J., Henson, J., Van Volkenburgh, E., Hoy, M., Wright, L., *et al.* (2008) Stress tolerance in plants via habitat-adapted symbiosis. *ISME Journal* 2, 404–416. doi:10.1038/ismej.2007.106

Rodriguez, R.J., White, J.F., Arnold, A.E. and Redman, R.S. (2009) Fungal endophytes: Diversity and functional roles. *New Phytologist* 182, 314–330. doi:10.1111/j.1469-8137.2009.02773.x

Rodríguez-Rajo, F.J., Iglesias, I. and Jato, V. (2005) Variation assessment of airborne *Alternaria* and *Cladosporium* spores at different bioclimatical conditions. *Mycological Research* 109, 497–507. doi:10.1017/s0953756204001777

Sachs, J.L., Mueller, U.G., Wilcox, T.P. and Bull, J.J. (2004) The evolution of cooperation. *Quarterly Review of Biology* 79, 135–160. doi:10.1086/383541

Sánchez Márquez, S., Bills, G.F., Herrero, N. and Zabalgogeazcoa, I. (2012) Non-systemic fungal endophytes of grasses. *Fungal Ecology* 5, 289–297. doi:10.1016/j.funeco.2010.12.001

Schoebitz, M., Lopez, M.D. and Roldan, A. (2013) Bioencapsulation of microbial inoculants for better soil-plant fertilization. A review. *Agronomy for Sustainable Development* 33, 751–765. doi:10.1007/s13593-013-0142-0

Scott, K.J. and Wills, R.B.H. (1972) Ethylene produced by plastics in sunlight. *HortScience* 7, 177.

Shahzad, R., Khan, A.L., Bilal, S., Asaf, S. and Lee, I.J. (2018) What is there in seeds? Vertically transmitted endophytic resources for sustainable improvement in plant growth. *Frontiers in Plant Science* 9, 24, doi:10.3389/fpls.2018.00024

Shi, Y.W., Li, C., Yang, H.M., Zhang, T., Gao, Y., *et al.* (2016) Endophytic fungal diversity and space-time dynamics in sugar beet. *European Journal of Soil Biology* 77, 77–85. doi:10.1016/j.ejsobi.2016.09.005

Suryanarayanan, T.S. (2013) Endophyte research: Going beyond isolation and metabolite documentation. *Fungal Ecology* 6, 561–568. doi:10.1016/j.funeco.2013.09.007

Thakur, A., Kaur, S., Kaur, A. and Singh, V. (2013) Enhanced resistance to *Spodoptera litura* in endophyte infected cauliflower plants. *Environmental Entomology* 42, 240–246. doi:10.1603/en12001

Vega, F.E. (2018) The use of fungal entomopathogens as endophytes in biological control: A review. *Mycologia* 110, 4–30. doi:10.1080/00275514.2017.1418578

Vega, F.E., Posada, F., Aime, M.C., Pava-Ripoll, M., Infante, F., *et al.* (2008) Entomopathogenic fungal endophytes. *Biological Control* 46, 72–82. doi:10.1016/j.biocontrol.2008.01.008

Vega, F.E., Goettel, M.S., Blackwell, M., Chandler, D., Jackson, M.A., *et al.* (2009) Fungal entomopathogens: New insights on their ecology. *Fungal Ecology* 2, 149–159. doi:10.1016/j.funeco.2009.05.001

Vidal, S. and Jaber, L.R. (2015) Entomopathogenic fungi as endophytes: Plant-endophyte-herbivore interactions and prospects for use in biological control. *Current Science* 109, 46–54.

Wani, Z.A., Ashraf, N., Mohiuddin, T. and Riyaz-Ul-Hassan, S. (2015) Plant-endophyte symbiosis, an ecological perspective. *Applied Microbiology and Biotechnology* 99, 2955–2965. doi:10.1007/s00253-015-6487-3

Wearn, J.A., Sutton, B.C., Morley, N.J. and Gange, A.C. (2012) Species and organ specificity of fungal endophytes in herbaceous grassland plants. *Journal of Ecology* 100, 1085–1092. doi:10.1111/j.1365-2745.2012.01997.x

Wink, M. (2010) Introduction: Biochemistry, physiology and ecological functions of secondary metabolites. *Annual Plant Reviews* 40, 1–19. doi:10.1002/9781444320503.ch1

Yan, J.F., Broughton, S.J., Yang, S.L. and Gange, A.C. (2015) Do endophytic fungi grow through their hosts systemically? *Fungal Ecology* 13, 53–59. doi:10.1016/j.funeco.2014.07.005

Yang, T., Ma, S. and Dai, C.C. (2014) Drought degree constrains the beneficial effects of a fungal endophyte on *Atractylodes lancea*. *Journal of Applied Microbiology* 117, 1435–1449. doi:10.1111/jam.12615

Zarate, J.T., Caasi-Lit, M.T., Cuevas, V.C. and Cocjin, G.T. (2015) Terrestrial arthropod profile and soil microbial population dynamics on cabbage cropping as affected by application of *Trichoderma* Microbial Inoculant (TMI) in Quezon, Sariaya, Philippines. *Journal of Environmental Science and Management* 18, 10–22.

Zhou, W.Q., Starr, J.L., Krumm, J.L. and Sword, G.A. (2016) The fungal endophyte *Chaetomium globosum* negatively affects both above- and belowground herbivores in cotton. *FEMS Microbiology Ecology* 92, fiw158. doi:10.1093/femsec/fiw1158

7 Improved Adaptation of Temperate Grasses through Mutualism with Fungal Endophytes

John R. Caradus[1,*] and Linda J. Johnson[2]

[1]*Grasslanz Technology Limited, Palmerston North, New Zealand;* [2]*AgResearch Limited, Grasslands Research Centre, Palmerston North, New Zealand*

Abstract

Plants provide a unique ecological niche for diverse communities of fungal endophytes that vary in their impact, positive to negative, on the host plant. Fungal endophytes colonize plants without any visible disease symptoms for at least part of their life history. These symbionts are critical components of natural and semi-natural ecosystems, as well as cultivated agricultural communities, dramatically influencing plant adaptation and evolution. Many temperate grass species are reliant on their mutualistic association with obligate fungal endophytes of the genus *Epichloë* for a variety of fitness benefits, such as persistence under both invertebrate and vertebrate grazing pressure and improved adaptation to abiotic stresses. In New Zealand, *Epichloë* endophytes have been estimated to contribute $200 m per year in increased animal production through improved pasture persistence and yield. Other fungal endophytes have also been studied in temperate grasses, although much less is understood about the biological and economic impacts of these plant–fungal interactions. Determining which fungal endophytes are mutualistic has been, and continues to be, a challenge, particularly as we begin to realize that endophytes interact with other endophytes and little is understood about what factors shape endophyte community structures.

7.1 Introduction

All plants live in symbiotic association with a diverse collection of microorganisms that can vary in their lifestyle and relationship. These close symbiotic relationships between two different biological species are complex, spanning a continuum from beneficial to detrimental, with the nature of the interaction changing with plant genotype and at different life cycle stages and/or in response to fluctuating environmental cues (Hartley and Gange, 2009; Newton *et al.*, 2010). Endophytes are a diverse class of microorganisms that live internally and asymptomatically within host tissues (Hartley and Gange, 2009). They exhibit lifestyles ranging from facultative to obligate, with various degrees of host specificity, in planta colonization patterns (i.e. systemic or localized point infections), and propagate by either vertical and/or horizontal transmission (Rodriguez *et al.*, 2004; Rodriguez and Redman, 2008; Card *et al.*, 2016). Here we will restrict our examination to endophytic fungi in temperate grasses that show mutualistic behaviours, where both the microbe and the higher plant benefit from the association (Hartley and Gange, 2009). While traditionally arbuscular mycorrhizal fungi (AMF) have not been considered endophytes because they do not reside entirely within plant tissues

* Corresponding author e-mail: john.caradus@grasslanz.com

(Hartley and Grange, 2009; Rodriguez *et al.*, 2009), they are mutualistic symbionts that form intimate associations with the vast majority of extant land plants (Hoysted *et al.*, 2018). Mycorrhizal associations are common on root tissues of temperate grasses, and in so doing, these mutualistic symbionts interact with other fungal endophytes, and this connection will be investigated here. Fossil records indicate associations of fungal symbionts with plants have existed since 400 million years ago, indicating their importance in early plant habitat transitions (Remy *et al.*, 1994; Krings *et al.*, 2007). A microbial endophyte will spend some or all of its life cycle within plant tissues without causing any sign of infection (Wilson, 1995) and as such provides the basis for the formation of a mutualistic association.

Determining the degree of mutualism expressed by some microbes associated with temperate grasses can be a significant challenge but will be pivotal for this review. Mutualism has been defined as a reciprocally beneficial relationship between organisms, although it can range from diffuse and indirect interactions to highly integrated and co-evolved associations between pairs of species (Herre *et al.*, 1999). Herre *et al.* (1999) went on to further identify factors that align mutualists' interests, which include the movement of propagules from parent to offspring (vertical transmission), genotypic uniformity of symbionts within individual hosts, and the spatial structure of populations leading to repeated interactions between would-be mutualists. However, they did indicate that not all mutualistic associations require vertical transmission, and vertical transmission alone does not automatically result in a benefit to either partner. Benefits accrued to the host plant by mutualistic endophytic microbes should manifest through positive effects on production and/or survival of both partners in the association. Although the magnitude of benefits may change with time, rarely do these associations shift to parasitism (Frederickson, 2017), nor do they always persist despite reciprocal benefits, and can break down when two species cease to cooperate (Werner *et al.*, 2018).

The most well-studied microbial mutualistic association of temperate grasses is that of *Epichloë* (family Clavicipitaceae) in ryegrass and fescue. These intimate associations are thought to have co-evolved for over 40 million years towards specialization and mutually beneficial cooperation (Schardl *et al.*, 2004; Schardl *et al.*, 2008; Saikkonen *et al.*, 2016). *Epichloë* species, which include the asexual species formally known as *Neotyphodium* spp. (Leuchtmann *et al.*, 2014), are filamentous fungi that form natural associations with many grass genera and tribes of the subfamily *Pooideae* (White, 1987; Schardl *et al.*, 2004). The associations formed with grass genera include those with ryegrasses that are agriculturally important to Australia and New Zealand (Easton, 2007; Easton and Fletcher, 2007) and tall fescue in the United States (Schardl and Phillips 1997; Bouton *et al.*, 2000; Aitken and Strickland, 2013). However, there are many other documented fungal symbionts of temperate grasses, but the nature of the interaction is often unknown. Saikkonen *et al.* (1998) summarized numerous interactions identified among endophyte-infected grasses and invertebrate herbivores. While the focus will be on the *Epichloë* endophyte association with temperate grasses, other possible mutualistic associations will also be described and discussed.

7.2 Fungal Endophytes Found in Temperate Grasses

7.2.1 Obligate, systemic mutualists located in aerial tissues – *Epichloë* endophytes

Epichloë species are obligate and systemic mutualists of grasses found in the intercellular spaces of aboveground plant parts (Clay, 1990; Schardl, 2001; Scott, 2001; Schardl *et al.*, 2004). This association has been described as a 'defensive mutualism' (Faeth, 2002). Fig. 7.1 depicts the vegetative life cycle of *Epichloë* in relation to the grass vegetative and sexual reproductive cycle. The infection process within developing host leaves occurs by a novel mechanism known as intercalary growth, whereby colonization is tightly controlled and synchronized with the development of the grass host (Christensen

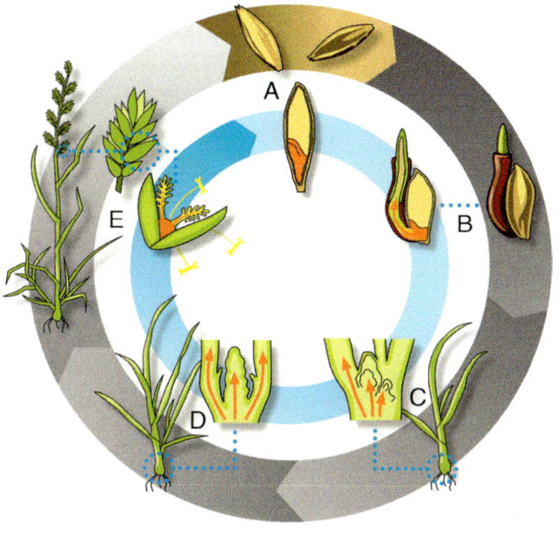

Fig. 7.1. The *Epichloë festucae* vegetative life cycle. The grass reproductive cycle (both vegetative and sexual) is presented in the outer ring, and the fungal transmission cycle (asexual only) is shown in the inner ring. (A) Hyphae (orange) are present in the embryo of the seed plus other structures including the scutellum and the 'infection layer' a region between the embryo and the endosperm (Philipson and Christey, 1986; Card *et al.*, 2011). During seed germination (B), hyphae in the embryo colonize the shoot apex of the developing seedling. (C) Hyphal tips in the shoot apex move into leaf primordia and, once there, extend with developing leaves, primarily through intercalary compartment growth (Christensen *et al.*, 2008). Hyphal growth ceases once the leaf matures. Axillary buds are colonized early, enabling hyphae to invade the newly forming daughter tillers. (D) Host reproductive tissues are colonized as the vegetative shoot apex becomes floral. Hyphae extend into floral tissues early in development, including the ovaries. (E) Hyphae penetrate the embryo and are incorporated into other seed structures (Sampson, 1933; Liu *et al.*, 2017; Zhang *et al.*, 2017a). (Figure from Gagic *et al.*, 2018.)

et al., 2008; Voisey, 2010). The reproduction strategies of *Epichloë* species spans a wide range from the strictly asexual types, to some capable of both vertical (via seed) and horizontal transmission (via ascospores), to truly sexual species that only transmit by ascospores (Leutchmann *et al.*, 2014). Importantly, the strictly asexual *Epichloë* species used in pastoral agriculture have lost the symptom-forming sexual phase characteristic of *Epichloë* species ('choke disease') and are uniquely seed-transmitted within their host populations (Siegel *et al.*, 1987). The centre of origin for *Epichloë* species symbiotic with *Lolium* and *Festuca* host plants is Eurasia and North Africa with numerous studies describing the association in native grass habitats (Latch *et al.*, 1987; Leuchtmann, 1994). *Epichloë* strains have also evolved in temperate grasses indigenous to North America (Clay and Leuchtmann, 1989), South America (Cabral *et al.*, 1999; Novas *et al.*, 2007), Asia (Nan and Li, 2000) and New Zealand and Australia (Moon *et al.*, 2002).

Discovering Epichloë *endophytes*

The importance of *Epichloë* endophytes was first identified because they were associated with animal welfare issues including fescue toxicosis in the United States (Bacon *et al.*, 1977; Schmidt *et al.*, 1982; Siegel *et al.*, 1984; Porter and Thompson, 1992; Hoveland, 1993) and ryegrass staggers in New Zealand (Fletcher and Harvey, 1981; Fletcher *et al.*, 1999). In the 1980s, it was established that some of the secondary metabolites produced by *Epichloë* in their host grasses were responsible for these animal disorders (Gallagher and Hawkes, 1985; Gallagher *et al.*, 1985; Bacon *et al.*, 1986), which also contributes to reduced animal liveweight gains and on-farm profitability (Paterson *et al.*, 1995; Fletcher, 1999; Ball *et al.*, 2007). The subsequent removal of the *Epichloë* endophyte to alleviate these ailments, however, resulted in poor pasture production and persistence (Barker *et al.*, 1985; Pottinger *et al.*, 1985). This led to the identification of *Epichloë* strains which provided benefits to the host plant without causing animal welfare issues (Siegel and Schardl, 1991; Latch, 1997, 1998). Methods of culture and inoculation (Latch and Christensen, 1985) of *Epichloë* into endophyte-free seedlings have allowed the commercial use of beneficial strains that do not cause serious animal health and welfare issues (Latch, 1989). Grass breeding initiatives, particularly in ryegrass and fescue, have been enhanced significantly by the introduction of novel *Epichloë* strains (Johnson *et al.*, 2013a). Managing endophytic seed appropriately is crucial for endophyte survival and application in grazed swards (Rolston *et al.*, 1986).

Step changes in *Epichloë* diversity have occurred repeatedly through hybridization events involving two or more *Epichloë* species (Tsai *et al.*, 1994; Moon *et al.*, 2000; Schardl and Wilkinson, 2000; Moon *et al.*, 2004). Selosse and Schardl (2007) have argued that the resulting meiotic sterility in these hybrids selects for vertical transmission and therefore mutualism. With rule changes to the nomenclature of pleomorphic fungi, *Epichloë* genera now includes almost all previously described *Neotyphodium* species and is comprised of 43 taxa made up of 34 described *Epichloë* species, as well as subspecies and varieties (Leuchtmann *et al.*, 2014). Of these, 24 are anamorph-typified species, many of which have been derived from the sexual progenitors (teleomorph-typified species) either as single species or more commonly through interspecific hybridization (Schardl *et al.*, 1994; Tsai *et al.*, 1994; Schardl and Wilkinson, 2000). Vertically transferred asexual *Epichloë* strains are reproductively isolated but can continue genetic differentiation by accumulation of random mutations (Clay, 1993).

Epichloë *endophyte chemistry*

Four major classes of bioactive secondary metabolites or alkaloids have been identified: the ergot alkaloids such as ergovaline and chanoclavine, the indole diterpenes (including lolitrem B and epoxy-janthitrem), pyrrolizidines, which includes lolines, and the pyrrolopyrazine metabolite, peramine (Lane *et al.*, 2000; Schardl *et al.*, 2013). It is these and other yet uncharacterized secondary metabolites that allow the host plant to tolerate biotic and abiotic stresses (Clay and Schardl, 2002; Schardl *et al.*, 2013; Panaccione *et al.*, 2014). Expression of these secondary metabolites is undetectable or low when the fungus is grown *in vitro* indicating that plant signalling is required to induce expression (Rowan, 1993; Young *et al.*, 2006). It is accepted that while the endophyte strain determines the types of alkaloid produced with the biosynthetic machinery encoded by the fungal genome, it is the host plant that regulates the quantitative expression of these compounds (Latch, 1994; Ball *et al.*, 1995; Lane *et al.*, 2000; Easton *et al.*, 2002; Spiering *et al.*, 2005; Schardl *et al.*, 2013). Iron-chelating siderophores are other important secondary metabolites synthesized by *Epichloë* (Koulman *et al.*, 2012; Johnson *et al.*, 2013b; Forester *et al.*, 2017), and it is unknown if the epichloëcyclins, multiple oligopeptides produced via ribosomal peptide synthesis (RiPS) (Johnson *et al.*, 2015), represent a new class of bioactive compounds or participate in another aspect of fungal biology.

Based on a meta-data analysis of published ecological studies, Bastías *et al.* (2017a) have proposed that *Epichloë* endophytes may improve antiherbivore defences via alkaloid-dependent as well as alkaloid-independent mechanisms. Alkaloid-independent mechanisms may be related to the plant immune system, specifically the jasmonic acid and salicylic acid pathways. It has been proposed that the endophyte presence modulates the plant immune system by promoting jasmonic acid and repressing the salicylic acid signalling pathways (Bastías *et al.*, 2017a). In agreement with this model, Bastías *et al.* (2018a) have shown that the plant hormone salicylic acid is reduced in endophyte-infected grasses compared to endophyte-free, and have linked artificially elevated salicylic acid (via exogenous hormonal treatment with salicylic acid) with reduced lolines that consequently reduced resistance to aphids. Similarly, in the case of the generalist chewing insect *Spodoptera frugiperda* (fall armyworm), induced plant levels of jasmonic acid decreased insect resistance, which was also attributed to a reduction in loline alkaloids (Bastías *et al.*, 2018b,c). Since jasmonic acid can activate host plant defences and re-allocate plant resources to the roots (possibly reducing plant nutrients for alkaloid production), both activities could negatively affect the growth of the endophyte and hence production of defensive alkaloids. Endophyte growth may also be affected by the phytohormone gibberellin, since ryegrass gibberellin biosynthetic genes are upregulated in response to *E. festucae* infection (Dupont *et al.*, 2015; Schmid *et al.*, 2017). There is also evidence that *Epichloë* endophytes can enhance establishment of perennial ryegrass by providing protection through elevating compounds that reduce oxidative stress levels caused by pathogenic fungi (Ma *et al.*, 2015).

Benefits to the host plant from Epichloë *endophyte associations*

While the specific characteristics of associations vary (Faeth and Saikkonen, 2007; Rudgers and Clay, 2007), *Epichloë* endophytes commonly enhance the fitness of the host (Clay, 1993). Host benefits include improved tolerance to water deficiency and to nutrient limitations (Bacon, 1993; West *et al.*, 1993; Elbersen and West, 1996; Saikkonen *et al.*, 1998; Malinowski and Belesky, 1999, 2000; Clay and Schardl, 2002; Malinowski *et al.*, 2005; Belesky and West, 2009; Nagabhyru *et al.*, 2013), elevated nutrient concentration in shoots (Vázquez-de-Aldana *et al.*, 2013), increased tolerance to heavy metals (Zhang *et al.*, 2010), increased tolerance to cold (Parsaeian *et al.*, 2006), increased competitiveness through allelopathic interactions (Matthews and Clay, 2001), and protection from insect and other invertebrate pests (Prestidge *et al.*, 1982; Barker, 1987; Cheplick and Clay, 1988; Clay, 1989; Latch, 1993; Breen, 1994; Vicari *et al.*, 2002; Popay and Bonos, 2005; Bastías *et al.* 2017b; Li *et al.*, 2007) and pathogenic fungi (Gwinn and Gavin, 1992; Burpee and Bouton, 1993; Christensen, 1996; Bonos *et al.*, 2005; Clarke *et al.*, 2006; Tian *et al.*, 2008; Ma and Nan, 2011; Panka *et al.*, 2013; Wiewiora *et al.*, 2015; Gorzynska *et al.*, 2017). These effects are mediated by bioactive compounds, some of which are very well characterized, such as peramine, lolines and ergovaline (Bush *et al.*, 1997; Lane *et al.*, 2000; Vázquez-de-Aldana *et al.*, 2013), and others more recently described, such as the epoxy-janthitrems (Johnson *et al.*, 2013a). The identification of approximately 40 secondary metabolite gene clusters in *E. festucae* genomes (Schardl *et al.*, 2013) and the discovery of several novel metabolites produced in endophyte-infected perennial ryegrass (Cao *et al.*, 2008) suggest that *Epichloë* species are capable of producing additional secondary metabolites with the potential to confer protective fitness benefits to plants.

Epichloë infection can affect plant development and metabolism. There is evidence that the presence of *Epichloë* endophyte can modify morphology by increasing tiller numbers (Arachevaleta *et al.*, 1989; Belesky *et al.*, 1989; Nan and Li, 2000), root architecture (Malinowski *et al.*, 1999; Crush *et al.*, 2004), and dry matter yield (Nan and Li, 2000). *Epichloë* infection may also affect host plant physiology including higher antioxidant enzyme activities and photosynthetic capability, but this is strain dependent (Richardson *et al.*, 1993; Zhang *et al.*, 2017b). Trichome formation and host cell wall thickness changes have also been observed with infection of ryegrass with the non-native endophyte *E. festucae* Fl1 (Dupont *et al.*, 2015). Endophyte infection has also been shown to cause changes in sugar and amino acid concentrations and, along with fungal compounds (mannitol and lolines), may enable endophyte-infected tall fescue to recover faster than endophyte-free controls under water-deficit conditions (Nagabhyru *et al.*, 2013). Pan and Clay (2004) also demonstrated that, in the clonal grass *Glyceria striata, Epichloë* infection can stimulate carbon translocation.

Other studies have shown that nitrogen and carbohydrate availability can differentially alter endophyte biomass and alkaloid production depending on the particular *Epichloë*–grass interaction (Rasmussen *et al.*, 2007; Ryan *et al.*, 2015). There is likely a metabolic cost to the host plant for maintaining infection, which may be tolerated when there is a net fitness benefit to the plant, but may be lost under non-stressed conditions. The degree of mutualism may be dependent on environmental factors such as nutrient availability (Saikkonen *et al.*, 2006). Control of nutrient exchange between symbiotic partners is therefore likely to be an important mechanism for determining how partners interact. Cheplick and Cho (2003) concluded that the direction of impact due to endophyte infection will be contingent on not only abiotic factors but also genotype-specific responses to endophytic fungi, herbivores and competitors.

Negative impacts of Epichloë *endophytes on host plant adaptation*

There are numerous examples in the literature where *Epichloë* endophytes have been inconsistent or failed to improve host adaptation to biotic or abiotic stresses (Lewis, 2004). Examples of zero impacts include plant tolerance to insects (Lewis and Clements, 1986) and pathogens (Welty *et al.*, 1991) or resilience against drought or nutrient stress (Lewis, 1992; Cheplick *et al.*, 2000; Cheplick 2004). It may be that the particular strain used did not produce the secondary metabolite required for the effect sought, or that there was no adaptive advantage in the *Epichloë* strains trialled. The host plant genotype can also alter the impact of the endophyte (Belesky and Fedders, 1996) through host genotype effects on expression of bioactive secondary metabolites (Johnson *et al.*, 2007; Schardl *et al.*, 2013).

While the association between *Epichloë* and temperate grasses is generally considered mutualistic and beneficial, there is some evidence to the contrary (Muller and Krauss, 2005). For example, the cost of endophytes may outweigh their benefits in resource-limited conditions including nutrients and/or moisture (Cheplick *et al.*, 1989; Ahlholm *et al.*, 2002; Cheplick, 2004). In these experiments, there was no invertebrate pest pressure exerted and so the direct effect of the endophyte on the plant growth and reproduction was measured. In the undomesticated American bunchgrass, Arizona fescue (*Festuca arizonica*), *Epichloë* infection may decrease plant fitness and susceptibility to vertebrates, invertebrates (Schulthess and Faeth, 1998; Saikkonen *et al.*, 1999) and plant pathogens (Faeth *et al.*, 2001). It has also been shown that beneficial effects of endophyte infection in wild grasses can vary for different grass species, *Brachypodium sylvaticum* with *Epichloë sylvatica* and *Bromus benekenii* with *Epichloë bromicola* (Brem and Leuchtmann, 2002). *E. bromicola* did increase the fitness of *B. benekenii*, which may explain the high infection rate observed in natural populations. However, this was not the case for *B. sylvaticum*, where other factors such as increased herbivore and pathogen resistance may be responsible for the very high incidence of this association in nature. This has led to the suggestion that *Epichloë* endophytes could have antagonistic as well as mutualistic effects on its host (Morse *et al.*, 2002), and as a result, many have argued that the relationship cannot be considered an obligate mutualism (Ravel *et al.*, 1995; Cheplick, 1997; Saikkonen *et al.*, 1998; Cheplick *et al.*, 2000; Faeth and Sullivan, 2003). Yet others have argued that mutualism becomes more apparent in stressful or limiting environments (Bacon 1993; Hill *et al.*, 1996).

Host responses to Epichloë *endophyte associations*

To investigate host responses induced by the presence of *Epichloë* symbionts of several grass hosts at the molecular level, differential gene expression and transcriptome studies have been conducted, but with contrasting outcomes (Johnson *et al.*, 2003; Khan *et al.*, 2010; Ambrose and Belanger, 2012; Dupont *et al.*, 2015; Dinkins *et al.*, 2017; Schmid *et al.*, 2017). In the fine fescue–*E. festucae* symbiosis, 209 plant genes were moderately affected by endophyte infection under non-stressed conditions (Ambrose and Belanger, 2012). Likewise, in the tall fescue–*E. coenophiala* symbiosis, minimal effects of *Epichloë* on the plant transcriptome during vegetative and reproductive host stages were observed (Dinkins *et al.*, 2017; Nagabhyru *et al.*, 2019). indicating that the endophyte does not cause major changes in the host transcriptome for symbiotic persistence and transmission through to the seeds. However, Dupont *et al.* (2015), using the model perennial ryegrass–*E. festucae* Fl1 system (a mutualistic symbiosis with a non-native endophyte), showed that endophyte infection induces large changes in ryegrass gene expression (including suppression of host defences), suggesting that endophyte infection can trigger reprogramming of host metabolism and substantially alter host development. In contrast, in the natural perennial ryegrass–*E. festucae* var. *lolii* symbiosis (Schmid *et al.*, 2017), the number of differentially expressed genes in the host were much reduced, with little correlation to the Dupont *et al.* (2015) study, with a host response supporting the hypothesis that the endophyte primes host defences. Different grass–endophyte symbioses therefore appear to lead to various levels of host response. Further research is required to understand the signalling mechanisms for symbiosis establishment and maintenance and, in particular, how the endophyte induces changes in host growth and function under specific environmental conditions.

Epichloë *strain and host compatibility*

Epichloë strains are generally host-specific but can be artificially induced to form symbioses with related host species (Christensen, 1995). Novel associations can be created by inoculation of fungal strains with plant species closely related to the natural host. In a functional (i.e. compatible) association, endophyte hyphae proliferate in a coordinated manner that does not overwhelm the host and is in synchrony with host plant development (Christensen *et al.*, 2008; Voisey, 2010). Interaction between grasses and *Epichloë* endophytes has been shown to be driven predominantly by compatibility between host plant genotype and fungal endophyte strain (Ryan *et al.*, 2015). Incompatibility is often evident when an *Epichloë* strain is artificially transferred from its natural host to an alternative host (Koga *et al.*, 1993; Christensen, 1995). Essential elements of stable functional associations are maintenance and coordinated proliferation of the fungus in planta, effective transmission of viable mycelium to seed, and production by the endophyte of its bioactive alkaloids. Compatible endophyte strains have been shown to produce reactive oxygen species that prevent them from aggressive and parasitic growth in their natural hosts, whereas parasitic strains lack the ability to produce those oxidants in incompatible hosts (Tanaka *et al.*, 2006).

While *Epichloë* studies have predominantly focused on associations with *Lolium* and *Festuca*, their occurrence is widespread across many genera in the temperate grass subfamily *Pooideae*. This includes *Bromus*, *Poa* (White and Cole, 1986; Cabral *et al.*, 1999), *Dactylis* (Sánchez Márquez *et al.*, 2007), *Elymus* (Nan and Li, 2000; Card *et al.*, 2014; Simpson *et al.*, 2014), *Echinopogon* (Moon *et al.*, 2002) and *Hordeum* (Wilson *et al.*, 1991; Wilson, 2007; Card *et al.*, 2014). Modern cereals such as cultivated wheat, barley and rye, however, do not naturally host *Epichloë* endophytes (Card *et al.*, 2014; Simpson *et al.*, 2014).

Epichloë *persistence in the host*

An important question is whether grass symbioses with *Epichloë* can persist without conferring improved plant production and persistence and hence improved reproductive advantage to their hosts. Gundel *et al.* (2008) have argued that as long as the transmission efficiency is high, endophytes will ultimately persist even when the benefit of infection to the host is minor. Saikkonen *et al.* (2002) concluded that even when a meta-population (i.e. a group of populations of the same species is separated by space, but interactions occur among them by individual members moving from one population to another) consists of qualitatively different patches, endophyte-infected plants may persist at the meta-population level even if both the vertical transmission is imperfect and the endophyte decreases the host grass fitness in certain environments. This may explain why, in Europe and Scandinavia, *Epichloë* infection in natural grasslands is often maintained at 40–50% despite its benefits being hard to define (Lewis *et al.*, 1997; Oldenburg, 1997; Oliveira *et al.*, 1997; Ravel *et al.*, 1997; Saikkonen *et al.*, 2000; Jensen and Roulund, 2004; Bazely *et al.*, 2007; Granath *et al.*, 2007). However, endophyte presence was higher at sites where drought stress was common (Valle Ribeiro *et al.*, 1996; Arroyo García *et al.*, 2002) or grazing pressure was high (Jensen and Roulund, 2004).

7.2.2 Facultative, systemic mutualists located in shoot and root tissue

In nature, both facultative and obligate endophytes can colonize the interior of plants, but facultative ones are also able to survive in the soil and/or on the plant surface as well (Hardoim *et al.*, 2015). A number of different types of facultative, systemic fungal endophytes have been found in shoots and roots (i.e. all non-embryonic tissues) of temperate grasses. The best characterized of these are the dark septate endophytes (DSE) and a diverse group of fungi classified as Class 2 endophytes by Rodriguez *et al.* (2009), including the *Phialophora*-like or p-endophytes and endophytic fungal entomopathogens. All of these facultative endophytes colonize the plant by forming extensive systemic hyphal networks.

The Class 2 endophytes share the common attributes of being transmitted via seed coats and/or rhizomes, with low abundance in the rhizosphere (Rodriguez *et al.*, 2009). Two significant examples of these are *Curvularia protuberata,* which colonizes the geothermal grass *Dichanthelium lanuginosum* and is required to allow the symbiosis to tolerate temperatures up to 65°C (Redman *et al.*, 2002; Márquez *et al.*, 2007), and *Fusarium culmorum,* which colonizes the coastal dune grass (*Leymus mollis*) and is required for the symbiosis to survive and the host plant to grow at high levels of salinity experienced in their native habitat (Rodriguez *et al.*, 2008). Of interest is the observation that isolates of *C. protuberata* and *F. culmorum,* obtained from plants in habitats devoid of heat or salt stress, respectively, do not confer either heat or salt tolerance but are still able to asymptomatically colonize these plants (Rodriguez *et al.*, 2008).

Entomopathogenic fungi can also be considered Class 2 endophytes, but they form a distinct group due to their pathogenicity of insects. They can also behave as mutualistic endophytes (for comprehensive reviews see Vidal and Jaber, 2015 and Jaber and Ownley, 2017) and include taxa like *Beauveria, Torrubiella, Metarhizium, Lecanicillium* and *Tolypocladium,* which have been frequently isolated as symptomless endophytes from grasses and other plant species (Sánchez Márquez *et al.*, 2008; Vega *et al.*, 2008; Ownley *et al.*, 2010; Sánchez Márquez *et al.*, 2010). While being well known for their control of insect pests, these fungi, as plant colonizers, have also been shown to suppress plant diseases (Goettel *et al.*, 2008; Ownley *et al.*, 2008).

Dark septate endophytes (DSE) are abundant fungal root colonizers, particularly, of woody plants, but are also found in temperate grasses (Barrow, 2003; Mandyam *et al.*, 2010), including perennial ryegrass (Skipp and Christensen, 1989), *Elymus farctus* (Sánchez Márquez *et al.*, 2008), *Poa alpigena* (Vare *et al.*, 1992), *Cyanodon dactylon, Digitaria cruciate* and *Poa annua* (Li *et al.*, 2005). DSE are a ubiquitous group of biotrophic fungi that form melanized, septate hyphae, colonizing root tissue without causing any obvious host response (Kernaghan and Patriquin, 2011) and are found in both roots and aboveground tissues (Jumpponen and Trappe, 1998). There is some evidence that DSE may enhance host tolerance to drought as well as other extreme environmental stresses (Barrow, 2003; Waller *et al.*, 2005), increase foliar concentrations of N and P (Haselwandter and Read, 1982; Jumpponen *et al.*, 1998; Newsham, 1999) and deter pathogens (Mandyam and Jumpponen, 2005), although their ecological function remains largely elusive. Transmission is most likely horizontal through mycelia and conidiospores (Jumpponen and Trappe, 1998). DSE include *Phialophora* spp., *Phialocephala* spp., *Chloridium paucisporum, Heteroconium chaetospira* and *Leptodontidium orchidicola* (Schulz, 2006).

Phialophora-like and *Gliocladium*-like endophytes (Philipson, 1989, 1991a,b), as yet not formally named, are known collectively as penicilliate or p-endophytes due to their penicilliate conidiophores. They are often co-symbiotic with *Epichloë* endophytes in a number of cool season grasses, including *Festuca* and *Lolium* species (Latch *et al.*, 1984; Philipson, 1991a; An *et al.*, 1993). Unlike *Epichloë,* which generally grow synchronously with the host, p-endophyte growth is more heavily branched and continues beyond leaf maturation (Christensen and Voisey, 2007). In Arizona fescue (*Festuca arizonica*), *Epichloë* has been shown to possibly inhibit the colonization and growth of other fungal endophytes (Schulthess and Faeth, 1998). However, Hayes and Faeth (2002) working with sleepy grass (*Achnatherum robustum*), a cool season grass that is native to high elevations in the Southwestern United States, believed that while *Epichloë* may benefit the host plant, it may incur a cost by allowing the co-symbiotic p-endophytes to bypass host defences. In this case, the authors referred to the behaviour of the p-endophyte as parasitic.

High concentrations of branched hyphae can be found in old leaves (Christensen *et al.*, 2002). The p-endophytes sporulate on leaves, and spores can move horizontally between host plants, but they can also be vertically transmitted in host seed (Latch and Christensen, 1985). These endophytes can be found throughout the plant including the roots (Latch *et al.*, 1984; Skipp and Christensen, 1989). However, although p-endophytes are symbionts, the question remains: are they mutualistic? They have been shown to provide antifungal properties (Siegel and Latch, 1991); however, to date, there is no documented evidence that p-endophytes provide any benefit directly to the host plant (Latch *et al.*, 1985; Christensen *et al.*, 2002).

7.2.3 Facultative, non-systemic endophytes located in shoot tissue

Non-systemic endophytes are a taxonomically diverse group of ubiquitous endophytes found in every plant species to date and are incapable of seed transmission (Canals *et al.*, 2014). Little is understood about their ecological function. Indeed, non-systemic facultative endophytes of grasses found in the aboveground plant parts are common, and species diversity is large and dominated by ascomycetes such as *Acremonium, Alternaria, Cladosporium, Epicoccum* and *Penicillium* species (Sánchez Márquez *et al.*, 2012). Varvas *et al.* (2013) identified ten different symbiotic fungal taxa in *Phleum pratense* (timothy), and they were dominated by *Epicoccum nigrum* with a 67% colonization frequency. Others occurring at lower frequency included *Alternaria arbusti, Lewia viburni, Apiospora montagnei, Aureobasidium pullulans, Fusarium* sp., *Gibberella avenacea, Monographella* sp., *Paraphaeosphaeria michotii* and *Phaeosphaeria herpotrichoides*. While many have been found to be symbiotic in behaviour, determination of their mutualism is open to debate. Sánchez Márquez *et al.* (2007) sampled 15 populations of cocksfoot (*Dactylis glomerata*) in Spain and collected 1400 fungal isolates from 120 plants of which 316 were selected as putative endophytes from leaf and root material. Ninety-one different species of fungi belonging to 63 genera were identified. Of these, 48 species were found in leaves, 22 species in roots, and 21 species from both above- and belowground parts. The genera most abundant in terms of the number of isolates collected were *Penicillium* (34 isolates), *Cladosporium* (21), *Acremonium* (20), *Helgardia* (18), *Podospora* (18), *Fusarium* (17), *Phaeosphaeria* (17), *Epicoccum* (15), *Alternaria* (7), *Chaetomium* (9) and *Lewia* (7). These genera represent diverse fungi, including *Epichloë* species that were reportedly isolated, thereby suggesting that these fungi can exhibit different microbial lifestyles and behaviours with their host grasses.

Morchella elata, an endophyte of *Bromus tectorum* or cheatgrass, has been shown to provide benefit to the host plant through simultaneously increasing both the probability of fire and survival of that event, via more fuel and a greater belowground seed bank, respectively (Baynes *et al.*, 2012). Cheatgrass is an invasive winter annual native to Eurasia, and it is postulated that this mutualism with fire-associated fungi may be contributing to its distribution across western North America.

7.3 Interactions between Endophytic Taxa in Temperate Grasses

Plants are colonized by more than one type of endophyte (Hardoim *et al.*, 2015), and how interactions between symbionts of temperate grasses affect plant fitness within grassland ecosystems is important to investigate. In particular, the effects of *Epichloë* species on the colonization frequency of AMF, co-symbionts of grasses, has been well studied. AMF, while colonizing root tissues, are not considered true endophytes due to their growth into the rhizosphere which provides the means for substantial fungus-to-plant nutrient transfer (Brundrett, 2002, 2004, 2006; Rodriguez *et al.*, 2009). As such, they behave as mutualistic symbionts providing an advantage to plants grown in nutrient-deficient (Asghari and Cavagnaro, 2011; Hoysted *et al.*, 2018; Li *et al.*, 2018) and drought-prone soils (Duan *et al.*, 1996; Augé, 2001; Allen, 2007). In return, AMF receive both carbohydrate and fatty acids from the host plant to enable their growth and reproduction (Pfeffer *et al.*, 1999; Roth and Paszkowski, 2017).

AMF have also been shown to provide resistance against soil pathogens (Newsham *et al.*, 1995; Azcon-Aguilar and Barea, 1996; Sikes *et al.*, 2009) and tolerances to soil toxicities due to contaminants such as heavy metals (Hildebrandt *et al.*, 2007). Disease protection by AMF may result from a number of mechanisms including improved nutrient acquisition, resulting in stronger plants, or through direct elicitation of a plant defence response (Azcon-Aguilar and Barea, 1996).

AMF are common in pastures (Spatafora *et al.*, 2016), and Berthelot *et al.* (2018) demonstrated that perennial ryegrass biomass and shoot Na, P, K and Mg concentrations significantly increased following AMF inoculation as compared to non-inoculated controls. AMF colonization also increased alkaline phosphatase activity and P mobility in the soil.

Mycorrhizal infection has been shown to be markedly reduced when the plant contains an *Epichloë* endophyte in annual ryegrass (Omacini *et al.*, 2006), perennial ryegrass (Muller 2003; Liu *et al.*, 2011) or tall fescue (Chu-Chou *et al.*, 1992; Guo *et al.*, 1992; Antunes *et al.*, 2008; Mack and Rudgers, 2008; Buyer *et al.*, 2011). Mack and Rudgers (2008) considered that the negative effects of endophytes on AMF might be due to inhibition via endophyte alkaloids, altered nutritional requirements of the host plant and/or temporal and spatial priority effects in the interactions among plants and multiple symbionts. However, in swards with both endophytic and non-endophytic plants, the impacts of *Epichloë* endophytes on AMF symbionts can be more dynamic. Omacini *et al.* (2006) observed that while endophyte-infected plants had lower levels of mycorrhizal colonization, the presence of endophyte-infected plants caused an increase in AMF colonization in non-endophyte-infected conspecific neighbours.

Furthermore, Kalosa-Kenyon *et al.* (2018) concluded that effects of *Epichloë* species on AMF co-symbionts are not easily generalized across plant–endophyte symbiota after comparing the interactions between *Epichloë* and AMF in three temperate grass species. They showed that *Epichloë* infection increased AMF colonization of roots in both *Poa alsodes* and *P. sylvestris*, although this effect was only significant for hyphal colonization in *P. sylvestris*, while *Epichloë* did not significantly alter AMF colonization in *S. arundinaceus*. Similarly, in observational field trials with native grasses, *Epichloë* presence has been associated with higher root AMF colonization in *Bromus setifolius* (Novas *et al.*, 2005; Novas *et al.*, 2011), *Poa bonariensis* (Novas *et al.*, 2009) and *Elymus hystrix* (Larimer *et al.*, 2012). In tall fescue, Slaughter *et al.* (2018) showed that *Epichloë* infection decreased arbuscule formation (>50% reduction), only in treatments without added precipitation, but had no effect on the total rate of AMF colonization (arbuscules + vesicles + hyphae) in tall fescue roots.

Few studies have accessed the effect of AMF symbiosis on the plant protective benefits conferred by *Epichloë*. In one study, AMF infection has been shown to reduce the level of resistance provided by *Epichloë* to ryegrass against Argentine stem weevil (Barker, 1987). However, the mechanism for this effect is unclear. Similarly, the beneficial effect of *Epichloë* in perennial ryegrass on resistance to larvae of the noctuid moth, *Phlogophora meticulosa*, was reduced, but not eliminated, by AMF (Vicari *et al.*, 2002). However, dual inoculation of perennial ryegrass with the mycorrhizal fungus *Claroideoglomus etunicatum* and *Epichloë festucae* var. *lolii* resulted in reduced incidence of leaf spot caused by *Bipolaris sorokiniana* (Li *et al.*, 2018).

Other belowground endophytes that colonize the roots of perennial ryegrass are the DSE. Berthelot *et al.* (2018) investigated the impact of dual inoculation with an AMF (*Funneliformis mosseae*) and a DSE (*Cadophora* sp.) and their interaction on perennial ryegrass grown in a trace element (Cd/Zn/Pb) polluted soil. Although DSE had no effect on AMF colonization, AMF colonization slightly decreased DSE frequency. Dual inoculation significantly decreased shoot Cd concentration. *Epichloë* presence in tall fescue has been shown to reduce root colonization by the DSE leading to the conclusion that *Epichloë* associations with tall fescue may have divergent long-term impacts on other host–symbiont interactions (Slaughter *et al.*, 2018). The contrasting outcomes obtained from analysing the effects of multi-symbiont interactions may be related to difference in species used, as well as host genetics and environmental conditions.

7.4 Economic Importance of Fungal Endophytes in Temperate Grasses

Irrespective of whether the endophytic microbes discussed here are facultative or obligate mutualists, they have been shown to provide advantages to their host plants by improving adaptation to abiotic stresses and 'resistance' to biotic pests and diseases. However, quantifying that impact in terms of financial returns to a farmer or the economy has only rarely been attempted.

Epichloë endophytes have been used extensively in commerce for the last 20 years in both New Zealand (Caradus *et al.*, 2013) and the United States (Bouton and Easton, 2005; Young *et al.*, 2013; Table 7.1). In New Zealand, the uptake by farmers was high (Milne, 2007; Caradus *et al.*, 2013), driven by need (a solution to a significant problem), ease of use/application and the resulting improvements in production and profitability. Almost all proprietary ryegrass cultivars in New Zealand are now sold with an accompanying *Epichloë* endophyte that provides improved pasture performance. The New Zealand seed industry has self-regulated *Epichloë* endophyte use to ensure that farmers are sold a product with at least 70% viable endophyte in seed at point of sale and an agreed rating system for all endophytes in commerce that is freely available to all end users. Endophyte strains are viewed as added value points of difference between ryegrass cultivars by proprietary seed companies.

Table 7.1. Commercialized *Epichloë* endophyte strains for temperate grasses used for ruminant grazing.

Strain	Trademark	Host species	Year of commercialization	Territories
AR5	Endosafe	Perennial ryegrass	1992	New Zealand
AR542	MaxQ	Tall fescue	2000	USA, Argentina, Uruguay
AR1	AR1	Perennial ryegrass	2001	New Zealand, Australia, Chile
AR542	MaxP	Tall fescue	2003	New Zealand, Australia
NEA2 and NEA6	NEA2	Perennial ryegrass	2003	New Zealand, Australia
AR5	Endo5	Perennial ryegrass	2006	Australia
AR37	AR37	Perennial and annual ryegrass	2007	New Zealand, Australia
E34	E34	Tall fescue	2007	USA
UArk4	ArkShield	Tall fescue	2011	USA
AR584	MaxQ II	Tall fescue	2011	USA, Argentina, Uruguay
AR584	MaxP	Tall fescue	2012	New Zealand, Australia
U2	U2	Festulolium	2012	New Zealand, Australia
647 Protek	Protek	Tall fescue	2014	USA, New Zealand
Edge	Edge	Perennial ryegrass	2016	New Zealand
Happe	Happe	Perennial ryegrass	2017	New Zealand
NEA2	NEA	Perennial and hybrid ryegrass	2017	New Zealand, Australia
NEA2 and NEA3	NEA4	Perennial ryegrass	2018	New Zealand, Australia

In New Zealand in the early 1980s, a significant animal health issue for farmers was that 'wild-type' or standard endophyte, responsible for causing ryegrass staggers (Fletcher *et al.*, 1999) and, in warm summers, heat stress (Easton *et al.*, 1996), was widespread throughout the country's pastures. Replacing the 'wild-type' endophyte with non-toxic endophyte strains such as AR1 and AR37 has been a major driver in perennial ryegrass sales over the last two decades. AR1 was delivered in response to the prevalence of staggers on ryegrass with the standard endophyte expressing lolitrem B. A peramine-expressing endophyte without producing lolitrem B was required to protect the ryegrass from Argentine stem weevil and remove the animal health issue. At the time of its release, AR1 was considered a significant breakthrough. The endophyte did not cause ryegrass staggers and also reduced heat stress, flystrike and faecal soiling of sheep. It increased animal liveweight gains, resulting in 22% returns over farms using standard endophyte pastures (Fletcher, 1999). A 9% increase in milk solids was measured in dairy systems (Bluett *et al.*, 2003).

When poor pasture persistence with AR1 was observed with the increasing impact of other pasture pests such as African black beetle, root aphid, porina and pasture mealy bug (Thom *et al.*, 2014), a new endophyte AR37 was released (Hume *et al.*, 2009). AR37 provided broad resistance to insect pests, increasing ryegrass persistence (Popay and Hume, 2011). Estimates of value to the New Zealand economy have shown that AR37 alone will contribute NZ$3.6 billion over the 20-year lifetime of its patent (ACIL Allen Consulting, 2017).

In the United States, the widespread use of Kentucky 31 tall fescue with an endophyte responsible for causing fescue toxicosis is estimated to cause a loss in productivity of over US$1.5 billion per year (Aiken and Strickland, 2013). The use of novel endophytes such as MaxQ, which do not produce the toxic alkaloid ergovaline, has meant that all toxic fescue pastures can now be replaced without any significant impact on pasture persistence and production (Latch, 1997; Latch *et al.*, 2000; Bouton *et al.*, 2002). It has been estimated that the average daily gain for cattle on tall fescue pastures decreases by 39 and 33 g/day, for each increase of 100 ppb of total ergot alkaloid concentration (Liebe and White, 2018). The significant financial benefits of MaxQ tall fescue have been demonstrated for beef cows, calves and feeder cattle systems (Bouton *et al.*, 2001; Duckett *et al.*, 2007; Biermacher and Beck, 2013) and sheep farming systems (Bouton *et al.*, 2000).

In Australia, perennial ryegrass toxicosis caused by ergovaline and lolitrem B expression, again in the wild-type endophyte, is estimated to cause losses of A$100 m per year (Leury *et al.*, 2014). *Epichloë* endophytes developed and used in New Zealand are now available for use in Australia and are reducing the impact of endophyte toxins.

7.5 The Future for Fungal Endophytes of Temperate Grasses

There is still much to learn about the impact and ecological benefits of fungal endophytes in temperate grasses. The biological role of non-systemic endophytes in temperate grasses requires better understanding. While we understand some of the direct impacts of particular insect-active secondary metabolites produced by *Epichloë* endophytes, less is known and understood about the impacts of these fungi on plant metabolism. The complexity of endophyte–grass interactions indicates that there is still much to explore in understanding their pivotal role in providing protection against wider multitrophic assemblages. For example, in *Dactylis glomerata*, over 100 different fungal species have been isolated and only five have corresponded to known pathogens of that grass (Sánchez Márquez *et al.*, 2007), so this raises the question as to the role and function of the other 90 plus fungi present.

Research exploring the types of interactions fungal endophytes form with temperate grasses reveals that, even for the well-characterized *Epichloë* genus, the symbioses can span a continuum from pathogenic/parasitic to mutualistic. Studies of these fungi in natural ecosystems enable the full range of symbiotic outcomes to be observed, as changes in environment, plant genotype and fungal genotype can be extremely varied.

This situation contrasts the strictly asexual *Epichloë* species used in agricultural settings, where the associations have been co-selected for optimum compatibility with selected elite germplasm. The specific characteristics selected for are 100% seed transmission, high endophyte viability in seed and high expression of selected alkaloids to provide bioactivity against pests, diseases and some abiotic stresses and a reduction or elimination of associated animal health and welfare issues.

The relationships symbiotic fungi form with their hosts are dependent on multiple factors, with the most influential being the reproduction mode of the fungus. Other dominant factors include the genotype of the host plant and endophyte and the environment (nutrient availability). Changes in resource availability are known to trigger shifts in functionality of fungi (Termorshuizen and Jeger, 2009). We are still learning whether the metabolic cost of the endophyte to the host plant may outweigh the benefits accrued in some situations. However, there is little doubt that in some well-researched associations, such as *Epichloë* in temperate grasses, the benefits are such that some pastoral agricultural systems would fail to function without the presence of these mutualistic endophytes.

Plant–fungal symbiotic associations are ubiquitously distributed in natural plant communities. Besides the well-studied mycorrhizal symbiosis and grass systemic clavicipitaceous endophytes, recently, non-systemic and horizontally transmitted fungal endophytes serving as plant symbionts have been increasingly recognized and provide new opportunities to proffer sustainable solutions for food production systems.

Acknowledgements

The authors thank Christine Voisey for the critical reading of this chapter.

References

ACIL Allen Consulting (2017) New Zealand's Science System: Case Studies, Report to the Ministry of Business, Innovation and Employment. Sydney, Australia.

Ahlholm, J.U., Helander, M., Lehtimaki, S., Wali, P. and Saikkonen, K. (2002) Vertically transmitted fungal endophytes: Different responses host-parasite systems to environmental conditions. *Oikos* 99, 173–183.

Aiken, G.E. and Strickland, J.R. (2013) Forages and Pastures Symposium: Managing the tall fescue-fungal endophyte symbiosis for optimum forage-animal production. *Journal of Animal Science* 91, 2369–2378.

Allen, M.F. (2007) Mycorrhizal fungi: Highways for water and nutrients in arid soils. *Vadose Zone Journal* 6, 291–297.

Ambrose, K.V. and Belanger, F.C. (2012) SOLiD-SAGE of endophyte-infected red fescue reveals numerous effects on host transcriptome and an abundance of highly expressed fungal secreted proteins. *PLOS ONE* 7, e53214.

An, Z.-Q., Siegel, M.R., Hollin, W., Tsai, H.-F., Schmidt, D. and Schardl, C.L. (1993) Relationships among non-*Acremonium* sp. fungal endophytes in five grass species. *Applied and Environmental Microbiology* 59, 1540–1548.

Antunes, P.M., Miller, J., Carvalho, L.M., Klironomos, J.N. and Newman, J.A. (2008) Even after death the endophytic fungus of *Schedonorus phoenix* reduces the arbuscular mycorrhizas of other plants. *Functional Ecology* 22, 912–918.

Arachevaleta, M., Bacon, C.W., Hoveland, C.S. and Radcliffe, D.E. (1989) Effect of the tall fescue endophyte on plant response to environmental stress. *Agronomy Journal* 81, 83–90.

Arroyo García R., Martínez Zapater J.M., García Criado B. and Zabalgogeazcoa I. (2002) Genetic structure of natural populations of the grass endophyte *Epichlöe festucae* in semiarid grasslands. *Molecular Ecology* 11, 355–364.

Asghari, H.R. and Cavagnaro, T.R. (2011) Arbuscular mycorrhizas enhance plant interception of leached nutrients. *Functional Plant Biology* 38, 219–226.

Augé, R.M. (2001) Water relations, drought and vesicular-arbuscular mycorrhizal symbiosis. *Mycorrhiza* 11, 3–42.

Azcón-Aguilar, C. and Barea, M.J. (1996) Arbuscular mycorrhizas and biological control of soil-borne plant pathogens – An overview of the mechanisms involved. *Mycorrhiza* 6, 457–464.

Bacon, C.W. (1993) Abiotic stress tolerances (moisture, nutrients) and photosynthesis in endophyte-infected tall fescue. *Agriculture, Ecosystems & Environment* 44, 123–141.

Bacon, C.W., Porter, J.K., Robbins, J.D. and Luttrell, E.S. (1977) *Epichloë typhina* from toxic tall fescue grasses. *Applied Environmental Microbiology* 34, 576–581.

Bacon, C.W., Lyons, P.C., Porter, J.K. and Robbins, J.D. (1986) Ergot toxicity from endophyte-infected grasses: A review. *Agronomy Journal* 78, 106–116.

Ball, D.M., Lacefield, G.D., Agee, C.S. and Hoveland, C.S. (2007) Introduction and acceptance of novel endophyte tall fescue in USA. In: Popay, A.J. and Thom, E.R. (eds) *6th International Symposium on Fungal Endophytes of Grasses. Grassland Research and Practice Series 13*. New Zealand Grassland Association, Dunedin, New Zealand, pp. 249–251.

Ball, O.J.-P., Prestidge, R.A. and Sprosen, J.M. (1995) Interrelationships between *Acremonium lolii*, peramine and lolitrem B in perennial ryegrass. *Applied and Environmental Microbiology* 61, 1527–1533.

Barker, G.M., Prestidge, R.A. and Pottinger, R.P. (1985) Strategies for Argentine stem weevil control: Effects of drought and endophyte. *Proceeding of New Zealand Grassland Association* 47, 107–114.

Barker, G.M. (1987) Mycorrhizal infection influences *Acremonium* induced resistance to argentine stem weevil in ryegrass. Proceeding of 40th New Zealand Weed and Pest Control Conference, Wellington, pp. 199–203.

Barrow, J.R. (2003) Atypical morphology of dark septate fungal root endophytes of *Bouteloua* in arid southwestern USA rangelands. *Mycorrhiza* 13, 239–247.

Bastías, D.A., Martínez-Ghersa, M.A., Ballaré, C.L. and Gundel, P.E. (2017a) *Epichloë* fungal endophytes and plant defenses: Not just alkaloids. *Trends in Plant Science* 22, 939–948. doi:10.1016/j.tplants.2017.08.005

Bastías, D.A., Ueno, A.C., Machado Asseth, C.R., Alvarez, A.E., Young, C.A., *et al.* (2017b) Metabolism or behaviour: Explaining the performance of aphids on alkaloids-producing fungal endophytes in annual ryegrass (*Lolium multiflorum*). *Oecologia* 185, 245–256.

Bastías, D.A., Alejandra Martínez-Ghersa, M., Newman, J.A., Card, S.D., Mace, W.J., *et al.* (2018a) The plant hormone salicylic acid Interacts with the mechanism of anti-herbivory conferred by fungal endophytes in grasses. *Plant, Cell & Environment* 41, 395–405.

Bastías, D.A., Martínez-Ghersa, A.M., Newman, J., Card, S.D., Mace, W.J., *et al.* (2018b) Jasmonic acid regulation of the anti-herbivory mechanism conferred by fungal endophytes in grasses. *Journal of Ecology* 106, 2365–2379.

Bastías, D.A., Martínez-Ghersa, A.M., Newman, J.A., Card, S.D., Mace, W.J., *et al.* (2018c) Data from: Jasmonic acid regulation of the anti-herbivory mechanism conferred by fungal endophytes in grasses. *Dryad Digital Repository*. doi:10.5061/dryad.k87m52h

Baynes, M., Newcombe, G., Dixon, L., Castlebury, L. and O'Donnell, K. (2012) A novel plant fungal mutualism associated with fire. *Fungal Biology* 116, 133–144.

Bazely, D.R., Ball, J.P., Vicari, M., Tanentzap, A.J., Berenger, M., *et al.* (2007) Broad-scale geographic patterns in the distribution of vertically transmitted, asexual endophytes in four naturally-occurring grasses. *Ecography* 30, 367–374.

Belesky, D.P. and Fedders, J.M. (1996) Does endophyte influence regrowth of tall fescue? *Annals of Botany* 78, 499–505.

Belesky, D.P. and West, C.P. (2009) Abiotic stresses and endophyte effects. In: Fribourg, H.A., Hannaway, D.B. and West, C.P. (eds) *Tall Fescue for the Twenty-First Century. Agronomy. Monograph No. 53*. American Society of Agronomy, Crop Science Society of America, and Soil Science Society of America, Madison, Wisconsin, pp. 129–149.

Belesky, D.P., Stringer, W.C. and Hill, N.S. (1989) Influence of endophyte and water regime upon tall fescue accessions. I. Growth characteristics. *Annals of Botany* 63, 495–503.

Berthelot, C., Blaudez, D., Beguiristain, T., Chalot, M. and Leyval, C. (2018) Co-inoculation of *Lolium perenne* with *Funneliformis mosseae* and the dark septate endophyte *Cadophora* sp. in a trace element-polluted soil. *Mycorrhiza* 28, 301–314.

Biermacher, J. and Beck, P. (2013) Study reveals legume-fescue mixture economics. *The Samuel Roberts Noble Foundation Ag News and Views*, 2 pp.

Bluett, S.J., Thom, E.R., Clark, D.A., Macdonald, K.A. and Minneé, E.M.K. (2003) Milksolids production from cows grazing perennial ryegrass containing AR1 or wild endophyte. *Proceedings of the New Zealand Grassland Association* 65, 83–90.

Bonos, S.A., Wilson, M.M., Meyer, W.A. and Funk, C.R. (2005) Suppression of red thread in fine fescues through endophyte-mediated resistance. *Applied Turfgrass Research* 2. doi:10.1094/ATS-2005-0725-01-RS

Bouton, J.H. and Easton, S. (2005) Endophytes in forage cultivars. In: Roberts, C.A., West, C.P. and Spiers, D.E. (eds) *Neotyphodium in Cool-Season Grasses*. Blackwell Publishing, Ames, Iowa, pp. 327–340.

Bouton, J.H., Hill, N.S., Hoveland, C.S., McCann, M.A., Thompson, F.N., *et al.* (2000) Performance of tall fescue cultivars infected with non-toxic endophytes. In: Paul, V.H. and Dapprich, P.D. (eds) *Proceedings of the 4th International Neotyphodium/Grass Interactions Symposium*. Soest, Germany, 179–185.

Bouton, J.H., Hill, N.S., Hoveland, C.S., McCann, M.A., Thompson, F.N., *et al.* (2001) Georgia State

Report to SERAIEG Group 8, Chapel Hill, Tennessee. November 5–6, 2001, pp. 9–20.

Bouton, J.H., Latch, G.C.M., Hill, N.S., Hoveland, C.S., McCann, M.A., *et al.* (2002) Reinfection of tall fescue cultivars with non-ergot alkaloid producing endophytes. *Agronomy Journal* 94, 567–574.

Breen, J.P. (1994) *Acremonium* endophyte interactions with enhanced plant resistance to insects. *Annual Review in Entomology* 39, 401–423.

Brem, B. and Leuchtmann, A. (2002) Intraspecific competition of endophyte infected vs uninfected plants of two woodland grass species. *Oikos* 96, 281–290.

Brundrett, M.C. (2002) Coevolution of roots and mycorrhizas of land plants. *New Phytology* 154, 275–304.

Brundrett, M.C. (2004) Diversity and classification of mycorrhizal associations. *Biological Review* 79, 473–495.

Brundrett, M.C. (2006) Understanding the roles of multifunctional mycorrhizal and endophytic fungi. In: Schulz, B., Boyle, C. and Sieber, T.N. (eds) *Soil Biology, vol. 9: Microbial Root Endophytes*. Springer, Berlin/Heidelberg, pp. 281–298.

Burpee, L.L. and Bouton, J.H. (1993) Effect of eradication of the endophyte *Acremonium coenophialum* on epidemics of *Rhizoctonia* blight in tall fescue. *Plant Disease* 77, 157–159.

Bush, L.P., Wilkinson, H.H. and Schardl, C.L. (1997) Bioprotective alkaloids of grass-fungal endophyte symbioses. *Plant Physiology* 114, 1–7.

Buyer, J.S., Zuberer, D.A., Nichols, K.A. and Franzluebbers, A.J. (2011) Soil microbial community function, structure, and glomalin in response to tall fescue endophyte infection. *Plant and Soil* 339, 401–412.

Cabral, D., Cafaro, M.J., Saidman, B., Lugo, M., Reddy, P.V., *et al.* (1999) Evidence supporting the occurrence of a new species of endophyte in some South American grasses. *Mycologia* 91, 315–325.

Canals, R.M., San-Emeterio, L., Sánchez-Márquez, S., Ruiz de los Mozos, I., Pujol, P., *et al.* (2014) Non-systemic fungal endophytes in *Carex brevicollis* may influence the toxicity of the sedge to livestock. *Spanish Journal of Agricultural Research* 12, 623–632.

Cao, M., Koulman, A., Johnson, L.J., Lane, G.A. and Rasmussen, S. (2008) Advanced data-mining strategies for the analysis of direct-infusion ion trap mass spectrometry data from the association of perennial ryegrass with its endophytic fungus, *Neotyphodium lolii*. *Plant Physiology* 146, 1501–1514.

Caradus, J.R., Lovatt, S. and Belgrave, B. (2013) Adoption of forage technologies. *Proceedings of New Zealand Grassland Association* 75, 39–44.

Card, S.D., Rolston, M., Park, Z., Cox, N. and Hume, D. (2011) Fungal endophyte detection in pasture grass seed utilising the infection layer and comparison to other detection techniques. *Seed Science and Technology* 39, 581–592.

Card, S.D., Faville, M.J., Simpson, W.R., Johnson, R.D., Voisey, C.R., *et al.* (2014) Mutualistic fungal endophytes in the *Triticeae* – Survey and description. *FEMS Microbiology Ecology* 88, 94–106.

Card, S.D., Johnson, L., Teasdale, S. and Caradus, J. (2016) Deciphering endophyte behaviour – The link between endophyte biology and efficacious biological control agents. *FEMS Microbiology Ecology* 92, pii: fiw114. doi:10.1093/femsec/fiw114

Cheplick, G.P. (1997) Effects of endophytic fungi on the phenotypic plasticity of *Lolium perenne* (Poaceae). *American Journal of Botany* 84, 34–40.

Cheplick, G.P. (2004) Recovery from drought stress in *Lolium perenne* (Poaceae): Are fungal endophytes detrimental? *American Journal of Botany* 91, 1960–1968.

Cheplick, G.P. and Cho, R. (2003) Interactive effects of fungal endophyte infection and host genotype on growth and storage in *Lolium perenne*. *New Phytologist* 158, 183–191.

Cheplick, G.P. and Clay, K. (1988) Acquired chemical defences in grasses: The role of fungal endophytes. *Oikos* 52, 309–318.

Cheplick, G.P., Clay, K. and Marks, S. (1989) Interactions between infection by endophytic fungi and nutrient limitation in the grasses *Lolium perenne* and *Festuca arundinacea*. *New Phytologist* 111, 89–97.

Cheplick, G.P., Perera, A. and Koulouris, K. (2000) Effect of drought on the growth of *Lolium perenne* genotypes with and without fungal endophytes. *Functional Ecology* 14, 657–667.

Christensen, M.J. (1995) Variation in the ability of *Acremonium* endophytes *of Lolium perenne, Festuca arundinacea* and *F. pratensis* to form compatible associations in the three grasses. *Mycological Research* 99, 466–470.

Christensen, M.J. (1996) Antifungal activity in grasses infected with *Acremonium* and *Epichloë* endophytes. *Australasian Plant Pathology* 25, 186–191.

Christensen, M.J., Bennett, R.J., Ansari, H.A., Koga, H., Johnson, R.D., *et al.* (2008) *Epichloë* endophytes grow by intercalary hyphal extension in elongating grass leaves. *Fungal Genetics and Biology* 45, 84–93.

Christensen, M.J., Bennet, R.J. and Schmid, J. (2002) Growth of *Epichloë*/ *Neotyphodium* and p-endophytes in leaves of grass hosts. *Mycological Research* 106, 93–106. doi:10.1017/S095375620100510X

Christensen. M.J. and Voisey, C.R. (2007) The biology of the endophyte/grass partnership. In: Popay A.J. and Thom E.R. (eds) *6th International Symposium on Fungal Endophytes of Grasses. Grassland Research and Practice Series 13.* New Zealand Grassland Association, Dunedin, New Zealand, pp. 123–133.

Chu-Chou, M., Guo, B., An, Z.-Q., Hendrix, J.W., Ferriss, R.S., *et al.* (1992) Suppression of mycorrhizal fungi in fescue by the *Acremonium coenophialium* endophyte. *Soil Biology and Biochemistry* 24, 633–637.

Clarke, B.B., White, J.F., Hurley, H., Torres, M.S., Sun, S., *et al.* (2006) Endophyte-mediated suppression of dollar spot disease in fine fescues. *Plant Disease* 90, 994–998.

Clay, K. (1989) Clavicipitaceous endophytes of grasses: Their potential as biocontrol agents. *Mycological Research* 92, 1–12.

Clay, K. (1990) Fungal endophytes of grasses. *Annual Review of Ecology and Systematics* 21, 275–295.

Clay, K. (1993) The ecology and evolution of endophytes. *Agriculture, Ecosystems & Environment* 44, 39–64.

Clay, K. and Leuchtmann, A. (1989) Infection of woodland grasses by fungal endophytes. *Mycologia* 81, 805–811.

Clay, K. and Schardl, C. (2002) Evolutionary origins and ecological consequences of endophyte symbiosis with grasses. *The American Naturalist* 160(Suppl), S99–S127.

Crush, J.R., Popay, A.J. and Waller, J. (2004) Effect of different *Neotyphodium* endophytes on root distribution of a perennial ryegrass (*Lolium perenne* L.) cultivar. *New Zealand Journal of Agricultural Research* 47, 345–349.

Dinkins, R.D., Nagabhyru, P., Graham, M.A., Boykin, D. and Schardl, C.L. (2017) Transcriptome response of *Lolium arun*dinaceum to its fungal endophyte *Epichloë coenophiala*. *New Phytologist* 213, 324–337.

Duan, X., Newman D.S., Reibee, J.M., Green, C.D., Saxton, A.M., *et al.* (1996) Mycorrhizal influence on hydraulic and hormonal factors implicated in the control of stomatal conductance during drought. *Journal of Experimental Botany* 47, 1541–1550.

Duckett, S.K., Lacy, R.C., Andrae, J.G., Hoveland, C.S., Bouton, J.H., *et al.* (2007) Animal performance, carcass quality and economics of cattle finished after grazing endophyte-infected, endophyte-free or non-ergot alkaloid-producing endophyte-infected tall fescue. In: Popay, A.J. and Thom, E.R. (eds) *6th International Symposium on Fungal Endophytes of Grasses. Grassland Research and Practice Series 13.* New Zealand Grassland Association, Dunedin, New Zealand, pp. 253–255.

Dupont, P.Y., Eaton, C.J., Wargent, J.J., Fechtner, S., Solomon, P., *et al.* (2015) Fungal endophyte infection of ryegrass reprograms host metabolism and alters development. *New Phytologist* 208, 1227–1240.

Easton, H., Latch, G., Tapper, B. and Ball, J. (2002) Ryegrass host genetic control of concentrations of endophyte-derived alkaloids. *Crop Science* 42, 51–57

Easton, H.S. (2007) Grasses and *Neotyphodium* endophytes: Co-adaptation and adaptive breeding. *Euphytica* 54, 295–306.

Easton, H.S. and Fletcher, L.R. (2007) The importance of endophyte in agricultural systems – Changing plant and animal productivity. In: Popay, A.J. and Thom, E.R. (eds) *6th International Symposium on Fungal Endophytes of Grasses. Grassland Research and Practice Series 13.* New Zealand Grassland Association, Dunedin, New Zealand, pp. 11–18.

Easton, H.S., Lane, G.A., Tapper, B.A., Keogh, R.G., Cooper, B.M., *et al.* (1996) Ryegrass endophyte-related heat stress in cattle. *Proceedings of the New Zealand Grassland Association* 57, 37–41.

Elbersen, H.W. and West, C.P. (1996) Growth and water relations of field-grown tall fescue as influenced by drought and endophyte. *Grass and Forage Science* 51, 333–342.

Faeth, S.H. (2002) Are endophytic fungi defensive plant mutualists? *Oikos* 98, 25–36.

Faeth, S.H. and Saikkonen, K. (2007) Variability is the nature of the endophyte-grass interaction. In: Popay, A.J. and Thom, E.R. (eds) *6th International Symposium on Fungal Endophytes of Grasses. Grassland Research and Practice Series 13.* New Zealand Grassland Association, Dunedin, New Zealand, pp. 37–48.

Faeth, S.H. and Sullivan, T.J. (2003) Mutualistic asexual endophytes in a native grass are usually parasitic. *The American Naturalist* 161, 310–325.

Faeth, S.H., Sullivan, T.J. and Hamilton, C.E. (2001) What maintains high levels of *Neotyphodium* endophytes in native grasses? A dissenting view and alternative hypotheses. In: Paul, V.H. and Dapprich P.D. (eds) *Proceedings of the Fourth International Neotyphodium/Grass Interactions Symposium.* Soest, Germany, pp. 65–69.

Fletcher, L.R. (1999) "Non-toxic" endophytes in ryegrass and their effect on livestock health and production. In: Woodfield, D.R. and Matthew, C. (eds) *Ryegrass Endophyte: An Essential New Zealand Symbiosis. Grassland Research and Practice Series 7.* New Zealand Grasslanz Association, Palmerston North, New Zealand, pp. 133–139.

Fletcher, L.R. and Harvey, C. (1981) An association of a *Lolium* endophyte with ryegrass staggers. *New Zealand Veterinary Journal* 29, 85–86.

Fletcher, L.R., Sutherland, B.L. and Fletcher, C.G. (1999) The impact of endophyte on the health and productivity of sheep grazing ryegrass-based pasture. In: Woodfield, D.R. and Matthew, C. (eds) *Ryegrass Endophyte: An Essential New Zealand Symbiosis. Grassland Research and Practice Series 7.* New Zealand Grasslanz Association, Palmerston North, New Zealand, pp. 11–17.

Forester, N.T., Lane, G.A., Steringa, M., Lamont, I.L. and Johnson, L.J. (2017) Contrasting roles of fungal siderophores in maintaining iron homeostasis in *Epichloë festucae*. *Fungal Genetics and Biology* 111, 60–72.

Frederickson, M.E. (2017) Mutualisms are not on the verge of breakdown. *Trends in Ecology and Evolution* 32, 727–733.

Gagic, M., Faville, M.J., Zhang, W., Forester, N.T, Rolston, P.M., *et al.* (2018) Seed transmission of *Epichloë* endophytes in *Lolium perenne* is heavily influenced by host genetics. *Frontiers in Plant Science* 9, 1580.

Gallagher, R.T. and Hawkes, A.D. (1985) Estimation of neurotoxin levels in perennial ryegrass by mouse bioassay. *New Zealand Journal of Agricultural Research* 28, 427–431.

Gallagher, R.T., Hawkes, A.D. and Stewart, J.M. (1985) Rapid determination of the neurotoxin lolitrem B in perennial ryegrass by high-performance liquid chromatography with fluorescence detection. *Journal of Chromatography* 321, 217–226.

Goettel, M.S., Koike, M., Kim, J.J., Aiuchi, D., Shinya, R., *et al.* (2008) Potential of *Lecanicillium* spp. for management of insects, nematodes and plant diseases. *Journal of Invertebrate Pathology* 98, 256–261.

Górzynska, K., Ryszka, P., Anielska, T., Turnau, K. and Lembicz, M. (2017) Effect of *Epichloë typhina* fungal endophyte on the diversity and incidence of other fungi in *Puccinellia distans* wild grass seeds. *Flora* 228, 60–64.

Granath, G., Vicari, M., Bazely, D.R., Ball, J.P., Puentes, A., *et al.* (2007) Variation in the abundance of fungal endophytes in fescue grasses along altitudinal and grazing gradients. *Ecography* 30, 422–430.

Gundel, P.E., Batista, W.B., Texeira, M., Martínez-Ghersa, M.A., Omacini, M., *et al.* (2008) *Neotyphodium* endophyte infection frequency in annual grass populations: Relative importance of mutualism and transmission efficiency. *Proceedings of the Royal Society B* 275, 897–905.

Guo, B.Z., Hendrix, J.W., An, Z.-Q. and Ferriss, R.S. (1992) Role of *Acremonium* endophyte of fescue on inhibition of colonization and reproduction of mycorrhizal fungi. *Mycologia* 84, 882–885.

Gwinn, K.D. and Gavin, A.M. (1992) Relationship between endophytic infection level of tall fescue seed lots and *Rhizoctonia zeae* seedling disease. *Plant Disease* 76, 911–914.

Hardoim, P.R., van Overbeek, L.S., Berg, G., Pirttilä, A.M., Compant, S., *et al.* (2015) The hidden world within plants: Ecological and evolutionary considerations for defining functioning of microbial endophytes. *Microbiology and Molecular Biology Reviews* 79, 293–320.

Hartley, S.E. and Gange, A.C. (2009) Impacts of plant symbiotic fungi on insect herbivores: Mutualism in a multitrophic context. *Annual Review of Entomology* 54, 323–342.

Haselwandter, K. and Read, D.J. (1982) The significance of root-fungus associations in two *Carex* species of high-alpine plant communities. *Oecologia* 53, 352–354.

Hayes, C.J. and Faeth, S.H. (2002) Shift of a *Neotyphodium* mutualism by a parasitic co-occurring endophyte. Annual Meeting of the Ecological Society of America, 8 August 2002, Tucson, Arizona.

Herre, E.A., Knowlton, N., Mueller, U.G. and Rehner, S.A. (1999) The evolution of mutualism: Exploring the paths between conflict and cooperation. *Trends in Ecology and Evolution* 14, 49–53.

Hildebrandt, U., Regvar, M. and Bothe, H. (2007) Arbuscular mycorrhiza and heavy metal tolerance. *Phytochemistry* 68, 139–146. doi:10.1016/j.phytochem.2006.09.023

Hill, N.S., Pachon, J.G. and, Bacon, C.W. (1996) *Acremonium coenophialum* mediated short and long-term drought acclimation in tall fescue. *Crop Science* 36, 665–672.

Hoveland, C.S. (1993) Importance and economic significance of the *Acremonium* endophytes to performance of animals and grass plant. *Agriculture, Ecosystems & Environment* 44, 3–12.

Hoysted, G.A., Kowal, J., Jacob, A., Rimington, W.R., Duckett, J.G., *et al.* (2018) A mycorrhizal revolution. *Current Opinion in Plant Biology* 44, 1–6.

Hume, D.E., Cooper, B.M. and Panckhurst, K.A. (2009) The role of endophyte in determining the persistence and productivity of ryegrass, tall fescue and meadow fescue in Northland. *Proceedings of the New Zealand Grassland Association* 71, 145–150.

Jaber L.R and Ownley, B.H. (2017) Can we use entomopathogenic fungi as endophytes for dual biological control of insect pests and plant pathogens? *Biological Control* 116, 36–45. doi:10.1016/j.biocontrol.2017.01.018

Jensen, A.M.D. and Roulund, N. (2004) Occurrence of *Neotyphodium* endophytes in permanent grassland with perennial ryegrass (*Lolium perenne*) in Denmark. *Agriculture, Ecosystems & Environment* 104, 419–427.

Johnson, L.J., Johnson, R.D., Schardl, C.L. and Panaccione, D.G. (2003) Identification of differentially expressed genes in the mutualistic association of tall fescue with *Neotyphodium coenophialum*. *Physiological and Molecular Plant Pathology* 63, 305–317.

Johnson, L.J., de Bonth, A.C.M., Briggs, L.R., Caradus, J.R., Finch, S.C., *et al.* (2013a) The exploitation of epichloae endophytes for agricultural benefit. *Fungal Diversity* 60, 171–188.

Johnson, L.J., Koulman, A., Christensen, M., Lane, G.A., Fraser, K., *et al.* (2013b) An extracellular siderophore is required to maintain the mutualistic interaction of *Epichloë festucae* with *Lolium Perenne*. *PLOS Pathogens* 9, e1003332. doi:10.1371/journal.ppat.1003332

Johnson, R., Voisey, C., Johnson, L., Pratt, J., Fleetwood, D., *et al.* (2007) Distribution of NRPS gene families within the *Neotyphodium/Epichloë* complex. *Fungal Genetics and Biology* 44, 1180–1190.

Johnson, R.D., Lane, G.A., Koulman, A., Cao, M., Fraser, K., *et al.* (2015) A novel family of cyclic oligopeptides derived from ribosomal peptide synthesis of an *in planta*-induced gene, *gigA*, in *Epichloë* endophytes of grasses. *Fungal Genetics and Biology* 85, 14–24.

Jumpponen, A. and Trappe, J.M. (1998) Dark septate endophytes: A review of facultative biotrophic root-colonizing fungi. *New Phytologist* 140, 295–310.

Jumpponen, A., Mattson, K.G. and Trappe, J.M. (1998) Mycorrhizal functioning of *Phialocephala fortinii*: Interactions with soil nitrogen and organic matter. *Mycorrhiza* 7, 261–265.

Kalosa-Kenyon, E., Slaughter, L.C., Rudgers, J.A. and McCulley, R.L. (2018) Asexual *Epichloë* endophytes do not consistently alter arbuscular mycorrhizal fungi colonization in three grasses. *The American Midland Naturalist* 179, 157–165. doi:10.1674/0003-0031-179.2.157

Kernaghan, G. and Patriquin, G. (2011) Host associations between fungal root endophytes and boreal trees. *Microbial Ecology* 62, 460–473.

Khan, A., Bassett, S., Voisey, C., Gaborit, C., Johnson, L., *et al.* (2010) Gene expression profiling of the endophytic fungus *Neotyphodium lolii* in association with its host plant perennial ryegrass. *Australasian Plant Pathology* 39, 467–476.

Koga, H., Christensen, M.J. and Bennett, R.J. (1993) Incompatibility of some grass/*Acremonium* endophyte associations. *Mycological Research* 97, 1237–1244.

Koulman, A., Lee, T.V., Fraser, K., Johnson, L., Arcus, V., *et al.* (2012) Identification of extracellular siderophores and a related peptide from the endophytic fungus *Epichloë festucae* in culture and endophyte-infected *Lolium perenne*. *Phytochemistry* 75, 128–139.

Krings, M., Taylor, T.N., Hass, H., Kerp, H., Dotzler, N., *et al.* (2007) Fungal endophytes in a 400-million-yr-old land plant: Infection pathways, spatial distribution, and host responses. *New Phytologist* 174, 648–657.

Lane, G.A., Christensen, M.J. and Miles, C.O. (2000) Coevolution of fungal endophytes with grasses: The significance of secondary metabolites. In: Bacon, C.W. and White, J.F. (eds) *Microbial Endophytes*. Marcel Dekker, New York, pp. 341–388.

Larimer, A.L., Bever, J.D. and Lay, K. (2012) Consequences of simultaneous interactions of fungal endophytes and arbuscular mycorrhizal fungi with a shared host grass. *Oikos* 121, 2090–2096.

Latch, G.C.M. (1989) Plant improvement using endophytic fungi. *Proceedings XVI International Grasslands Congress*. The French Grassland Society, Nice, France, pp. 345–346.

Latch, G.C.M. (1993) Physiological interactions of endophytic fungi and their hosts. Biotic stress tolerance imparted to grasses by endophytes. *Agriculture, Ecosystems & Environment* 44, 143–156.

Latch, G.C.M. (1994) Influence of *Acremonium* endophytes on perennial ryegrass improvement. *New Zealand Journal of Agricultural Research* 37, 311–318.

Latch, G.C.M. (1997) An overview of neotyphodium/grass interactions. In: Bacon, C.W. and Hill, N.S. (eds) *Neotyphodium/Grass Interactions*. Plenum Press, New York, pp. 1–11.

Latch, G.C.M. (1998) Grass endophytes as a model. *Sydowia Horn* 50, 213–228.

Latch, G.C.M. and Christensen, M.J. (1985) Artificial infections of grasses with endophytes. *Annals of Applied Biology* 107, 17–24.

Latch, G.C.M., Christensen, M.J. and Samuels, G-.J. (1984) Five endophytes of *Lolium* and *Festuca* in New Zealand. *Mycotaxon* 20, 535–550.

Latch, G.C.M., Hunt, W.F and Musgrave, D.R. (1985) Endophytic fungi affect growth of perennial ryegrass. *New Zealand Journal of Agricultural Research* 28, 165–168.

Latch, G.C.M., Christensen, M.J., Tapper, B.A., Easton, H.S., Hume, D.E., *et al.* (2000) Tall fescue

endophytes. *US Patent* 6,111,170. Date issued: 29 August 2000.
Latch, G.C.M., Potter, L.R. and Tyler, B.R. (1987) Incidence of endophytes in seeds from collections of *Lolium* and *Festuca* species. *Annals of Applied Biology* 111, 59–64.
Leuchtmann, A. (1994) Isozyme relationships of *Acremonium* endophytes from twelve *Festuca* species. *Mycological Research* 98, 25–33.
Leuchtmann, A., Bacon, C.W., Schardl, C.L., White, J.F. and Tadych, M. (2014) Nomenclatural realignment of *Neotyphodium* species with genus *Epichloë*. *Mycologia* 106, 202–215.
Leury, B., Henry, M., Kemp. S. and Webb-Ware, J. (2014) Developing Increased Understanding, Awareness and Potential Mitigation Strategies for Perennial Ryegrass Toxicosis in Sheep Production Systems. Project B.AHE.0039. Meat and Livestock Australia Limited, North Sydney.
Lewis, G.C. (1992) Effect of ryegrass endophyte in mixed swards of perennial ryegrass and white clover under two levels of irrigation and pesticide treatment. *Grass and Forage Science* 47, 302–305.
Lewis, G.C. (2004) Effects of biotic and abiotic stress on the growth of three genotypes of *Lolium perenne* with and without infection by the fungal endophyte *Neotyphodium lolii*. *Annals of Applied Biology* 144, 53–63.
Lewis, G.C. and Clements, R.O. (1986) A survey of ryegrass endophyte (*Acremonium loliae*) in the UK and its apparent ineffectuality on a seedling pest. *Journal Agricultural Science* 107, 633–638.
Lewis, G.C., Ravel, C., Naafa, W., Astier, C. and Charmet, G. (1997) Occurrence of *Acremonium*-endophytes of wild populations of *Lolium* spp. in European countries and a relationship between level of infection and climate in France. *Annals of Applied Biology* 130, 227–238.
Li, C., Zhang, X., Li, F., Nan, Z. and Schardl, C.L. (2007) Disease and pest resistance of endophyte infected and non-infected drunken horse grass. In: Popay, A. and Thom, E.R. (eds) *Proceedings of the 6th International Symposium on Fungal Endophytes of Grasses*, New Zealand Grassland Association, Dunedin, New Zealand, pp. 111–114.
Li, F., Guo, Y.E., Christensen, M.J., Gao P, Li, Y.Z. and Duan, T.Y. (2018) An arbuscular mycorrhizal fungus and *Epichloë festucae* var. *lolii* reduce *Bipolaris sorokiniana* disease incidence and improve perennial ryegrass growth. *Mycorrhiza* 28, 59–169.
Li, L., Yang, A. and Zhao, Z. (2005) Seasonality of arbuscular mycorrhizal symbiosis and dark septate endophytes in a grassland site in southwest China. *FEMS Microbiology Ecology* 5, 367–373.
Liebe, D.M. and White, R.R. (2018) Meta-analysis of endophyte-infected tall fescue effects on cattle growth rates. *Journal of Animal Science* 96, 1350–1361.
Liu, J., Nagabhyru, P. and Schardl, C.L. (2017) *Epichloë festucae* endophytic growth in florets, seeds, and seedlings of perennial ryegrass (*Lolium perenne*). *Mycologia* 109, 691–700.
Liu, Q., Parsons, A.J., Xue, H., Fraser, K., Ryan, G.D., *et al.* (2011) Competition between foliar *Neotyphodium lolii* endophytes and mycorrhizal *Glomus* spp. fungi in *Lolium perenne* depends on resource supply and host carbohydrate content. *Functional Ecology* 25, 910–920.
Ma, M.Z. and Nan, Z.B. (2011) Effect of fungal endophytes against rust disease of perennial ryegrass *(Lolium perenne)* on growth and physiological indices. *Acta Prataculturae Sinica* 20, 150–156.
Ma, M.Z., Christensen, M.J. and Nan, Z.B. (2015) Effects of the endophyte *Epichloë festucae* var. *lolii* of perennial ryegrass (*Lolium perenne*) on indicators of oxidative stress from pathogenic fungi during seed germination and seedling growth. *European Journal of Plant Pathology* 141, 571–583.
Mack, K.M.L. and Rudgers, J.A. (2008) Balancing multiple mutualists: Asymmetric interactions among plants, arbuscular mycorrhizal fungi, and fungal endophytes. *Oikos* 117, 310–320.
Malinowski, D.P. and Belesky, D.P. (1999) Infection with *Neotyphodium coenophialum*-endophyte may affect the ability of tall fescue to use sparingly available phosphorus. *Journal of Plant Nutrition* 22, 835–853.
Malinowski, D.P. and Belesky, D.P. (2000) Adaptations of endophyte-infected cool-season grasses to environmental stresses: Mechanisms of drought and mineral stress tolerance. *Crop Science* 40, 923–940.
Malinowski, D.P., Brauer, D.K. and Belesky, D.P. (1999) The endophyte *Neotyphodium coenophialum* affects root morphology of tall fescue grown under phosphorus deficiency. *Journal of Agronomy and Crop Science* 183, 53–60.
Malinowski, D.P., Belesky, D.P. and Lewis, G.C. (2005) Abiotic stresses in endophytic grasses. In: Roberts, C.A., West, C.P. and Spiers, D.E. (eds) *Neotyphodium in Cool-Season Grasses*. Blackwell Publishing, Ames, Iowa, pp. 187–199.
Mandyam, K. and Jumpponen, A. (2005) Seeking the elusive function of the root-colonising dark septate endophyte. *Studies in Mycology* 53, 173–189.
Mandyam, K., Loughin, T. and Jumpponen, A. (2010) Isolation and morphological and metabolic characterization of common endophytes in annually burned tallgrass prairie. *Mycologia* 102, 813–821.

Márquez, L.M., Redman, R.S., Rodriguez, R.J. and Roossinck, M.J. (2007) A virus in a fungus in a plant – Three way symbiosis required for thermal tolerance. *Science* 315, 513–515.

Matthews, J.W. and Clay, K. (2001) Influence of fungal endophyte infection on plant-soil feedback and community interactions. *Ecology* 82, 500–509.

Milne, G.D. (2007) Technology transfer of novel ryegrass endophytes in New Zealand. In: Popay, A. and Thom, E.R. (eds) *Proceedings of the 6th International Symposium on Fungal Endophytes of Grasses*. New Zealand Grassland Association, Dunedin, New Zealand, pp. 237–239.

Moon, C.D., Scott, B., Schardl, C.L. and Christensen, M.J. (2000) The evolutionary origins of *Epichloë* endophytes from annual ryegrasses. *Mycologia* 92, 1103–1118.

Moon, C.D., Miles, C.O., Järlfors, U. and Schardl, C.L. (2002) The evolutionary origins of the three new *Neotyphodium* endophyte species from grasses indigenous to the Southern Hemisphere. *Mycologia* 94, 694–711.

Moon, C.D., Craven, K.D., Leuchtmann, A., Clement, S.L. and Schardl, C.L. (2004) Prevalence of interspecific hybrids amongst asexual fungal endophytes of grasses. *Molecular Ecology* 13, 1455–1467.

Morse, L.J., Day, T.A. and Faeth, S.H. (2002) Effect of *Neotyphodium* endophyte infection on growth and leaf gas exchange of Arizona fescue under contrasting water availability regimes. *Environmental and Experimental Botany* 48, 257–268.

Muller, C.B. and Krauss, J. (2005) Symbiosis between grasses and asexual fungal endophytes. *Current Opinions in Plant Biology* 8, 450–456.

Muller, J. (2003) Artificial infection by endophytes affects growth and mycorrhizal colonisation of *Lolium perenne*. *Functional Plant Biology* 30, 419–424.

Nagabhyru, P., Dinkins, R.D., Wood, C.L., Bacon, C.W. and Schardl, C.L. (2013) Tall fescue endophyte effects on tolerance to water-deficit stress. *BMC Plant Biology* 13, 127.

Nagabhyru, P., Dinkins, R.D. and Schardl, C.L. (2019) Transcriptomics of *Epichloë*-grass symbioses in host vegetative and reproductive stages. *Molecular Plant Microbe Interactions* 32(2), 194–207.

Nan, Z.B. and Li, C.J. (2000) *Neotyphodium* in native grasses in China and observations on endophyte/host interactions. *Proceedings of 4th International Neotyphodium/Grass Interactions Symposium*. University Paderborn, Soest, Germany, pp. 41–50.

Newsham, K.K. (1999) *Phialophora graminicola*, a dark septate fungus, is a beneficial associate of the grass *Vulpia ciliata* ssp. *ambigua*. *New Phytologist* 144, 517–524.

Newsham, K.K., Fitter, A.H. and Watkinson, A.R. (1995) Multi-functionality and biodiversity in arbuscular mycorrhizas. *Trends in Ecology and Evolution* 1, 407–411.

Newton, A.C., Fitt, B.D.L., Atkins, S.D., Walters, D.R. and Daniell, T.J. (2010) Pathogenesis, parasitism and mutualism in the trophic space of microbe–plant interactions. *Trends in Microbiology* 18, 365–373.

Novas, M.V., Cabral, D. and Godeas, A. (2005) Interaction between grass endophytes and mycorrhizas in *Bromus setifolius* from Patagonia, Argentina. *Symbiosis* 40, 23–30.

Novas, M.V., Collantes, M. and Cabral, D. (2007) Environmental effects on grass-endophyte associations in the harsh conditions of south Patagonia. *FEMS Microbiology Ecology* 61, 164–173.

Novas, M.V., Iannone, L.J., Godeas, A.M. and Cabral, D. (2009) Positive association between mycorrhiza and foliar endophytes in *Poa bonariensis*, a native grass. *Mycological Progress* 8, 75–81.

Novas, M.V., Iannone, L.J., Godeas, A.M. and Scervino, J.M. (2011) Evidence for leaf endophyte regulation of root symbionts: Effect of *Neotyphodium* endophytes on the pre-infective state of mycorrhizal fungi. *Symbiosis* 55, 19–28.

Oldenburg, E. (1997) Endophytic fungi and alkaloid production in perennial ryegrass in Germany. *Grass and Forage Science* 52, 425–431.

Oliveira, J.A., Rottinghaus, G.E., Collar, J. and Castro, P. (1997) Perennial ryegrass endophytes in Galicia, northwest Spain. *Journal of Agricultural Science* 129, 173–177.

Omacini, M., Eggers, T., Bonkowski, M., Gange, A.C. and Jones, T.H. (2006) Leaf endophytes affect mycorrhizal status and growth of co-infected and neighbouring plants. *Functional Ecology* 20, 226–232.

Ownley, B.H., Griffin, M.R., Klingeman, W.E., Gwinn, K.D., Moulton, J.K., *et al.* (2008) *Beauveria bassiana*: Endophytic colonization and plant disease control. *Journal of Invertebrate Pathology* 3, 267–270.

Ownley, B.H., Gwinn, K.D. and Vega, F.E. (2010) Endophytic fungal entomopathogens with activity against plant pathogens: Ecology and evolution. *BioControl* 55, 113–128.

Pan, J.J. and Clay, K. (2004) *Epichloë* glyceriae infection affects carbon translocation in the clonal grass *Glyceria striata*. *New Phytologist* 164, 467–475.

Panaccione, D.G., Beaulieu, W.T. and Cook, D. (2014) Bioactive alkaloids in vertically transmitted fungal endophytes. *Functional Ecology* 28, 299–314.

Panka, D., Piesik, D., Jeske, M. and Baturo-Cieśniewska, A. (2013) Production of phenolics and the emission of volatile organic compounds by perennial ryegrass (*Lolium perenne* L.)/*Neotyphodium lolii* association as a response to infection by *Fusarium poae*. *Journal of Plant Physiology* 170, 1010–1019.

Parsaeian, M., Mirlouhi, A.F., Almajid, R.A. and Baradaran, M. (2006) The morphological changes affected by endophytes on induction of cold tolerance in two species of *Festuca*. *Iranian Journal of Crop Science* 8, 139–152.

Paterson, J., Forcherio, C., Larson, B., Samford, M. and Kerley, M. (1995) The effects of fescue toxicosis on beef cattle productivity. *Journal of Animal Science* 73, 889–898.

Pfeffer, P.E., Douds, D.D. Jr, Becard, G. and Shachar-Hill, Y. (1999) Carbon uptake and the metabolism and transport of lipids in an arbuscular mycorrhiza. *Plant Physiology* 120, 587–598.

Philipson, M.N. (1989) A symptomless endophyte of ryegrass (*Lolium perenne*) that spores on its host-a light microscope study. *New Zealand Journal of Botany* 27, 513–519.

Philipson, M.N. (1991a) Ultrastructure of the *Gliocladium*-like endophyte of perennial ryegrass (*Lolium perenne* L.). I. Vegetative phase and leaf blade sporulation. *New Phytologist* 117, 271–280.

Philipson, M.N. (1991b) Ultrastructure of the *Gliocladium*-like endophyte of perennial ryegrass (*Lolium perenne* L.). II. Sporulation in the leaf sheath. *New Phytologist* 117, 281–288.

Philipson, M.N. and Christey, M.C. (1986) The relationship of host and endophyte during flowering, seed formation, and germination of *Lolium perenne*. *New Zealand Journal of Botany* 24, 125–134.

Popay, A.J. and Bonos, S.A. (2005) Biotic responses in endophytic grasses. In: Roberts, C.A., West, C.P. and Spiers, D.E. (eds) *Neotyphodium in Cool-Season Grasses*. Blackwell Publishing, Ames, Iowa, pp. 163–185.

Popay, A.J. and Hume, D.H. (2011) Endophytes improve ryegrass persistence by controlling insects. New Zealand Grassland Association Pasture Persistence Symposium. *Grassland Research and Practice Series* 15, 149–156.

Porter, J.K. and Thompson, F.N. Jr. (1992) Effects of fescue toxicosis on reproduction in livestock. *Journal of Animal Science* 70, 1594–1603.

Pottinger, R.P., Barker, G.M. and Prestidge, R.A. (1985) A review of the relationships between endophytic fungi of grasses (*Acremonium* spp.) and Argentine stem weevil (*Listronotus bonariensis* Kuschel). In: *Proceedings 4th Australasian Conference of Grassland Invertebrate Ecology*. Lincoln College, Canterbury, pp. 322–331.

Prestidge, R.A, Pottinger, R.P. and Barker, G.M. (1982) An association of *Lolium* endophyte with ryegrass resistance to Argentine stem weevil. In: *Proceedings of the 35th New Zealand Weed and Pest Control Conference*, vol. 35, pp. 199–222.

Rasmussen, S., Parsons, A.J., Bassett, S., Christensen, M.J., Hume, D.E., *et al.* (2007) High nitrogen supply and carbohydrate content reduce fungal endophyte and alkaloid concentration in *Lolium perenne*. *New Phytologist* 173, 787–797.

Ravel, C., Charmet, G. and Balfourier, F. (1995) Influence of the fungal endophyte *Acremonium lolii* on agronomic traits of perennial ryegrass in France. *Grass and Forage Science* 50, 75–80.

Ravel, C., Michalakis, Y. and Charmet, G. (1997) The effect of imperfect transmission on the frequency of mutualistic seed-borne endophytes in natural populations of grasses. *Oikos* 80, 18–24.

Redman, R.S., Sheehan, K.B., Stout, R.G., Rodriguez, R.J. and Henson, J.M. (2002) Thermotolerance conferred to plant host and fungal endophyte during mutualistic symbiosis. *Science* 298, 1581.

Remy, W., Taylor, T.N., Hass, H. and Kerp, H. (1994) Four hundred-million-year-old vesicular arbuscular mycorrhizae. *Proceedings of the National Academy of Sciences of the United States of America* 91, 11841–11843.

Richardson, M.D., Hoveland, C.S. and Bacon, C.W. (1993) Photosynthesis and stomatal conductance of symbiotic and non-symbiotic tall fescue. *Crop Science* 33, 145–149.

Rodriquez, R.J. and Redman, R. (2008) More than 400 million years of evolution and some plants still can't make it on their own: Plant stress tolerance via fungal symbiosis. *Journal of Experimental Botany* 59, 1109–1114.

Rodriguez, R.J., Redman, R.S. and Henson, J.M. (2004) The role of fungal symbioses in the adaptation of plants to high stress environments. *Mitigation and Adaptation Strategies for Global Change* 9, 261–272.

Rodriguez, R.J., Henson, J., Van Volkenburgh, E., Hoy, M., Wright, L., *et al.* (2008) Stress tolerance in plants via habitat-adapted symbiosis. *International Society of Microbial Ecology* 2, 404–416.

Rodriguez, R.J., White, J.F., Arnold, A.E. and Redman, R.S. (2009) Fungal endophytes: Diversity and functional roles. *New Phytologist* 182, 314–330.

Rolston, M.P., Hare, M.D., Moore, K.K. and Christensen, M.J. (1986) Viability of *Lolium* endophyte fungus in seed stored at different seed moisture contents and temperature. *New Zealand Journal of Experimental Agriculture* 14, 297–300.

Roth, R. and Paszkowski, U. (2017) Plant carbon nourishment of arbuscular mycorrhizal fungi. *Current Opinion in Plant Biology* 39, 50–56.

Rowan D. (1993) Lolitrems, peramine and paxilline: Mycotoxins of the ryegrass/endophyte interaction. *Agricultural Ecosystems and Environment* 44, 103–122.

Rudgers, J.A. and Clay, K. (2007) Community and ecosystem consequences of endophyte symbiosis with tall fescue. *Grassland Research and Practice Series* 13, 19–35.

Ryan, G.D., Rasmussen, S., Parsons, A.J. and Newman, J.A. (2015) The effects of carbohydrate supply and host genetic background on *Epichloë* endophyte and alkaloid concentrations in perennial ryegrass. *Fungal Ecology* 18, 115–125.

Saikkonen, K., Faeth, S.H., Helander, M. and Sullivan, T.J. (1998) Fungal endophytes: A continuum of interactions with host plants. *Annual Review of Ecology and Systematics* 29, 319–343.

Saikkonen, K., Helander, M., Faeth, S.H., Schulthess, F. and Wilson, D. (1999) *Neotyphodium* endophytes in native grass populations: Against herbivory-based defensive mutualism. *Oecologia (Berlin)* 121, 411–420.

Saikkonen, K., Ahlholm, J., Helander, M., Lehtimäki, S. and Niemelainen, O. (2000) Endophytic fungi in wild and cultivated grasses in Finland. *Ecography* 23, 360–366.

Saikkonen, K., Ion, D. and Gyllenberg, M. (2002) The persistence of vertically transmitted fungi in grass metapopulations. *Proceedings of Royal Society B* 269, 1397–1403.

Saikkonen, K., Lehtonen, P., Helander, M., Koricheva, J. and Faeth, S.H. (2006) Model systems in ecology: Dissecting the endophyte-grass literature. *Trends in Plant Science* 11, 428–433.

Saikkonen, K., Young, C.A., Helander, M. and Schardl, C.L. (2016) Endophytic *Epichloë* species and their grass hosts: From evolution to applications. *Plant Molecular Biology* 90, 665–675.

Sampson, K. (1933) The systemic infection of grasses by *Epichloe typhina* (Pers.) Tul. *Transactions of the British Mycological Society* 18, 30–47.

Sánchez Márquez, S., Bills, G.F. and Zabalgogeazcoa, I. (2007) The endophytic mycobiota of the grass *Dactylis glomerata*. *Fungal Diversity* 27, 171–195.

Sánchez Márquez, S., Bills G.F. and Zabalgogeazcoa, I. (2008) Diversity and structure of the fungal endophytic assemblages from two sympatric coastal grasses. *Fungal Diversity* 33, 87–100.

Sánchez Márquez, S., Bills, G.F., Domínguez Acuña, L. and Zabalgogeazcoa, I. (2010) Endophytic mycobiota of leaves and roots of the grass *Holcus lanatus*. *Fungal Diversity* 41, 115–123.

Sánchez Márquez, S., Bills, G.F., Herrero, N. and Zabalgogeazcoa, I. (2012) Non-systemic fungal endophytes of grasses. *Fungal Ecology* 5, 289–297.

Schardl, C.L. (2001) *Epichloë festucae* and related mutualistic symbionts of grasses. *Fungal Genetics and Biology* 33, 69–82.

Schardl, C.L. and Phillips, T.D. (1997) Protective grass endophytes: Where are they from and where are they going? *Plant Disease* 81, 430–438.

Schardl, C.L. and Wilkinson, H.H. (2000) Hybridization and co-speciation hypotheses for the evolution of grass endophytes. In: Bacon, C.W. and White, J.F. (eds) *Microbial Endophytes*. Marcel Dekker, New York, pp. 63–83.

Schardl, C.L., Leuchtmann, A., Tsai, H.F., Collett, M.A., Watt, D.M., *et al.* (1994) Origin of a fungal symbiont of perennial ryegrass by interspecific hybridization of a mutualist with the ryegrass choke pathogen, *Epichloë typhina*. *Genetics* 136, 1307–1317.

Schardl, C.L., Leuchtmann, A. and Spiering, M.J. (2004) Symbioses of grasses with seedborne fungal endophytes. *Annual Review of Plant Biology* 55, 315–340.

Schardl, C.L., Craven, K.D., Speakman, S., Stromberg, A., Lindstrom, A., *et al.* (2008) A novel test for host-symbiont codivergence indicates ancient origin of fungal endophytes in grasses. *Systematic Biology* 57, 483–498.

Schardl, C.L., Young, C.A., Hesse, U., Amyotte, S.G., Andreeva, K., *et al.* (2013) Plant symbiotic fungi as chemical engineers: Multi-genome analysis of the *Clavicipitaceae* reveals dynamics of alkaloid loci. *PLOS Genetics* 9(2), e1003323. doi:10.1371/journal.pgen.1003323

Schmid, J., Day, R., Zhang, N., Dupont, P.Y., Cox, M.P., *et al.* (2017) Host tissue environment directs activities of an *Epichloë* endophyte, while it induces systemic hormone and defense responses in its native perennial ryegrass host. *Molecular Plant-Microbe Interactions* 30, 138–149.

Schmidt, S.P., Hoveland, C.S., Clark, E.M., David, N.D., Smith, L.A., *et al.* (1982) Association of an endophytic fungus with fescue toxicity in steers fed Kentucky 31 tall fescue seed or hay. *Journal of Animal Science* 55, 1259–1263.

Schulthess, F.M. and Faeth, S.H. (1998) Distribution, abundances, and associations of the endophytic fungal community of Arizona fescue (*Festuca arizonica*). *Mycologia* 90, 569–578.

Schulz, B. (2006) Mutualistic interactions with fungal root endophytes. In: Schulz, B., Boyle, C. and Sieber, T.N. (eds) *Microbial Root Endophytes*. Springer, Berlin/Heidelberg, pp. 261–279.

Scott, B. (2001) Epichloë endophytes: Fungal symbionts of grasses. *Current Opinion in Microbiology* 4, 393–398.

Selosse, M.-A. and Schardl, C.L. (2007) Fungal endophytes of grasses: Hybrids rescued by vertical transmission? An evolutionary perspective. *New Phytologist* 173, 452–458.

Siegel, M.R., Johnson, M.C., Varney, D.R., Nesmith, W.C., Buckner, R.C., *et al.* (1984) A fungal endophyte in tall fescue: Incidence and dissemination. *Phytopathology* 74, 932–937.

Siegel, M.R., Latch, G.C.M. and Johnson M.C. (1987) Fungal endophytes of grasses. *Annual Review of Phytopathology* 25, 293–315.

Siegel, M.R. and Latch, G.C.M. (1991) Expression of antifungal activity in agar culture by isolates of grass endophytes. *Mycologia* 83, 529–537.

Siegel, M.R. and Schardl, C.L. (1991) Fungal endophytes of grasses: Detrimental and beneficial associations. In: Andrew, J.H. and Hirano, S.S. (eds) *Microbial Ecology of Leaves*. Springer, Berlin, pp. 198–221.

Sikes, B.A., Cottenie, K. and Klironomos, J.N. (2009) Plant and fungal identity determines pathogen protection of plant roots by arbuscular mycorrhizas. *Journal of Ecology* 97, 1274–1280.

Simpson, W.R., Faville, M.J., Moraga, R.A., Williams, W.M., McManus, M.T., *et al.* (2014) *Epichloë* fungal endophytes and the formation of synthetic symbioses in Hordeeae (=Triticeae) grasses. *Journal of Systematics and Evolution* (Special Issue: Systematics, evolution, and breeding of the wheat tribe – Proceedings of the 7th International Triticeae Symposium) 52, 794–806.

Skipp R.A. and Christensen, M.J. (1989) Fungi invading roots of perennial ryegrass (*Lolium perenne* L.) in pasture. *New Zealand Journal of Agricultural Research* 32, 423–431.

Slaughter, L.C., Nelson, J.A., Carlisle, E., Bourguignon, M., Dinkins, R.D., *et al.* (2018) Climate change and *Epichloë coenophiala* association modify belowground fungal symbioses of tall fescue host. *Fungal Ecology* 31, 37–46.

Spatafora J.W., Chang Y, Benny G.L., Lazarus K, Smith M.E., *et al.* (2016) Aphylum-level phylogenetic classification of zygomycete fungi based on genome-scale data. *Mycologia* 108, 1028–1046.

Spiering, M.J., Lane, G.A., Christensen, M.J. and Schmid, J. (2005) Distribution of the fungal endophyte *Neotyphodium lolii* is not a major determinant of the distribution of fungal alkaloids in *Lolium perenne* plants. *Phytochemistry* 66, 195–202.

Tanaka, A., Christensen, M.J., Takemoto, D., Park, P. and Scott, B. (2006) Reactive oxygen species play a role in regulating a fungus-perennial ryegrass mutualistic interaction. *The Plant Cell* 18, 1052–1066.

Termorshuizen, A.J. and Jeger, M.J. (2009) Strategies of soilborne plant pathogenic fungi in relation to disease suppression. *Fungal Ecology* 1, 108–114.

Thom, E.R., Popay, A.J., Waugh, C.D. and Minneé, E.M. (2014) Impact of novel endophytes in perennial ryegrass on herbage production and insect pests from pastures under dairy cow grazing in northern New Zealand. *Grass Forage Science* 69, 191–204.

Tian, P., Nan, Z., Li, C. and Spangenberg, G. (2008) Effect of the endophyte *Neotyphodium lolii* on susceptibility and host physiological response of perennial ryegrass to fungal pathogens. *European Journal of Plant Pathology* 122, 593–602.

Tsai, H-F., Liu, J-S., Staben, C., Christensen, M.J., Latch, G.C.M., *et al.* (1994) Evolutionary diversification of fungal endophytes of tall fescue grass by hybridization with *Epichloë* species. *Proceedings of the National Academy of Sciences of the United States of America* 91, 2542–2546.

Valle Ribeiro, M.A.M., Carnide, V. and Bush, L.P. (1996) *Interaccao endofito/graminea em ecotipos de Lolium perenne L. colhidos em Tras-os-Montes. Resumos da XVII Reuniao de primavera da Sociedades Portuguesa de Pastagens e Forragens*, p. 8. Vila Real, Portugal.

Vare, H., Vestberg, M. and Eurola, S. (1992) Mycorrhizae and root-associated fungi in Spitsbergen. *Mycorrhiza* 1, 93–104.

Varvas, T., Kasekamo, K. and Kulman, B. (2013) Preliminary study of endophytic fungi in timothy (*Phleum pratense*) in Estonia. *Acta Mycologica* 48, 41–49.

Vázquez-de-Aldana, B.R., García-Ciudad, A., García-Criado, B., Vicente-Tavera, S. and Zabalgogeazcoa, I. (2013) Fungal endophyte (*Epichloë festucae*) alters the nutrient content of *Festuca rubra* regardless of water availability. *PLOS ONE* 8(12), e84539. doi:10.1371/journal.pone.0084539

Vega, F.E., Posada, F., Aime, M.C., Pava-Ripoll, M., Infante, F., *et al.* (2008) Entomopathogenic fungal endophytes. *Biological Control* 46, 72–82.

Vicari, M., Hatcher, P.E. and Ayres, P.G. (2002) Combined effect of foliar and mycorrhizal endophytes on an insect herbivore. *Ecology* 83, 2452–2464. doi:10.1890/0012-9658(2002)083[2452:CEOFAM]2.0.CO;2

Vidal, S and Jaber L.R. (2015) Entomopathogenic fungi as endophytes: Plant-endophyte herbivore interactions and prospects for use in biological control. *Current Science* 109, 46–54.

Voisey, C.R. (2010) Intercalary growth in hyphae of filamentous fungi. *Fungal Biology Reviews* 24, 123–131.

Waller, F., Achatz, B., Baltruschat, H., Fodor, J., Becker, K., *et al.* (2005) The endophytic fungus *Piriformospora indica* reprograms barley to salt-stress tolerance, disease resistance, and higher yield. *Proceedings of the National Academy of Sciences of the United States of America* 102(38), 13386–13391. doi:10.1073/pnas.0504423102

Welty, R.E., Barker, R.E. and Azavedo, M.D. (1991) Reaction of tall fescue infected and non-infected by *Acremonium coenophialium* to *Puccina graminis* subsp *graminicola*. *Plant Diseases* 75, 883–886.

Werner, G.D.A., Cornelissen, J.H.C., Cornwell, W.K., Soudzilovskaia, N.A., Kattge, J., *et al.* (2018) Symbiont switching and alternative resource acquisition strategies drive mutualism breakdown. *Proceedings of the National Academy of Sciences of the United States of America* 115, 5229.

West, C.P., Izekor, E., Turner, K.E. and Elmi, A.A. (1993) Endophyte effects on growth and persistence of tall fescue along a water-supply gradient. *Agronomy Journal* 85, 264–270. doi:10.2134/agronj1993

White, J.F. Jr. (1987) Widespread distribution of endophytes in the Poaceae. *Plant Disease* 71, 340–342.

White, J.F. Jr and Cole, G.T. (1986) Endophyte-host associations in forage grasses. V. Occurrence of fungal endophytes in certain species of Bromus and Poa. *Mycologia* 78, 846–850.

Wiewiora, B., Zurek, G. and Zurek, M. (2015) Endophyte-mediated disease resistance in wild populations of perennial ryegrass (*Lolium perenne*). *Fungal Ecology* 15, 1–8.

Wilson, A., Clement, S., Kaiser, W. and Lester, D. (1991) First report of clavicipitaceous anamorphic endophytes in *Hordeum* species. *Plant Disease* 75, 215.

Wilson, A.D. (2007) Clavicipitaceous anamorphic endophytes in *Hordeum* germplasm. *Plant Pathology Journal* 6, 1–113.

Wilson, D. (1995) Endophyte: The evolution of a term and a clarification of its use and definition. *Oikos* 73, 274–276.

Young, C.A., Felitti, S., Shields, K., Spangenberg, G., Johnson, R.D., *et al.* (2006) A complex gene cluster for indole-diterpene biosynthesis in the grass endophyte *Neotyphodium lolii*. *Fungal Genetics and Biology* 43, 679–693.

Young, C.A., Hume, D.E. and McCulley, R.L. (2013) Fungal endophytes of tall fescue and perennial ryegrass: Pasture friend or foe? *Journal of Animal Science* 91, 2379–2394.

Zhang, W., Card, S.D., Mace, W.J., Christensen, M.J., McGill, C.R., *et al.* (2017a) Defining the pathways of symbiotic *Epichloë* colonization in grass embryos with confocal microscopy. *Mycologia* 109, 153–161.

Zhang, U., Zhou, Y., Zhang, X, Duan, T. and Nan, Z. (2017b) Effects of *Epichloë* endophyte on antioxidant enzymes activities, photosynthesis and growth of three ecotypes of *Elymus dahuricus*. *Frontiers of Agricultural Science and Engineering* 5, 148–158.

Zhang, X.X., Li, C.J. and Nan, Z.B. (2010) Effects of cadmium stress on growth and anti-oxidative systems in *Achnatherum inebrians* symbiotic with *Neotyphodium gansuense*. *Journal of Hazardous Materials* 175, 703–709.

8 Interactive Effects of Co-occurring Epichloid Endophytes, Rhizobia and Arbuscular Mycorrhizal Fungi Modulating Their Benefits to Grasses and Legumes

Pablo Adrián García-Parisi[1,2]* and Marina Omacini[1,3]

[1]IFEVA, Facultad de Agronomía, Universidad de Buenos Aires, CONICET, Buenos Aires; [2]Cátedra de Forrajicultura, Departamento de Producción Animal, Facultad de Agronomía, Universidad de Buenos Aires; [3]Cátedra de Ecología, Departamento de Recursos Naturales y Ambiente, Facultad de Agronomía, Universidad de Buenos Aires

Abstract

Symbiotic interactions are very extended in nature and their multiple co-occurrence among plants, fungi and bacteria is highly likely within a community. Thus, a single plant can harbour different strains, species or types of symbionts. Furthermore, at the neighbourhood level, the co-occurrence of grasses and legumes is a frequent event associated with their specific (endophyte, rhizobia) or generalist (arbuscular mycorrhizal fungi) symbionts. The simultaneous presence of two symbionts may induce additive or interactive effects (i.e. synergisms or antagonisms) both at plant and neighbourhood level. In this chapter, we explore the responses of plants when two symbionts are present within the same host or within neighbouring plants. Here we review studies researching the effects of epichloid endophytes and systemic and asymptomatic fungal symbionts of grasses on colonization of arbuscular mycorrhizal fungi (AMF) and their consequences on host plant performance. Also, we explore the current knowledge related to the presence of epichloid endophyte and AMF on the host grass affecting legume–rhizobia interaction, whether coexisting in the neighbourhood or growing after grass dies. Interestingly, endophyte effects go beyond the host and impact on other symbioses, either within the host or established in co-occurring plants in the neighbourhood. Endophytes either increase or impair AMF colonization within the host, and both symbionts can have interactive or additive effects on host performance, depending on the grass, endophyte and AMF species, and on the abiotic and biotic environment. Furthermore, endophyte presence on the host grass can affect different attributes at community level through altering the performance of a neighbouring legume, or one that grows after grass dies. As an outcome, the effects of these specific symbionts can result in potential public benefits for non-host plants through the propagation of interactive effects of several symbionts to the whole symbiosis influence area, the symbiosphere. The benefits give rise to the appearance of agroecosystem processes or services that could favour their sustainability.

8.1 Introduction

Since the term symbiosis first came into use in 1879, it has shifted its focus from naming a 'rarity or curiosity' in nature to referring to a widespread phenomenon with great biological relevance at different organizational levels (Douglas, 2010). Symbiosis can be

* Corresponding author e-mail: pgarcia@agro.uba.ar

defined as a persistent mutualism: an interaction in which the host and the microbial symbiont are in close and permanent contact and benefit each other (Douglas, 2010). The interaction should increase the reproductive success of both members, otherwise the relationship would not be evolutionarily stable (Denison, 2000; Kiers *et al.*, 2003; Kiers and van der Heijden, 2006; Gundel *et al.*, 2008; Kiers *et al.*, 2011; Gundel *et al.*, 2012). In part, the stability of mutualisms is provided by the multi-functionality of the interaction (Walder and van der Heijden, 2015), which means that a symbiont can provide multiple types of benefits to the host during its life cycle (Newsham *et al.*, 1995). Then, these multiple, although sometimes minor, benefits may ensure the mutualistic outcome in complex or varying contexts (Douglas, 2010). However, under certain biotic or abiotic environmental conditions, or during some stages of each partner's life cycle, the benefits disappear and the interaction may even be negative, at least for one of them, which is known as context dependency (Thrall *et al.*, 2007; Cheplick and Faeth, 2009; Gundel *et al.*, 2012).

Despite the existence of multi-functionality, the symbiotic microorganisms of plants can be classified as protectors or providers according to their main functional role within the host (Thrall *et al.*, 2007). Protective symbionts are those that reduce the attack of herbivores, parasites and pathogens (White and Torres, 2010). For example, epichloid fungal endophytes are foliar symbionts, specific for several cool-season grasses, which protect the host against various groups of herbivores (Clay and Schardl, 2002). The provider microorganisms, on the other hand, are those that enhance their host acquisition of scarce or inaccessible resources (Douglas, 2010). These symbionts are, in general, microorganisms that are horizontally transmitted belowground and whose propagules remain in the soil after the host dies. For example, nitrogen-fixing bacteria (rhizobia) establish symbiosis with legumes in root structures called nodules in which atmospheric nitrogen (N) is converted into mineral N, which can be harnessed by the plant (Sprent, 2007). Also classified as providers, arbuscular mycorrhizal fungi (AMF) are soil fungi that establish associations with the plant roots. This symbiosis develops internal and external hyphae that explore the soil and supply the plant with nutrients, mainly phosphorus (P) (Johnson *et al.*, 1997).

Symbiotic interactions are so widespread in nature that the co-occurrence of several symbioses among plants, fungi and bacteria is highly likely within a community (Stanton, 2003). A single plant can harbour different strains, species or types of symbionts. For example, a legume plant may be holding one or more species of mycorrhizal fungi and one or more rhizobia genotypes simultaneously (Denison, 2000; Larimer *et al.*, 2010). Similarly, grasses can be associated with their specific endophyte and also support one or several species of mycorrhizal fungi (Omacini *et al.*, 2006, Mack and Rudgers, 2008). At neighbourhood level, the co-occurrence of grasses and legumes, associated to their specific symbionts, is highly frequent. They can either share or impair their benefits. Furthermore, AMF are relatively more promiscuous and can be associated with different plants, forming a complex network of subterranean hyphae connecting several plants, known as common mycorrhizal networks (Smith and Read, 1997; Hodge *et al.*, 2001). In terrestrial communities, the different hosts and symbionts form a complex system of associations. This fact has triggered a growing interest in understanding the interactions between different symbioses, both within the same plant and among neighbouring plants. However, the number of studies involving several symbioses is still scarce.

At different hierarchical levels, the outcome of the simultaneous presence of two symbionts can be the result of additive or interactive effects (Fig. 8.1). Additive effects occur when the plant response to two (or more) symbionts acting together is the same as the cumulative effects of each symbiont with no resulting adverse effects. Interactive effects on the other hand occur when the combined effects on plant or community performance are greater (synergisms) or smaller (antagonisms) than the cumulative effects of each symbiont. Depending on the variable considered, it is sometimes impossible to determine which symbiont presents

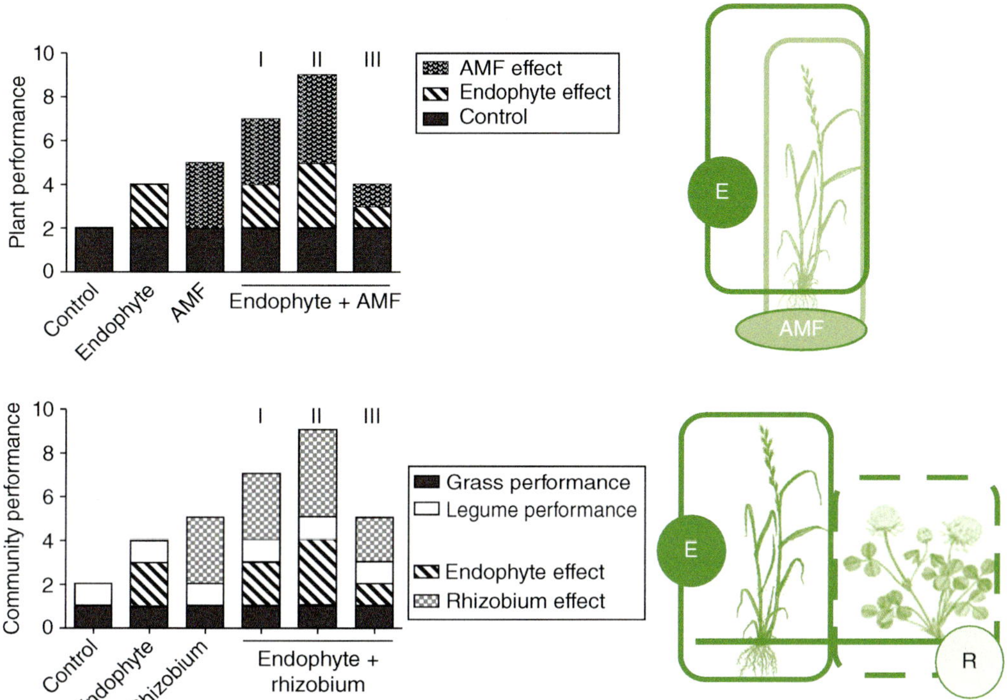

Fig. 8.1. Hypothesized additive and interactive effects of two distinct symbionts' co-occurrence on plant (e.g. biomass production, growth rate, nutrient acquisition) or community (primary productivity, nutrient dynamics, diversity) attributes. Additive (I), synergic (II) or antagonistic (III) effects may appear when both symbionts are present. The upper panel illustrates different possible results on host plant performance when manipulating the occurrence of endophytes and arbuscular mycorrhizal fungi (AMF) in the same host grass. Performance of host grass when they are absent (control treatment in black) is represented along with the contribution of endophyte presence only (hatched), AMF presence only (dotted) or that of both co-occurring symbionts. The lower panel illustrates the outcome of plant community performance when manipulating the occurrence of endophyte and rhizobia in neighbouring plants. The performance of the community is represented, divided into the contribution of grass (black bar) or legume (white bar) performance and the effect of endophyte only (hatched bar) or rhizobia only (chequered bars) or of both co-occurring.

an increased effect (e.g. when measuring biomass). In other cases, when it is possible to determine a specific amount of the effect caused by a symbiont, the interaction could be due to a change in the effects of both symbionts (as in Fig. 8.1, upper panel) or due to a change in the effect of only one of them. In general, interactive effects are studied at host level (Fig. 8.1, upper panel). However, interactive effect can appear at community level when symbionts which are specific for different host plants co-occur (Fig. 8.1, lower panel). In this case, the total effect could be driven by the independent effects of endophyte and rhizobia on the grass and legume, respectively (as shown in Fig. 8.1, lower panel, with an increase of each host biomass by the presence of their specific symbionts). However, interactive effects can be the result of each symbiont's effect on a neighbouring non-host plant (i.e. extended benefits). The inclusion of the third type of symbiont in the analysis is interesting and challenging. Unfortunately, there are very few studies including the three functionally distinct symbionts.

In this chapter, we explore the responses of plants when at least two symbionts co-occur within the same host or neighbouring plants. First, we briefly introduce the epichloid endophytes, systemic and asymptomatic fungal symbionts of temperate grasses. Second, we review studies that investigate the effects of epichloid endophytes on AMF colonization and their consequences on host plant performance. Then, as a third point of concern, we explore the current knowledge related to the presence of endophyte and AMF on the host grass affecting legume–rhizobia interaction, whether coexisting in the neighbourhood or growing after grass dies. Finally, we highlight specific concepts in the light of symbiotic interactions with the goal of gaining deeper knowledge about the implications of multiple symbiotic systems for their agronomic management and for future research.

8.2 Epichloid Endophytes of Grasses – a Private Symbiont with Multiple Effects within the Host Neighbourhood

In most terrestrial ecosystems there are grasses associated with asexual endophytic fungi of the genus *Epichloë* (formerly *Neotyphodium*) (Ascomycetes: Clavicipitaceae) (Clay and Schardl, 2002; Leuchtmann *et al*., 2014; Semmartin *et al*., 2015). These fungi intercellularly grow in sheaths and seeds of many cool-season grasses, including the agronomic grasses *Schenodorus phoenix* (tall fescue), *Lolium perenne* (ryegrass perennial) and *Lolium multiflorum* (annual ryegrass) (Clay and Schardl, 2002; Leuchtmann *et al*., 2014). These epichloid endophytes are considered protector symbionts because their presence can confer the host plant with protection against different invertebrate or vertebrate herbivores and pathogens (Clay *et al*., 1993; Bush *et al*., 1997; Clay and Schardl, 2002; Pérez *et al*., 2013, 2016). Although the production of alkaloids has been considered responsible for the effects observed in plants, other putative mechanisms could contribute to the anti-herbivory defences (Rasmussen *et al*., 2007; García-Parisi *et al*., 2014; Dupont *et al*., 2015; Bastias *et al*., 2017). For example, changes in volatile compound production of endophyte-associated plants could be involved in the observed protection (Yue *et al*., 2001; García-Parisi *et al*., 2014; Li *et al*., 2014). In particular, the common strains of the endophyte associated with *L. perenne* and *S. phoenix* produce intoxication in domestic livestock and markedly decrease the preference for these grasses. By contrast, the endophyte associated with *L. multiflorum* and some modified strains associated with *L. perenne* and *S. phoenix* do not generate toxicity to livestock (Gundel *et al*., 2009; Shiba *et al*., 2011) and only protect against the attack of certain insect species (Omacini *et al*., 2001).

In addition to the protective function of the endophyte, plants in symbiosis usually show greater tolerance and resistance to different situations of abiotic stress, including drought (Malinowski and Belesky, 2000; Davitt *et al*., 2011; but see Gundel *et al*., 2016) or nutrient deprivation (Malinowski and Belesky, 2000; Belesky *et al*., 2008; García-Parisi *et al*., 2015, 2017). In particular, increased nutrient uptake in host by endophyte presence could be due to changes in the root morphology (Malinowski and Belesky, 2000) or changes in microbial processes of soil related to the availability of nutrients (Franzluebbers *et al*., 1999; Franzluebbers, 2006; Bowatte *et al*., 2011; Casas *et al*., 2011; Iqbal *et al*., 2013). Furthermore, although the endophytes colonize only aerial tissues, multiple belowground effects were observed in the presence of the endophyte (Omacini *et al*., 2012). Changes have been observed in different soil nutrient reservoirs (Franzluebbers *et al*., 1999; Franzluebbers, 2006) and soil biota involved in decomposition processes and mineralization of nutrients (Bowatte *et al*., 2011; Casas *et al*., 2011; Iqbal *et al*., 2013). Then, the changes that the epichloid endophyte produces in its host are propagated to community level, through affecting the performance of other members of the aboveground or belowground neighbourhood (Omacini *et al*., 2005; Omacini, 2014). Because of these multiple changes in the neighbourhood, we propose a new term 'symbiosphere' to refer to the whole area in the neighbourhood that is affected by the presence of symbionts. In the following sections, we will analyse the current knowledge of the effects of epichloid endophytes on plant belowground symbionts in the symbiosphere and their joint effects on their host plants and other plants or symbioses within the neighbourhood.

8.3 Co-occurrence of Functionally Distinct Symbionts that Can Share the Same Host: Epichloid Endophytes and AMF

The simultaneous presence of AMF and epichloid endophytes within a common host has been analysed through alterations in symbiosis traits (e.g. AM spore production, degree of AM root colonization, endophyte hyphal length, density or biomass, alkaloid production) or through their impact on host plant performance or ecosystem processes (e.g. biomass production, fitness, nutrient acquisition, primary productivity, nutrient dynamics). Two reviews that were conducted based on works about endophyte effects on host root colonization by AMF found contrasting results (Larimer *et al.*, 2010; Omacini *et al.*, 2012). Similarly, the impact of both symbionts on the performance of their shared host plant also showed a great diversity of results, ranging from synergism to antagonism (see Larimer *et al.*, 2010). In this section, we incorporate recent studies on endophyte effects on AMF and host attributes into the discussion of the potential factors and processes introducing variability in the outcomes of symbiont co-occurrence, including changes in plant or microorganism species, in the abiotic conditions or in the neighbourhood complexity.

Endophyte presence is thought to impair host root colonization by AMF (Omacini *et al.*, 2012) although the strength and direction of its effects are apparently dependent on the origin or identity of the partners and abiotic conditions. The first studies were developed resorting to agronomic grasses such as *L. perenne* and *S. phoenix*, showing a significant negative endophyte effect on AMF colonization and spore production (e.g. Chu-Chou *et al.*, 1992; Guo *et al.*, 1992; Müller, 2003; Omacini *et al.*, 2006; Mack and Rudgers, 2008; Liu *et al.*, 2011). However, when surveying fields with non-agronomic grass species, the opposite effect was detected: endophyte presence was associated with higher AMF colonization (Novas *et al.*, 2005, 2009). The authors suggested that the identity of the host grass, and in particular being agronomic or not, defined the outcome of the tripartite interaction. Later on, new studies were developed with agronomic and non-agronomic grasses manipulating the origin or composition of AMF inoculum. In non-agronomic grasses, two studies detected that endophyte effect on AMF colonization may range from positive to negative according to AM species that were included or to abiotic context (Larimer *et al.*, 2010, 2012; Bell-Dereske *et al.*, 2017, but see Zhou *et al.*, 2016, 2018). For agronomic grasses, recent studies also showed that an improved (or less harsh) environmental condition may shift the effect of the endophyte on AMF colonization from negative to neutral (Li *et al.*, 2018; Slaughter *et al.*, 2018). However, other studies on the agronomic grasses *S. phoenix* and *L. perenne* did not find this response (Guo *et al.*, 1992; Mack and Rudgers, 2008; Liu *et al.*, 2011).

The host performance, when there is simultaneous presence of epichloid endophytes and AMF, also depends on the origin of the plant species, on the AMF species and on the context. Out of nine studies analysing the effect of simultaneous presence of both endophyte and AMF on the host outcome, five used agronomic grasses (*L. perenne*, *L. multiflorum* and *S. phoenix*), while four used non-agronomic grasses (*Elymus hystrix*, *Bromus auleticus* and *Achanatherum sibricum*). Most of these studies found non-interactive effects for dual symbiotic associations with both groups of plants (Mack and Rudgers, 2008; Larimer *et al.*, 2012; Vignale *et al.*, 2016). However, antagonistic effects on the host growth were found by Liu *et al.* (2011) and by Guo *et al.* (2017) in agronomic grasses. The common strain of endophyte *E. lolii* (and not the AR1 modified strain) and AM fungus *Rhizophagus intraradices* (and not the *Funneliformis mosseae*) antagonistically affected the performance of two cultivars of *L. perenne*, but only under high P availability (Liu *et al.*, 2011). Similarly, Guo *et al.* (2017) found antagonistic interactive effect on *L. perenne* biomass, under both low and high P availability. In both studies, endophyte or

AMF separately showed no effect on the host biomass, but when both microorganisms were present, a decrease was observed. Furthermore, synergic responses to endophyte and AMF presence were observed in two experiments, one using agronomic and the other using non-agronomic grasses. First, Li *et al.* (2018) found that even when the single association with endophyte or AMF produced no effect on host growth, their dual presence increased the *L. perenne* host growth, but only when the pot was inoculated with a pathogen. In the second one, Zhou *et al.* (2016) tested the dual association of AMF and epichloid endophytes in the non-agronomic *A. sibricum*, using two species of AMF (*Glomus etunicatum* and *F. mosseae*), under varying availability of N and P. They found a synergic effect only under sufficient N and P availability, explained by a positive effect of the endophyte only when inoculated with *F. mosseae*. Indeed, without endophyte, the inoculation with *F. mosseae* was negative for the host grass.

Under more complex scenarios, manipulating the neighbourhood of the host plants, new results emerged, showing that the biotic context may also change the outcome of the tripartite interaction. Omacini *et al.* (2006) studied the effect of a plant associated with endophyte on AM colonization of a neighbouring plant without endophytes. They found that, even when the endophyte was negative on the endophyte-associated grass, it was positive for the neighbouring endophyte-free grass of the same agronomic species. Recently, Vignale *et al.* (2018) and Zhou *et al.* (2018) tested the dual association in native grasses (*B. auleticus* and *A. sibiricum*, respectively). While no effect of the endophyte presence on the neighbouring plant was observed in *B. auleticus* (Vignale *et al.*, 2018), the effect of the AM fungi was different if the plant was growing with or without competition (Zhou *et al.*, 2018). Mycorrhiza showed no effect on host growth when growing without competition, but it either increased or decreased its performance when growing in competition with a non-endophytic grass, depending on the AMF species (Zhou *et al.*, 2018). Furthermore, the origin of the AMF inoculum may also affect the outcome of the tripartite symbiosis, as it was shown in an experiment developed with the agronomic *L. multiflorum* grass, the endophyte *E. occultans* and three species of AMF. When AMF were previously multiplied in non-endophytic plants (*Bromus unioloides*, *Plantago lanceolata* and *Lotus tenuis*), they exerted a negative effect on endophyte-associated host performance. When multiplied in endophyte-associated *L. multiflorum* plants, the effect was neutral and it was positive when multiplied in endophyte-associated *L. multiflorum* plants. No effect of AMF presence was observed in endophyte-free *L. multiflorum* plants irrespective of the inoculum origin (Fig. 8.2; García-Parisi and Omacini, 2017). These results suggest that the history of the AMF inoculum used in the experiments may influence symbiosis and host responses: when it was previously multiplied in endophyte-free grasses, endophyte effect on AMF colonization was negative (Liu *et al.*, 2011), whereas when multiplied in endophyte-associated grass, the effect of the AMF and endophyte on host performance was positive (García-Parisi and Omacini, 2017).

8.4 Co-occurrence of Functionally Distinct Symbioses that Can Share the Neighbourhood: Grass-Endophyte, Legume-Rhizobia and Mycorrhiza

Cattle production systems such as pastures and grasslands usually include two functional groups of plants associated with specific and generalist symbionts co-occurring in the neighbourhood. Similarly, in agricultural rotations where the interspersing monocultures of grasses with monocultures of legumes is very frequent, the legacy of both groups of plants and their symbionts may positively or negatively affect the performance of the next generation of plants (Klironomos, 2002; Bever, 2003; Antunes *et al.*, 2008; Omacini *et al.*, 2009; Cripps *et al.*, 2013; van der Putten *et al.*, 2013; García-Parisi *et al.*, 2017). Considering that the association with different types of symbionts is quite usual in both groups of plants, the lack of studies including the presence of epichloid endophytes, N-fixing

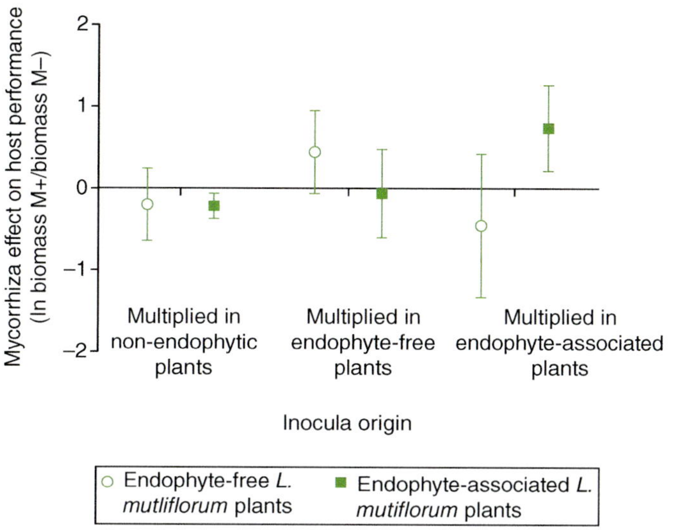

Fig. 8.2. Mycorrhizal effect on biomass production of endophyte-free (open symbols) or endophyte-associated (closed squares) *L. multiflorum* plants, according to the origin of the arbuscular mycorrhizal fungi (AMF) inoculum. The three AMF species (*Simiglomus hoi*, *Rhizophagus irregularis* and *Funneliformis mosseae*) were previously multiplied in three non-endophytic plant species (*Lotus tenuis*, *Bromus unioloides* and *Plantago lanceolata*), or in the endophytic *L. multiflorum*, either associated to the endophyte *E. occultans* or not. Values are the log response ratio and the confidence interval (CI: 95%). When CI includes 0, there is no effect of the AMF. (Modified from García-Parisi and Omacini, 2017)

bacteria and AMF is surprising. Indeed, in our previous meta-analyses conducted to evaluate the effect of endophyte presence on host grass on different components of soil biota, we could not analyse the effect of the endophyte on rhizobia or free-living N-fixing bacteria due to the lack of information (Omacini *et al.*, 2012). In this section, we analyse the current knowledge about the impact of the grass–endophyte symbiosis on the association between legumes and both types of provider symbionts, either when grasses and legumes coexist in the same or subsequent crops (Cripps *et al.*, 2013; García-Parisi *et al.*, 2014, 2015; Slaughter *et al.*, 2016; García-Parisi *et al.*, 2017).

Endophyte presence in grass may affect the neighbouring legume in several ways: it can either increase its nodulation (Eerens *et al.*, 1998) or decrease it without affecting the amount of N fixed within the host plants (García-Parisi *et al.*, 2015). Indeed, in mixed stands of grasses and legumes, the presence of both symbionts showed additive effects on both plant species productivity and N acquisition at neighbourhood level (García-Parisi *et al.*, 2015). Symbiosis with endophytes increased the host grass growth and N acquisition from the soil, while symbiosis with rhizobia increased legume and N acquisition from the atmosphere (García-Parisi *et al.*, 2015). Similarly, Slaughter *et al.* (2016) found that endophyte presence in tall fescue did not affect the N fixation rate of the neighbouring red clover–rhizobia symbiosis, but increased the amount of fixed N transfer to the host grass. Then, the symbiosis with rhizobia may not only induce complementarity in N use between legumes and grasses but also improve the performance of grasses due to N transfer (Slaughter *et al.*, 2016). Furthermore, endophyte presence can also induce benefits from grass to legume through associational protection against insect herbivory (García-Parisi *et al.*, 2014). As this protection occurs only when legume is highly associated to rhizobium, synergic effects arise from the simultaneous presence of both plant species and both symbionts.

Legacy effects of endophytic grasses over legumes have been observed in studies mimicking the agricultural rotations, which include the interspersing monocultures of grasses with monocultures. White clover grown in soils previously conditioned by endophyte-associated *L. perenne* or *L. multiflorum* plants improved their performance (Cripps *et al.*, 2013; García-Parisi *et al.*, 2017). Interestingly, the changes in the soil induced by endophyte-associated grass impacted on the legume–rhizobia symbiosis establishment and functioning (García-Parisi *et al.*, 2017). Indeed, even when the endophyte presence in a grass showed a negative effect or none at all on legume nodulation, its imprint in the soil improved plant growth and nodule efficiency, by increasing the amount of N fixed per nodule in plants (García-Parisi *et al.*, 2017). Then, even after host death, endophyte presence can affect the following legume–rhizobia symbiosis. Putative mechanisms can be related with changes either in the soil conditions/microbiota or in the shoot or root host litter deposited after the host dies.

Comprehensive studies including epichloid endophytes, AMF and rhizobia are really scarce. In a recent study, which was designed similar to those by García-Parisi *et al.* (2014, 2015), we detected that endophyte presence in *L. multiflorum* grass impaired AMF colonization in neighbouring legumes, irrespective of rhizobia presence (Fig. 8.3). Then, when analysing how endophyte and AMF presence on a grass affects legume growth after the grass dies, no interactive effects were found between both fungal symbionts (García-Parisi *et al.*, 2017). Indeed, AMF increased legume establishment independently of the presence of other symbionts either in the previous grass or in the legume. Further studies are needed to unravel the role of AMF linking both symbioses. The common mycorrhizal networks established between plants can act as a highway transferring nutrients and defensive signals (Barto *et al.*, 2012). In this sense, it is probable that AMF networks can be the channel to transfer N from the legume to the grass (e.g. Slaughter *et al.*, 2016) or the defences against herbivory (García-Parisi *et al.*, 2014). Through socializing the private symbiont benefits, the AMF networks could represent an extremely important component, considering the great impact on multiple interactions in the neighbourhood.

8.5 Concluding Remarks: Multisymbioses Public Benefits that Impact on the Symbiosphere

The multiple studies reviewed in this chapter suggest that even when the presence of specific symbionts is restricted to an individual host plant, their benefits are not necessarily private but public. Examples of public benefits are the associational protection conferred by the endophytic grass to the neighbouring legume or the fixed N transfer from the legume to the grass. If endophyte presence in a grass enhances rhizobia N fixation, the greater inputs of N in the system could benefit the grass. Furthermore, common mycorrhizal networks could transfer private symbiont benefits (such as fixed N or antiherbivore defences) from symbiotic to non-symbiotic

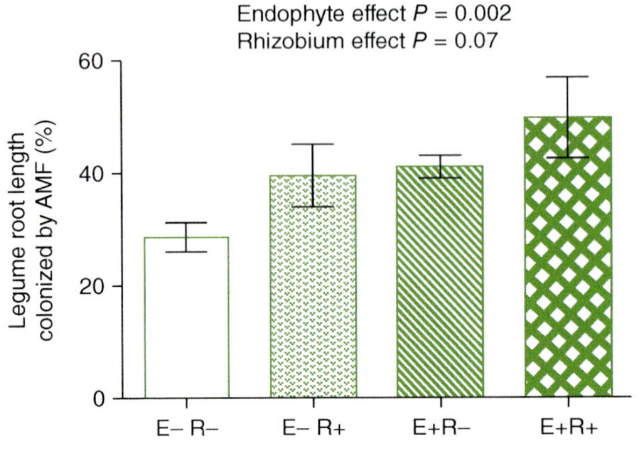

Fig. 8.3. Mycorrhizal colonization (root length colonization, %) measured in 30-day-old *Trifolium repens* plants that are inoculated (R+) or not (R–) with rhizobia, growing surrounded by endophyte-free (E–) or endophyte-associated (E+) *Lolium multiflorum* plants. Values are mean ±SE (n = 6). ANOVA statistical results (P value) of simple effects of endophyte and rhizobium are shown. No significant interaction was detected between factors.

plants (e.g. Ayres *et al*., 2007). Then, it would be difficult to determine where a symbiosis ends and where the other one starts. In this sense, any change observed in some plant or community attributes is no longer the effect of one or another symbiont but an emergent property arisen from their simultaneous presence.

The interactive effects of several symbionts could be propagated to the whole symbiosphere, the symbiosis influence area, giving rise to the appearance of processes or services of the agroecosystems that could favour their sustainability (Fig. 8.4; van der Heijden *et al*., 2008; Tikhonovich and Provorov, 2011; Wagg *et al*., 2014; Bender *et al*., 2016). When studied separately, the presence of endophytes, rhizobia and AMF has been observed to affect different above- and belowground ecosystem components – for example, other plants, herbivores and their consumers. The results of their interactions at the neighbourhood level might have an impact at the community level, affecting the species composition and diversity, the number of trophic levels, nutrient cycling and productivity. The ideas discussed in this chapter suggest that by affecting both herbivory relationships and nutrient dynamics, the simultaneous presence of these symbionts in the field could generate synergisms between regulatory (pest regulation) and support services (provision of nutrients), with consequences on primary and secondary productivity. Then, the knowledge about the impact of the interactions between plants and beneficial microorganisms in managed systems allows finding biological strategies to increase the protection of the plants against invertebrate herbivores or to reduce the use of pesticides and fertilizers (van der Heijden *et al*., 2008; Gianinazzi *et al*., 2010; Andrews *et al*., 2011; Tikhonovich and Provorov, 2011; Rillig *et al*., 2016). These studies are highly necessary to design and propose management strategies to enhance the

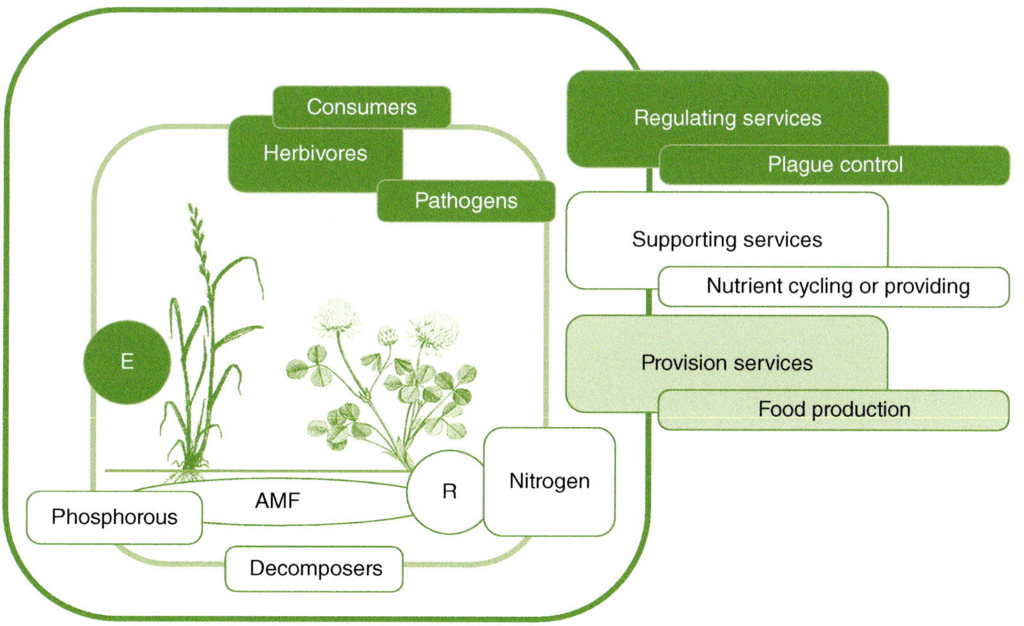

Fig. 8.4. Symbiosphere: Symbioses influence area at different levels. In the internal square are the biotic and abiotic ecosystem components that may respond to the presence of a symbiotic plant in the neighbourhood. In the external square are the ecosystem services that could be influenced by a symbiosis presence in agroecosystems. Populations bound to protector symbionts and regulating services (e.g. plague control) are represented by dark green boxes, whereas resources associated to provider symbionts and supporting services (e.g. nutrient cycling) are represented by white boxes. Both services contribute to provision services (e.g. food production, light green boxes).

synergies between services in agroecosystems that maximize their sustainability.

In conclusion, epichloid endophytes of grasses have effects that go beyond the host and impact on other symbioses either within the host or established with co-occurring plants in the neighbourhood. Endophytes can be either positive or negative for AMF colonization within the host, and both symbionts can have multiple interactive or additive effects on host performances, depending on the grass, endophyte and AMF species, and on the abiotic and biotic environment. In general, studies are focused on host performance or fitness, but no study evaluates the effects of several symbionts on the fitness of the endophytes or rhizobia. Indeed, the effect of a symbiont on other symbionts' public benefits can be considered part of its multi-functionality. Furthermore, endophytes' presence on the host grass can affect the community through the impact on the performance of a neighbouring legume or a legume that grows after the grass dies. Then, the effects of these specific symbionts can derive into public benefits for non-host plants that could be managed to increase sustainability of agroecosystems.

Acknowledgements

We are grateful to the Symbiosphere lab group. Most ideas presented here were developed through our helpful discussions on the issues dealt with in this essay. We wish to thank Agustin Grimoldi, Magdalena Druille, Luciano Boyero, Fernando Lattanzi and Hans Schnyder for thought-provoking conversations and helpful comments on earlier versions. Preparation of this chapter was facilitated by grants from the University of Buenos Aires (UBA), the National Research Council (CONICET) and the National Scientific and Technological Promotion (FONCYT).

References

Andrews, M., Edwards, G.R., Ridgway, H.J., Cameron, K.C., Di, H.J. and Raven, J.A. (2011) Positive plant microbial interactions in perennial ryegrass dairy pasture systems. *Annals of Applied Biology* 159, 79–92.

Antunes, P.M., Miller, J., Carvalho, L.M., Klironomos, J.N. and Newman, J.A. (2008) Even after death the endophytic fungus of *Schedonorus phoenix* reduces the arbuscular mycorrhizas of other plants. *Functional Ecology* 22, 912–918.

Ayres, E., Dromph, K.M., Cook, R., Ostle, N. and Bardgett, R.D. (2007) The influence of below-ground herbivory and defoliation of a legume on nitrogen transfer to neighbouring plants. *Functional Ecology* 21, 256–263.

Barto, E.K., Weidenhamer, J.D., Cipollini, D. and Rillig, M.C. (2012) Fungal superhighways: Do common mycorrhizal networks enhance below ground communication? *Trends in Plant Science* 17, 633–637.

Bastias, D.A., Martínez-Ghersa, M.A., Ballaré, C.L. and Gundel, P.E. (2017) *Epichloë* fungal endophytes and plant defenses: Not just alkaloids. *Trends in Plant Science* 22, 939–948.

Belesky, D.P., Burner, D.M. and Ruckle, J.M. (2008) Does endophyte influence resource acquisition and allocation in defoliated tall fescue as a function of microsite conditions? *Environmental and Experimental Botany* 63, 368–377.

Bell-Dereske, L., Takacs-Vesbach, C., Kivlin, S.N., Emery, S.M. and Rudgers, J.A. (2017) Leaf endophytic fungus interacts with precipitation to alter belowground microbial communities in primary successional dunes. *FEMS Microbiology Ecology* 93.

Bender, S.F., Wagg, C. and van der Heijden, M.G.A. (2016) An underground revolution: Biodiversity and soil ecological engineering for agricultural sustainability. *Trends in Ecology & Evolution* 31, 440–452.

Bever, J.D. (2003) Soil community feedback and the coexistence of competitors: Conceptual frameworks and empirical tests. *New Phytologist* 157, 465–473.

Bowatte, S., Barrett, B., Luscombe, C., Hume, D.E., Luo, D., Theobald, P. and Newton, P.C. (2011) Effect of grass species and fungal endophyte on soil nitrification potential. *New Zealand Journal of Agricultural Research* 54, 275–284.

Bush, L., Wilkinson, H.H. and Schardl, C.L. (1997) Bioprotective alkaloids of grass-fungal endophyte symbioses. *Plant Physiology* 114, 1–7.

Casas, C., Omacini, M., Montecchia, M.S. and Correa, O.S. (2011) Soil microbial community responses to the fungal endophyte *Neotyphodium* in Italian ryegrass. *Plant and Soil* 340, 347–355.

Cheplick, G.P. and Faeth, S.H. (2009) *Ecology and Evolution of the Grass-Endophyte Symbiosis*. Oxford Scholarship Online Monographs.

Chu-Chou, M., Guo, B., An, Z.-Q., Hendrix, J.W., Ferriss, R.S., *et al.* (1992) Suppression of mycorrhizal fungi in fescue by the *Acremonium coenophialum* endophyte. *Soil Biology and Biochemistry* 24, 633–637.
Clay, K. and Schardl, C.L. (2002) Evolutionary origins and ecological consequences of endophyte symbiosis with grasses. *The American Naturalist* 160, S99–S127.
Clay, K., Marks, S. and Cheplick, G.P. (1993) Effects of insect herbivory and fungal endophyte infection on competitive interactions among grasses. *Ecology* 74, 1767–1777.
Cripps, M.G., Edwards, G.R. and McKenzie, S.L. (2013) Grass species and their fungal symbionts affect subsequent forage growth. *Basic and Applied Ecology* 14, 225–234.
Davitt, A.J., Chen, C. and Rudgers, J.A. (2011) Understanding context-dependency in plant-microbe symbiosis: The influence of abiotic and biotic contexts on host fitness and the rate of symbiont transmission. *Environmental and Experimental Botany* 71, 137–145.
Denison, R.F. (2000) Legume sanctions and the evolution of symbiotic cooperation by rhizobia. *The American Naturalist* 156, 567–576.
Douglas, A.E. (2010) *The Symbiotic Habitat*. Princeton University Press, Princeton, New Jersey.
Dupont, P.-Y., Eaton, C.J., Wargent, J.J., Fechtner, S., Solomon, P., *et al.* (2015) Fungal endophyte infection of ryegrass reprograms host metabolism and alters development. *New Phytologist* 208, 1227–1240.
Eerens, J.P.J., Lucas, R.J., Easton, H.S. and White, J.G.H. (1998) Influence of the ryegrass endophyte (*Neotyphodium lolii*) in a cool-moist environment III. Interaction with white clover. *New Zealand Journal of Agricultural Research* 41, 201–207.
Franzluebbers, A.J. (2006) Short-term responses of soil C and N fractions to tall fescue endophyte infection. *Plant and Soil* 282, 153–164.
Franzluebbers, A.J., Nazih, N., Stuedemann, J.A., Fuhrmann, J.J., Schomberg, H.H. and Hartel, P.G. (1999) Soil carbon and nitrogen pools under low- and high-endophyte-infected tall fescue. *Soil Science Society of America* 63, 1687–1694.
García-Parisi, P.A. and Omacini, M. (2017) Arbuscular mycorrhizal fungi can shift plant-soil feedback of grass-endophyte symbiosis from negative to positive. *Plant and Soil* 419, 13–23.
García-Parisi, P.A., Grimoldi, A.A. and Omacini, M. (2014) Endophytic fungi of grasses protect other plants from aphid herbivory. *Fungal Ecology* 9, 61–64.
García Parisi, P.A., Lattanzi, F.A., Grimoldi, A.A. and Omacini, M. (2015) Multi-symbiotic systems: Functional implications of the coexistence of grass-endophyte and legume-rhizobia symbioses. *Oikos* 124, 553–560.
García-Parisi, P.A., Lattanzi, F.A., Grimoldi, A.A., Druille, M. and Omacini, M. (2017) Three symbionts involved in interspecific plant-soil feedback: Epichloid endophytes and mycorrhizal fungi affect the performance of rhizobia-legume symbiosis. *Plant and Soil* 412, 151–162.
Gianinazzi, S., Gollotte, A., Binet, M.-N., van Tuinen, D., Redecker, D. and Wipf, D. (2010) Agroecology: The key role of arbuscular mycorrhizas in ecosystem services. *Mycorrhiza* 20, 519–530.
Gundel, P.E., Batista, W.B., Texeira, M., Martinez-Ghersa, M.A., Omacini, M. and Ghersa, C.M. (2008) Neotyphodium endophyte infection frequency in annual grass populations: Relative importance of mutualism and transmission efficiency. *Proceedings of the Royal Society B* 275, 897–905.
Gundel, P.E., Garibaldi, L.A., Tognetti, P.M., Aragón, R., Ghersa, C.M. and Omacini, M. (2009) Imperfect vertical transmission of the endophyte *Neotyphodium* in exotic grasses in grasslands of the Flooding Pampa. *Microbial Ecology* 57, 740–748.
Gundel, P.E., Martínez-Ghersa, M.A., Omacini, M., Cuyeu, R., Pagano, E., Ríos, R. and Ghersa, C.M. (2012) Mutualisms effectiveness and vertical transmission of symbiotic fungal endophytes in response to host genetic background. *Evolutionary Applications* 5, 838–849.
Gundel, P.E., Irisarri, J.G.N., Fazio, L., Casas, C. and Pérez, L.I. (2016) Inferring field performance from drought experiments can be misleading: The case of symbiosis between grasses and *Epichloë* fungal endophytes. *Journal of Arid Environments* 132, 60–62.
Guo, B.Z., Hendrix, J.W., An, Z.-Q. and Ferriss, R.S. (1992) Role of *Acremonium* endophyte of fescue on inhibition of colonization and reproduction of mycorrhizal fungi. *Mycologia* 84, 882–885.
Guo, Y.-E., Wang, X.-Y., Gao, P. and Duan, T.-Y. (2017) Effects of *Glomus mosseae* and grass endophytes on the growth of *Lolium perenne* under phosphorus addition. *Acta Prataculturae Sinica* 26, 160–169.
Hodge, A., Campbell, C.D. and Fitter, A.H. (2001) An arbuscular mycorrhizal fungus accelerates decomposition and acquires nitrogen directly from organic material. *Nature* 413, 297–299.
Iqbal, J., Nelson, J.A. and McCulley, R.L. (2013) Fungal endophyte presence and genotype affect plant diversity and soil-to-atmosphere trace gas fluxes. *Plant and Soil* 364, 15–27.
Johnson, N.C., Graham, J.H. and Smith, F.A. (1997) Functioning of mycorrhizal associations along

the mutualism-parasitism continuum. *New Phytologist* 135, 575–585.
Kiers, E.T. and van der Heijden, M.G.A. (2006) Mutualistic stability in the arbuscular mycorrhizal symbiosis: Exploring hypothesis of evolutionary cooperation. *Ecology* 87, 1627–1636.
Kiers, E.T., Duhamel, M., Beesetty, Y., Mensah, J.A., Franken, O., *et al.* (2011) Reciprocal rewards stabilize cooperation in the mycorrhizal symbiosis. *Science* 333, 880–882.
Kiers, T., Rousseau, R.A., West, S.A. and Denison, R.F. (2003) Host sanctions and the legume–rhizobium mutualism. *Nature* 425, 1095–1098.
Klironomos, J.N. (2002) Feedback with soil biota contributes to plant rarity and invasiveness in communities. *Nature* 417, 67–70. doi:10.1038/417067a
Larimer, A.L., Bever, J.D. and Clay, K. (2010) The interactive effects of plant microbial symbionts: A review and meta-analysis. *Symbiosis* 51, 139–148.
Larimer, A.L., Bever, J.D. and Clay, K. (2012) Consequences of simultaneous interactions of fungal endophytes and arbuscular mycorrhizal fungi with a shared host grass. *Oikos* 121, 2090–2096.
Leuchtmann, A., Bacon, C.W., Schardl, C.L., White, J.F. and Tadych, M. (2014) Nomenclatural realignment of *Neotyphodium* species with genus *Epichloë*. *Mycologia* 106, 202–215.
Li, F., Christensen, M.J., Gao, P., Li, Y. and Duan, T. (2018) An arbuscular mycorrhizal fungus and *Epichloë festucae* var. *lolii* reduce *Bipolaris sorokiniana* disease incidence and improve perennial ryegrass growth. *Mycorrhiza* 28, 159–169.
Li, T., Blande, J.D., Gundel, P.E., Helander, M.L. and Saikkonen, K.T. (2014) *Epichloë* endophytes alter inducible defences in host grasses. *PLOS ONE* 9, e101331.
Liu, Q., Parsons, A.J., Xue, H., Fraser, K., Ryan, G.D., Newman, J. A. and Rasmussen, S. (2011) Competition between foliar *Neotyphodium lolii* endophytes and mycorrhizal *Glomus* spp. fungi in *Lolium perenne* depends on resource supply and host carbohydrate content. *Functional Ecology* 25, 910–920.
Mack, K.M.L. and Rudgers, J.A. (2008) Balancing multiple mutualists: Asymmetric interactions among plants, arbuscular mycorrhizal fungi, and fungal endophytes. *Oikos* 117, 310–320.
Malinowski, D.A.P. and Belesky, D.P. (2000) Adaptations of endophyte-infected cool-season grasses to environmental stresses: Mechanisms of drought and mineral stress tolerance. *Crop Science* 40, 923–940.
Müller, J. (2003) Artificial infection by endophytes affects growth and mycorrhizal colonisation of *Lolium perenne*. *Functional Plant Biology* 30, 419–424.
Newsham, K.K., Fitter, A.H. and Watkinson, A.R. (1995) Multi-functionality and biodiversity in arbuscular mycorrhizas. *Oikos* 10, 407–411.
Novas, M.V., Cabral, D. and Godeas, A.M. (2005) Interaction between grass endophytes and mycorrhizas in *Bromus setifolius* from Patagonia, Argentina. *Symbiosis* 40, 23–30.
Novas, M.V., Iannone, L.J., Godeas, A.M. and Cabral, D. (2009) Positive association between mycorrhiza and foliar endophytes in *Poa bonariensis*, a native grass. *Mycological Progress* 8, 75–81.
Omacini, M. (2014) Asexual endophytes of grasses: Invisible symbionts, visible imprints in the host neighbourhood. In: Verma, V.C. and Gange, A.C. (eds) *Advances in Endophytic Research*. Springer, India, pp. 143–157.
Omacini, M., Chaneton, E.J., Ghersa, C.M. and Muller, C.B. (2001) Symbiotic fungal endophytes control insect host-parasite interaction webs. *Nature* 409, 78–81.
Omacini, M., Chaneton, E.J. and Ghersa, C.M. (2005) A hierarchical framework for understanding the ecosystem consequences of endophyte-grass symbioses. In: Roberts, C., West, C.P. and Spiers, D.E. (eds), *Neotyphodium in Cool-Season Grasses Current Research Applications*. Blackwell Publishing, Ames, IA, USA, pp. 141–161.
Omacini, M., Eggers, T., Bonkowski, M., Gange, A.C. and Jones, T.H. (2006) Leaf endophytes affect mycorrhizal status and growth of co-infected and neighbouring plants. *Functional Ecology* 20, 226–232.
Omacini, M., Chaneton, E.J., Bush, L. and Ghersa, C.M. (2009) A fungal endosymbiont affects host plant recruitment through seed- and litter-mediated mechanisms. *Functional Ecology* 23, 1148–1156.
Omacini, M., Semmartin, M., Perez, L.I. and Gundel, P.E. (2012) Grass-endophyte symbiosis: A neglected aboveground interaction with multiple belowground consequences. *Applied Soil Ecology* 61, 273–279.
Pérez, L.I., Gundel, P.E., Ghersa, C.M. and Omacini, M. (2013) Family issues: Fungal endophyte protects host grass from the closely related pathogen *Claviceps purpurea*. *Fungal Ecology* 6, 379–386.
Pérez, L.I., Gundel, P.E. and Omacini, M. (2016) Can the defensive mutualism between grasses and fungal endophytes protect non-symbiotic neighbours from soil pathogens? *Plant and Soil* 405, 289–298.
Rasmussen, S., Parsons, A.J., Bassett, S., Christensen, M.J., Hume, D.E., *et al.* (2007) High nitrogen supply and carbohydrate content reduce fungal endophyte and alkaloid concentration in *Lolium perenne*. *New Phytologist* 173, 787–797.

Rillig, M.C., Sosa-Hernández, M.A., Roy, J., Aguilar-Trigueros, C.A., Vályi, K. and Lehmann, A. (2016) Towards an integrated mycorrhizal technology: Harnessing mycorrhiza for sustainable intensification in agriculture. *Frontiers in Plant Science* 7, 1625.

Semmartin, M., Omacini, M., Gundel, P.E. and Hernández-Agramonte, I.M. (2015) Broad-scale variation of fungal-endophyte incidence in temperate grasses. *Journal of Ecology* 103, 184–190.

Shiba, T., Sugawara, K. and Arakawa, A. (2011) Evaluating the fungal endophyte *Neotyphodium occultans* for resistance to the rice leaf bug, *Trigonotylus caelestialium*, in Italian ryegrass, *Lolium multiflorum*. *Entomologia Experimentalis et Applicata* 141, 45–51.

Slaughter, L.C., Carlisle, A.E., Nelson, J.A. and McCulley, R.L. (2016) Fungal endophyte symbiosis alters nitrogen source of tall fescue host, but not nitrogen fixation in co-occurring red clover. *Plant and Soil* 405, 243–256.

Slaughter, L.C., Nelson, J.A., Carlisle, E., Bourguignon, M., Dinkins, R.D., Phillips, T.D. and McCulley, R.L. (2018) Climate change and *Epichloë coenophiala* association modify belowground fungal symbioses of tall fescue host. *Fungal Ecology* 31, 37–46.

Smith, S.E. and Read, D.J. (1997) *Mycorrhizal Symbiosis*, 2nd edn. Academic Press, Inc., London.

Sprent, J.I. (2007) Evolving ideas of legume evolution and diversity: A taxonomic perspective on the occurrence of nodulation. *New Phytologist* 174, 11–25.

Stanton, M. (2003) Interacting guilds: Moving beyond the pairwise perspective on mutualisms. *The American Naturalist* 162(4 Suppl.), S10–S23.

Thrall, P.H., Hochberg, M.E., Burdon, J.J. and Bever, J.D. (2007) Coevolution of symbiotic mutualists and parasites in a community context. *Trends in Ecology and Evolution* 22, 120–126.

Tikhonovich, I.A. and Provorov, N. (2011) Microbiology is the basis of sustainable agriculture: An opinion. *Annals of Applied Biology* 159, 155–168.

van der Heijden, M.G.A., Bardgett, R.D. and Van Straalen, N.M. (2008) The unseen majority: Soil microbes as drivers of plant diversity and productivity in terrestrial ecosystems. *Ecology Letters* 11, 296–310.

van der Putten, W.H., Bardgett, R.D., Bever, J.D., Bezemer, T.M., Casper, B.B., *et al.* (2013) Plant-soil feedbacks: The past, the present and future challenges. *Journal of Ecology* 101, 265–276.

Vignale, M.V., Iannone, L.J., Pinget, A.D., De Battista, J.P., Novas, M.V. (2016) Effect of epichloid endophytes and soil fertilization on arbuscular mycorrhizal colonization of a wild grass. *Plant and Soil* 405, 279–287.

Vignale, M.V., Iannone, L.J., Scervino, J.M. and Novas, M.V. (2018) *Epichloë* exudates promote in vitro and in vivo arbuscular mycorrhizal fungi development and plant growth. *Plant and Soil* 422, 267–281.

Wagg, C., Bender, S.F., Widmer, F. and van der Heijden, M.G.A. (2014) Soil biodiversity and soil community composition determine ecosystem multifunctionality. *Proceedings of the National Academy of Sciences of the United States of America* 111, 5266–5270.

Walder, F. and van der Heijden, M.G.A. (2015) Regulation of resource exchange in the arbuscular mycorrhizal symbiosis. *Nature Plants* 1, 15159.

White, J.F. and Torres, M.S. (eds) (2010) *Defensive Mutualism in Microbial Symbiosis. Mycology*. CRC Press, Boca Raton, Florida.

Yue, Q., Wang, C., Gianfagna, T.J. and Meyer, W.A. (2001) Volatile compounds of endophyte-free and infected tall fescue (*Festuca arundinacea* Schreb.). *Phytochemistry* 58, 935–941.

Zhou, Y., Li, X., Qin, J., Liu, H., Chen, W., *et al.* (2016) Effects of simultaneous infections of endophytic fungi and arbuscular mycorrhizal fungi on the growth of their shared host grass *Achnatherum sibiricum* under varying N and P supply. *Fungal Ecology* 20, 56–65.

Zhou, Y., Li, X., Gao, Y., Liu, H., Gao, Y.-B., van der Heijden, M.G.A. and Ren, A.-Z. (2018) Plant endophytes and arbuscular mycorrhizal fungi alter plant competition. *Functional Ecology* 32, 1168–1179.

9 Saving Resources: The Exploitation of Endophytes by Plants for the Biosynthesis of Multi-functional Defence Compounds

Alexander Schouten*

Laboratory of Phytopathology, Wageningen University & Research, Wageningen, The Netherlands

Abstract

Plants are effective in defending themselves against herbivores, parasites and pathogens. To accomplish this, they employ various strategies, one of them being the synthesis of antimicrobial and antiherbivory compounds. To reduce the amount of energy spent, such compounds can be efficiently synthesized in multi-enzyme complexes and may have multiple roles in plant life. The synthesis can further be economized when the plant exploits associated microorganisms for the synthesis of these 'plant' compounds. Due to the potential multi-functionality of plant compounds, it is often difficult to establish what their roles are in the plant's physiology and ecology, particularly because these various roles can be quite unrelated. The research on endophytes, their synthetic abilities and their role in the ecology of the plant may, however, shed light on this issue. Indeed, it was found that particular compounds produced by endophytes, which are considered phytohormones, have additional activities, being toxic for nematodes.

9.1 Plants Are a Poor Food Source

Most plants are photoautotrophic and therefore capable of converting light energy into organic compounds, like sugars, adenosine-5'-triphosphate (ATP) and nicotinamide adenine dinucleotide phosphate ($NADPH_2$). Owing to this, next to chemoautotrophic organisms, plants serve as the primary energy source for heterotrophic organisms, i.e. parasitic plants, herbivores and (parasitic) microorganisms. Herbivores and parasites are primarily hostile to plants, although in some cases mutualism or even symbiosis through animal feeding, like seed dispersal and pollination, can be recognized. And while there are rare exceptions, leading to massive reductions in plant growth and reproduction and increases in plant mortality, herbivores only consume, on average, 10–20% of the annual net plant biomass production, which is the reason why the terrestrial world has been primarily 'green' (Hartley and Jones, 1997). Herbivores are apparently kept in check, not only through predators but also through the fact that terrestrial plants are rather inedible, sometimes even poisonous (Whittaker and Feeny, 1971; Hartley and Jones, 1997). Actually, plants turn out to be a difficult food source. Compared to herbivores, the nitrogen content in plants is low, and it was estimated that, for example, herbivorous insects need to consume at least 20 times their final body weight to reach adulthood,

* Corresponding author e-mail: sander.schouten@wur.nl

providing that all plant nitrogen can be extracted and utilized, which is usually not the case in practice (Hartley and Jones, 1997). The consumption of such high amounts of plant material is further complicated by the differences in elemental stoichiometry between plants and herbivores (Hartley and Jones, 1997). To acquire the required nitrogen amounts, the herbivore will simultaneously accumulate an excess of carbohydrates and material which must be excreted again, as exemplified by sap-sucking insects, which readily expel excess sugars through the production of honey dew (Feeny, 1992).

In addition to this, plants can be inedible or inaccessible because of the presence of complex morphological structures, such as spikes, thorns, glandular and stinging hairs and a nearly impenetrable bark, and the production of an array of compounds that are not considered essential for the primary metabolism, i.e. the compounds that are necessary for plant growth and development (Wink *et al.*, 1998; Wink, 1999; Wink and Schimmer, 1999). These secondary metabolites evolve from compounds of the primary metabolism, are generally highly complex and are rich in diversity, comprising tannins, coumarins, quinones, flavonoids, waxes, polyketides, terpenes, alkaloids, cyanogenic glycosides and glucosinolates (Reznik, 1960; Hartmann, 2007). Already in 2007, over 200,000 chemical constituents had been characterized (Hartmann, 2007), and new compounds are being discovered virtually every day. When first described, these constituents were initially considered metabolic waste or detoxification products (Peach, 1950; Reznik, 1960). However, from very early on there were indications that secondary metabolites served an important role from an ecological perspective, providing chemical defence against slugs and snails (Stahl, 1888; Fraenkel, 1959). This view really caught on in the 1970s, and it is currently widely accepted that secondary metabolites are fundamental in the interactions with herbivores, microbes and other plants and in coping with abiotic stress situations (Hartmann, 2007). Overall, secondary metabolites often play a crucial, and sometimes ambivalent, role not only in defence against herbivores and pathogenic microorganisms and the competition with other plants but also in the attraction of beneficial insects, vertebrates and microorganisms and the initiation and maintenance of mutualistic or symbiotic interactions (Hartmann, 2007).

9.2 Plants and the Costs of Secondary Metabolism

A plant has to continuously be aware of its abiotic and biotic environment in order to efficiently allocate its resources between defence chemistry to protect the parts that are already laid down against herbivory and diseases and new growth (Harborne, 1997). Secondary metabolism is considered a costly process in terms of resources, as it is dependent on primary metabolism, both as energy source for synthesis and the necessary precursors (e.g. amino acids, carbohydrates, acetyl coenzyme A). It was estimated that the synthesis of 1 g of phenolic resin, a cyanogenic glycoside and an alkaloid would consume 2.6, 2.8 and 5.0 g of photosynthetic CO_2, respectively. For comparison, to create 1 g of leaf, the necessary CO_2 consumption was estimated at 2.0–2.7 g. The accumulation of these secondary metabolites can vary significantly, ranging from 0.2 to 29% of the total leaf weight (Gulmon and Mooney, 1986). However, next to defence, secondary metabolites serve many other roles (Seigler and Price, 1976; Seigler, 1977), such as the attraction of pollinators, protection from UV light, structural support, temporary nutrient storage, phytohormone management, resisting drought, assisting in nutrient uptake, tolerating acidic and reducing environments, and mediating associations with beneficial microorganisms (Herms and Mattson, 1992; Neilson *et al.*, 2013). Individual secondary metabolites may be multifunctional as well (Wink and Schimmer, 1999), such as emodin, benzoxazinoids, strigolactones, ferulic acid and caffeine.

Emodin. The anthraquinone emodin is not only considered a feeding deterrent to herbivores (insects, birds and small mammals), but also has antibacterial, antifungal and allelochemical properties, can act as a UV protectant and attracts particular seed-dispersing fruit consumers (Whittaker and Feeny, 1971; Izhaki, 2002).

Benzoxazinoids. Several species within the Poaceae, maize (*Zea mays*), wheat (*Triticum aestivum*), rye (*Secale cereale*) and the wild barley *Hordeum lechleri* (Kremer and Ben-Hammouda, 2009; Niemeyer, 2009), can produce various benzoxazinoids (Niemeyer, 1988; Niemeyer, 2009; Du Fall and Solomon, 2011; Schulz *et al*., 2013; Makowska *et al*., 2015; Wouters *et al*., 2016). Benzoxazinoids are best known for their allelochemical activity, preventing competing plants from developing (Barnes and Putnam, 1987; Niemeyer, 1988; Pérez and Ormeño-Núñez, 1993; Bertin *et al*., 2003; Niemeyer, 2009; Schulz *et al*., 2013), but they were also associated with the detoxification of herbicides (Shimabukuro, 1967; Ioannou *et al*., 1980) and the protection against insect herbivores (Klun and Brindley, 1966; Klun and Robinson, 1969; Long *et al*., 1978; Guthrie *et al*., 1986; Thackray *et al*., 1990; Gianoli and Niemeyer, 1998; Niemeyer, 1988), plant pathogenic bacteria (Hartman *et al*., 1975; Corcuera *et al*., 1978; Niemeyer, 1988) and fungi (Virtanen and Hietala, 1955; Wahlroos and Virtanen, 1959; Couture *et al*., 1971; Baker and Smith, 1977; Long *et al*., 1978; Søltoft *et al*., 2008). Insights in the role of the individual benzoxazinoid hydroxamic acids are nevertheless fragmented, an issue which is caused by: (i) the numerous benzoxazinoids that have so far been identified in the different Poaceae species, all with their different activities and different bioactive derivatives that may evolve from each of them, (ii) the variations in spatial and timed in planta accumulation of benzoxazinoids, (iii) additional induced plant defence mechanisms or metabolites, (iv) the beneficial microbial conversion of benzoxazinoids and (v) the ability of fungal pathogens to handle (e.g. detoxify) benzoxazinoids (Schouten, 2018).

Strigolactones. Strigolactones, an ancient group of plant molecules, have an important role in plant life, both internally and externally. Externally, they function as cues for the germination of arbuscular mycorrhiza (AM) spores in the rhizosphere and to initiate a symbiotic interaction with the plant (Parniske, 2008). This interaction, which is probably very ancient (Remy *et al*., 1994; Taylor *et al*., 1995), may be crucial for the host plant because this association supports the host plant to efficiently acquire nutrients from the soil through the AM hyphae. Strigolactones are also cues for devastating parasitic weeds of the genus *Striga* to colonize the host plant, hence the name strigolactones (Parniske, 2008). Internally, strigolactones have more recently been shown to perform as plant hormones as well, controlling shoot branching, root growth, root-hair elongation, adventitious rooting, secondary growth, photomorphogenesis, seed germination and nodulation (Foo and Reid, 2013).

Ferulic acid. Ferulic acid (4-hydroxy-3-methoxycinnamic acid) is comparatively amply present in plants, predominantly in cereals, although it is principally conjugated with mono-, oligo- and polysaccharides, polyamines and lipids. Ferulic acid can, together with all hydroxycinnamic acid derivatives, be synthesized from the amino acid phenylalanine. For the plant, it is regarded valuable for its antioxidative and antimicrobial activity (Ou and Kwok, 2004). Besides antibacterial activity, ferulic acid showed antifungal activity toward *Sclerotinia sclerotiorum*, *Fusarium oxysporum*, *Alternaria* sp., *Botrytis cinerea* and *Penicillium digitatum* (Lattanzio *et al*., 1994; Ou and Kwok, 2004). Chitosan treatment of seeds of two spring wheat cultivars promoted germination and the accumulation of phenolic acids, particularly ferulic acid, and lignin in the primary leaf. This moderated seed-borne *F. graminearum* infection and suppressed the spreading of the pathogen to the primary roots of the germinating seedlings as well (Reddy *et al*., 1999). The accumulation of ferulic acid synthesis in the period from anthesis until approximately 20 days after anthesis seemed to correlate with resistance to *Fusarium* in various wheat cultivars (McKeehen *et al*., 1999). Wheat genotypes with the highest ferulic acid contents in wheat spikes turned out to be most resistant towards Fusarium head blight (FHB disease in a field situation

(Martin *et al.*, 2017). Ferulic acid could, at high concentrations, inhibit the growth of the watermelon (*Citrullus lanatus*) pathogen *Fusarium oxysporum* f. sp. *niveum* in *in vitro* experiments although the mycotoxin production by the fungus was increased more than twofold. (Wu *et al.*, 2010). This increase in mycotoxin production was surprising, as ferulic acid was shown to effectively inhibit Tri-gene transcription, in a dose-dependent manner, thus reducing mycotoxin type B trichothecene biosynthesis, including deoxynivalenol (DON) and its acetylated forms, 3-acetyl-4-deoxynivalenol and 15-acetyl-4-deoxynivalenol (3- and 15-ADON), and nivalenol (NIV) and its acetylated form, 4-acetylnivalenol (fusarenone X), in *F. culmorum* and *F. graminearum* strains (Boutigny *et al.*, 2009).

Caffeine. Depending on the level of accumulation, the alkaloid caffeine can act as an insecticide, a feeding deterrent for insects or an attractant for insect pollinators (Wink, 1998; Wright *et al.*, 2013). When caffeine concentrations exceeded 0.19 mg/ml, honeybees were deterred from drinking sucrose solutions (Wright *et al.*, 2013) and caffeine accumulation in vegetative parts and seeds can reach 24 mg/ml in coffee. However, the caffeine content within nectar of both coffee and citrus did not exceed 0.058 mg/ml, and these low levels had a rewarding effect on memorizing the floral scent, increasing the foraging on those plants and, consequently, enhancing the pollination by honeybees (Wright *et al.*, 2013). A clever allocation of secondary metabolites can thus have multiple and sometimes contradictory functions in plant ecology.

The aforementioned examples illustrate the diversity of secondary metabolites with individual versatile roles in plant life. This versatility has also been described for flavonoids and anthocyanins, glycoalkaloids, glucosinolates, cyanogenic glucosides, which, apart from their function in physiology or in structural maintenance, may serve in defence against microbes and herbivores, as signal compounds to attract pollinators or seed-dispersing animals (Wink, 1998; Wink *et al.*, 1998; Wink and Schimmer, 1999).

9.2.1 Streamlining the biosynthesis of secondary metabolites

Costs of secondary metabolism can be further reduced, because biosynthesis of the individual secondary compounds can be efficiently mediated by a limited number of enzymes, organized in a multi-enzyme complex, the metabolon, which uses up less energy to generate the secondary compounds (Jørgensen *et al.*, 2005; Neilson *et al.*, 2013). A typical example is cytochrome P450, CYP720B4, from the Sitka spruce (*Picea sitchensis*), which can use eight different olefins as a substrate and convert them in three consecutive oxidation steps into eight different diterpene resin acids (Hamberger *et al.*, 2011; Neilson *et al.*, 2013).

9.2.2 Sequestering: lowering the nutritional value of plants

In secondary metabolites, free amino acids, sugars or other nutrients are sequestered in an unpalatable form, making them unavailable to herbivores, lowering the nutritional value of the plant tissue in consequence. Yet, if needed, they can be turned over rather quickly, being reallocated into the primary metabolism again for growth and development (Herms and Mattson, 1992; Neilson *et al.*, 2013). Thus, secondary metabolite biosynthesis may be a mechanism of the plant to store precious resources, not only sugars but also elements like sulfur and nitrogen, in a way that they are, effectively, inaccessible for herbivores. This may also explain why the trade-off function between growth and secondary metabolism is non-linear and that the accumulation of secondary metabolites can be neutral for or even positively correlated with plant growth (Herms and Mattson, 1992; King *et al.*, 2006; Neilson *et al.*, 2013).

When these additional functions and benefits are taken into account, secondary metabolism may not be as costly as generally envisaged (Herms and Mattson, 1992; Neilson *et al.*, 2013).

9.3 Outsourcing of Secondary Metabolism through Endophytes

Plants can further economize on their resource expenses when the synthesis of secondary metabolites is transferred, fully or in part, to coexisting microorganisms, like endophytes. However, in practice, a close association between plant and microorganism makes it sometimes difficult to determine who is responsible for producing a particular metabolite, the host plant, an (inconspicuous) associated microorganism or both. Within the dicotyledonous bindweed family (Convolvulaceae), several species of the genera *Argyreia*, *Stictocardia*, *Rivea*, *Ipomea* and *Turbina* were, for a long time, suspected of producing ergometrine and related alkaloids (Buckingham, 1996; Wink, 1998; Wink and Schimmer, 1999; Steiner *et al*., 2006), which have anti-insect and antivertebrate activities (Wink, 1998) and are well known to be produced by species of *Claviceps*, a fungal genus that is commonly associated with the monocotyledonous Poaceae, Cyperaceae and *Juncaceae* families (Steiner *et al*., 2006). Several explanations were put forward for these findings: (i) ergoline alkaloid biosynthesis is quite an ancestral trait, which was generally lost over time and only retained in species belonging to Clavicipitaceae and Convolvulaceae, (ii) through convergent evolution, the pathways for ergoline alkaloid synthesis may have evolved independently, (iii) organisms may have acquired the coding genes for the relevant synthetic enzymes through horizontal gene transfer or (iv) particular Convolvulaceae species associate with one or more alkaloid synthesizing (fungal) microorganisms (Steiner *et al*., 2006). By fungicide treatment, ergoline alkaloid accumulation was abolished from *Ipomoea asarifolia* (Kucht *et al*., 2004), and *Ipomoea asarifolia* and *Turbina corymbosa* could associate with an epibiotic clavicipitaceous fungus, which is transmitted through seeds and seems to uniquely produce these compounds (Steiner *et al*., 2006).

The origin of constituents can be further obscured because, by combining the synthetic pathways present in both endophyte and plant, the accumulation of constituents can be increased and, more importantly, derivatives can be synthesized, which cannot be synthesized by the individual plant or endophyte. Moreover, the endophyte can initiate or stimulate particular synthesis pathways in the plant host and vice versa (Ludwig-Müller, 2015; see also Chapter 12, this volume). Such mechanisms may have significant consequences. The metabolites 2-benzoxazolin-2-one (BOA) and 6-methoxy-2-benzoxazolin-2-one (MBOA), released by various species of Poaceae, can be derivatized by soil microbes into the aminophenoxazines 2-amino-phenoxazin-3-one (APO) and 2-amino-7-methoxyphenoxazin-3-one (AMPO), respectively, which are both more stable in the soil environment (Niemeyer, 2009). APO was shown to be toxic to *Fusarium verticillioides*, but this fungus could prevent the accumulation of APO, by converting BOA, through a 2-aminophenol intermediate, into *N*-(2-hydroxyphenyl) malonamic acid (HPMA) and 2-acetamidophenol (BOA-X), which are both less toxic for the fungus (Glenn *et al*., 2003). An endophytic bacterium *Bacillus mojavensis*, which is not sensitive to APO, can, however, intrude with this fungal detoxification process by converting the 2-aminophenol intermediate into APO, thus restoring the defence against *Fusarium verticillioides* in maize and simultaneously reducing the accumulation of mycotoxins (Bacon *et al*., 2007). Endophytes from mayapple (*Podophyllum hexandrum*) and the common juniper (*Juniperus communis*), identified as *Trametes hirsuta* and *Aspergillus fumigatus*, respectively, were able to synthesize the insecticidal (Gao *et al*., 2004) and antimicrobial (Yu *et al*., 2017) secondary metabolites podophyllotoxin (Eyberger *et al*., 2006; Puri *et al*., 2006) and deoxypodophyllotoxin (Kusari *et al*., 2009a). Both plant and endophyte were reported as being capable of individually synthesizing these compounds (Kusari *et al*., 2013).

An endophyte, which seemed to be related to *Chaetomium globosum* on the basis of the sequence of the genomic coding region for the large subunit (LSU) rRNA, could, like its host plant, the medicinal herb

Hypericum perforatum (St. John's Wort), produce the multi-functional emodin and its antiviral, antimicrobial and antioxidative (Carpenter and Kraus, 1991; Radulović *et al.*, 2007) derivative, hypericin (Kusari *et al.*, 2008; Kusari *et al.*, 2009b).

An endophyte, identified as *Eupenicillium parvum* by sequence analysis of the rDNA ITS region of rDNA, isolated from the Indian neem tree, *Azadirachta indica*, could synthesize the azadirachtins A and B, which serve as insect antifeedants and insect growth regulators. It was postulated that host plant and endophyte were capable of producing azadirachtins (Kusari *et al.*, 2012).

Paclitaxel, which is also known as Taxol and has been in the spotlight from a medical perspective as a potent anticancer drug (Schiff *et al.*, 1979; Rowinsky *et al.*, 1990; Soliman *et al.*, 2011), is produced in various *Taxus* species and has antimicrobial properties against oomycetes, *Aphanomyces eochioides, Pythium splendens, Pythium ultimum, Pythium aphenadidermatum, Pythium myriotylum, Pythium irregulare, Phytophthora capsica, Phytophthora cactorum, Phytophthora citricola* and *Saprolegnia ferax*, the slime mould *Physarum polycephalum* and the basidiomycetes *Uromyces phaseoli* (bean rust fungus), *Stereum purpureum* (causal agent of silver leaf disease), *Heterobasidion annosum, Phaeolus schweinitzii* and *Perenniporia subacida* (Young *et al.*, 1992; Elmer *et al.*, 1994; Soliman *et al.*, 2015). The latter three are known as wood-decaying pathogens in conifers, with *P. schweinitzii* being pathogenic to taxus, thus suggesting a role of paclitaxel in plant defence. Other fungi were reported relatively insensitive to paclitaxel, such as the basidiomycete *Lentinus lepideus* and the ascomycetes *Pyricularia oryzae, Fusarium roseum, F. oxysporum, F. proliferatum, Botrytis cinerea, Monilinia fructicola, Rhizoctonia solani, Diaporthe sojae, Ceratocystis ulmi, Aspergillus nidulans, Verticillium dahliae, Venturia inaequalis* and the endophytic *Pestalotiopsis microspora* and *Paraconiothyrium* sp. (Young *et al.*, 1992; Elmer *et al.*, 1994; Mu *et al.*, 1999; Soliman *et al.*, 2015). Paclitaxel was not only found to be produced by the tree itself but by several *Taxus*-associated endophytes as well, like the ascomycetes *Taxomyces andreanae, Pestalotiopsis microspora, Tubercularia* sp., *Sporormia minima, Trichothecium* sp. and *Paraconiothyrium* sp. (Stierle *et al.*, 1993; Wang *et al.*, 2000; Shrestha *et al.*, 2001; Staniek *et al.*, 2009; Soliman *et al.*, 2015). *Paraconiothyrium* isolate SSM001 was shown to migrate to the potential entry points for *P. schweinitzii*, particularly branch cracks, where it released paclitaxel from intracellular hydrophobic bodies, in which it is stored in a sequestered form, through exocytosis upon induction by the pathogen. *In vitro* studies showed that paclitaxel synthesis is upregulated in the presence of *P. schweinitzii* or chloromethane, a compound that is produced by the pathogen during wood decay (Soliman *et al.*, 2015). This indicates that the SSM001 defends its habitat, i.e. the host plant, against a *Taxus* pathogen in a coordinated way. These observations would indicate that an endophyte can produce metabolites, which are identical to those produced by the host plant. However, the biosynthesis of paclitaxel is rather complex, and as discussed in Chapter 12, despite all the research, it is still not well established whether or not the host plant and endophyte each possess the complete paclitaxel synthetic pathway.

Two plants that were used in traditional medicine, *Nothapodytes nimmoniana* and *Camptotheca acuminata*, produce the alkaloid camptothecin. From *N. nimmoniana*, two endophytes were isolated, *Entrophospora infrequens* and *Neurospora crassa*, which have been reported to synthesize camptothecin as well (Puri *et al.*, 2005; Amna *et al.*, 2006; Rehman *et al.*, 2008). From *C. acuminata* an endophytic strain *Fusarium solani*, isolated from the inner bark, was capable of producing camptothecin and its derivatives 9-methoxycamptothecin and 10-hydroxycamptothecin (Kusari *et al.*, 2011). Two endophytic *F. solani* strains, isolated from white pear (*Apodytes dimidiata*), could also synthesize camptothecin, 9-methoxycamptothecin and 10-hydroxycamptothecin, although it is unclear whether or not the host plant can individually produce these metabolites (Shweta *et al.*, 2010).

9.3.1 Phytohormones produced by microorganisms

Contrary to the previous examples, it is well established that many soil-dwelling, epiphytic and endophytic microorganisms can produce phytohormones, such as indole acetic acid (IAA), ethylene, abscisic acid (ABA), cytokinins and gibberellins (Tan and Zou, 2001; Tudzynski and Sharon, 2002; Zhang *et al*., 2006; Aly *et al*., 2011; Redman *et al*., 2011; Hinsch *et al*., 2015) and the salicylic acid (SA) derivative 4-hydroxybenzoic acid (4-HBA) (Bogner *et al*., 2017). With the exception of 4-HBA, the hormonal activity of the individual compounds in plants has been firmly established, and although they are recognized as phytohormones, some of those may have additional functions in the ecology of plants.

Indole acetic acid (IAA)

The auxin indole-3-acetic acid (IAA) is the dominant auxin in nature (Tsavkelova *et al*., 2006; Duca *et al*., 2014). It has been postulated that 80% of the culturable bacteria residing in the rhizosphere can synthesize IAA (Přikryl *et al*., 1985; Loper and Schroth, 1986; Fuentes-Ramirez *et al*., 1993; Leinhos and Vacek, 1994). Among those are both plant pathogenic species, such as *Agrobacterium tumefaciens*, *A. rhizogenes*, *Erwinia herbicola*, *Pseudomonas syringae* pv. *savastanoi* and *P. syringae* pv. *syringae*, and plant beneficial species, such as *Alcaligenes faecalis*, *Enterobacter cloacae*, *Acetobacter diazotrophicus*, *Bradyrhizobium japonicum* and species within the genera of *Azospirillum*, *Pseudomonas*, *Xanthomonas* and *Rhizobium*. Moreover, as discussed further in more detail, several fungal species are known to produce IAA as well.

Microorganisms can synthesize IAA through various pathways (Spaepen *et al*., 2007): the indole-3-acetamide (IAM) pathway, the Indole-3-pyruvate (IPyA) pathway, the tryptamine (TAM) pathway, the tryptophane side chain oxidasae (TSO pathway), the indole-3-acetonitrile (IAN) pathway and a rather obscure tryptophane-independent pathway (Fig. 9.1). For IAA synthesis in fungi, the TAM pathway was identified in *Fusarium* sp. and *Colletotrichum gloeosporioides* (Gruen, 1959; Robinson *et al*., 1998; Tsavkelova *et al*., 2012), but auxins can also be produced through the IPyA pathway in other genera, such as *Ustilago* (Reineke *et al*., 2008) and *Rhizoctonia* (Furukawa *et al*., 1996), and the tryptophane-independent pathway (Chanclud and Morel, 2016). In plants, the IPyA pathway is considered the major IAA biosynthesis pathway, although in *Arabidopsis thaliana* the IAM pathway together with the tryptophane-independent pathway are present as well (Spaepen *et al*., 2007). Because of these redundancies in IAA synthesis pathways, generating mutants in either plants or bacteria, which are completely compromised in IAA synthesis, has thus far failed. For reasons currently unknown, it seems that phytopathogenic bacteria incline to employ the IAM pathway, whereas beneficial bacteria incline to employ the IPyA pathway (Spaepen *et al*., 2007). In plants, IAA can be stored in inactive forms by being sequestered to sugars, amino acids and proteins (Dörffling, 1982). Microorganisms, too, can store IAA in an inactive form by conjugation to amino acids. Strains of *Pseudomonas syringae* pv. *savastanoi* isolated from oleander showed IAA lysine synthetase activity, conjugating IAA to the amino group of lysine, to form 3-indole-acetyl-L-lysine (IAA-lysine) (Glass and Kosuge, 1986). Although the *IaaL* gene, coding for IAA lysine synthetase, seems to be commonly present in *P. syringae* strains, *IaaL*-related genes could not be detected in other bacteria, such as *Erwinia herbicola*, *Agrobacterium tumefaciens*, *Alcaligenes faecalis*, *Rhizobium meliloti*, *Psuedomonas fluorescens*, *P. putida*, *P. corrugata* and *P. stutzeri* (Glickmann *et al*., 1998).

As phytohormone, IAA has an important function in virtually every aspect of plant growth and development, such as cell division and elongation, initiating the development of roots, leaves, flowers, the vascular system, cambium and fruits, and also senescence (Duca *et al*., 2014). With respect to root development, IAA particularly induces lateral root formation in dicots and adventitious root formation in monocots (McSteen, 2010). Auxins are required for establishing

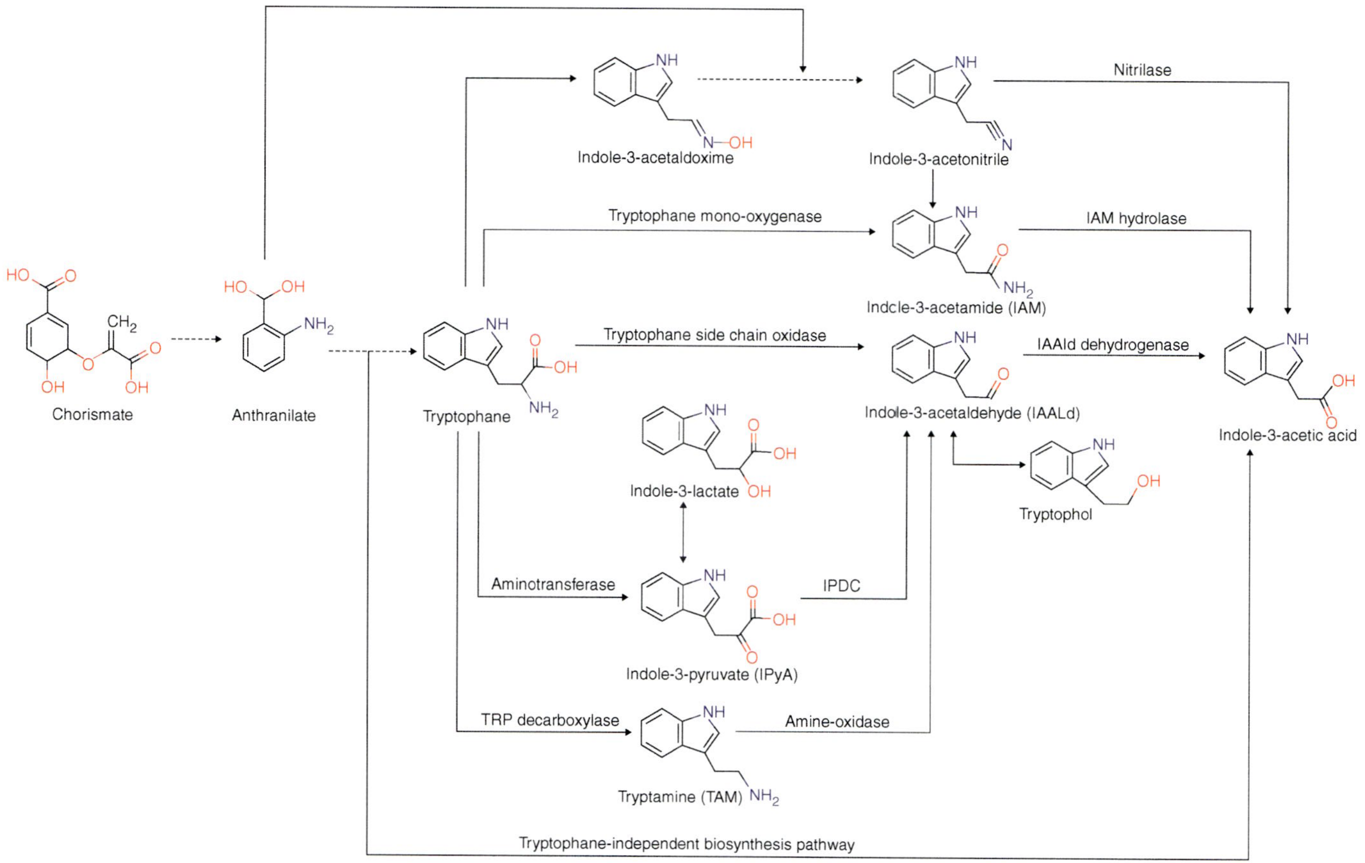

Fig. 9.1. Pathways by which indole acetic acids can be synthesized in microorganisms and plants. The IPyA, TAM and IAM pathways run over indole-3-pyruvate (IPyA), tryptamine (TAM) and indole-3-acetamide (IAM), respectively. All pathways can be found in bacteria, whereas only the TAM and the tryptophane-independent pathway have thus far been detected in fungi. In plants, the IPyA pathway is considered the primary indole acetic acid (IAA) biosynthesis pathway, although the IAM pathway together with the tryptophane-independent pathway seem to be present as well (IPDC, indole-3-pyruvate decarboxylase). (Modified from Spaepen *et al.*, 2017)

symbiotic interactions between plants and microorganisms. For both the initiation of nodule formation in the root by nitrogen-fixative bacteria (Hirsch and Fang, 1994) and the invasion of mycorrhizal fungi (Hanlon and Coenen, 2011; Etemadi *et al.*, 2014), auxins are necessary. For example, auxin-overproducing mutants of the ecto-mycorrhizal fungi *Hebeloma cylindrosporum* showed an improved capability of invading root tissues of *Pinus pinaster* (Gay *et al.*, 1994; Laurans *et al.*, 2001), although no difference in growth was observed between plants colonized by one of the mutants and those colonized by the wild type. This indicates that the fungal auxin facilitated host invasion without promoting host development. Generally, plants associated with mycorrhizal fungi hold higher auxin quantities than those that do not (Barker and Tagu, 2000; Meixner *et al.*, 2005), although it is thus far not clear from which organism these auxins originate, the host plant or the mycorrhizal fungus.

Despite the findings with *H. cylindrosporum*, the production of IAA is regarded as one of the plant growth-promoting traits microorganisms can possess. This was illustrated by analysing various *Arabidopsis* mutants that are auxin-resistant and produce fewer lateral roots (*axr1*), defective in acropetal and basispetal auxin transport at the root tip (*aux1*), defective in the proper translocation of particular auxin transport proteins (*doc1*) or lacking the auxin transporter AtPIN2 (*eir1*) (Contreras-Cornejo *et al.*, 2009). The endophytes *Trichoderma virens* and *Trichoderma atroviride* increased biomass and accelerated root development in the wild-type Arabidopsis seedlings, whereas these effects remained absent in all these tested mutants. *In vitro* experiments with *Arabidopsis* showed an increase in root and shoot proliferation in the presence of *F. oxysporum* Fo162 (Martinuz *et al.*, 2015), an isolate that is also capable of synthesizing IAA in liquid cultures (Bogner *et al.*, 2017).

The percentage of endophytic bacteria isolated from clover that were positive for the production of IAA, ACC deaminase and siderophores was higher than that isolated from the rhizosphere, suggesting that the plant allows growth-promoting bacteria to access the root system (Etesami *et al.*, 2014). Several of these endophytic isolates could be transferred to sterile rice seedlings, although endophytic colonization was cultivar dependent. Establishment of the endophytes in rice led to growth promotion and an increased resistance against pathogens, the latter being attributed to the induction of defence responses by the plant (Etesami *et al.*, 2014), a phenomenon that has been frequently observed when endophytes colonize roots (for references see Chapter 2).

Despite these observations, the role of microbial synthesized IAA in plant–host interactions is still rather ambivalent, because even pathogens seem to employ IAA to enter their host plants. IAA can initiate the softening of the cell wall and the opening of stomata, thus allowing easier access (Duca *et al.*, 2014, and references therein). Infection of susceptible plants with *Pseudomonas syringae* subsp. *savastanoi*, deficient in the biosynthesis of IAA, shows that IAA is necessary to produce disease symptoms in oleander (Comai and Kosuge, 1982; Smidt and Kosuge, 1978), while infection with mutants with an increased level of auxin production incites the formation of larger galls (Glass and Kosuge, 1986). For plant pathogenic necrotrophic fungi such as *Colletotrichum gloeosporioides* f. sp. *aeschynomene* and several pathogenic *Fusarium* species, the in planta expression of fungal IAA synthesis genes was reported to be highest during host colonization (Maor *et al.*, 2004; Tsavkelova *et al.*, 2012). The overexpression of IAA-biosynthesis genes rendered plants with elevated IAA levels, which were more susceptible to the pathogen *Pseudomonas syringae* DC 3000 (Mutka *et al.*, 2013; Duca *et al.*, 2014). SA represses the plant auxin-signalling pathway as part of the plant defence responses (Wang *et al.*, 2007).

These findings may not be confusing, as endophytes, like pathogens, need to gain access to the plant interior as well. This would illustrate that, rather than entering the plant host itself, the events after entering determine whether or not a plant–microbe interaction goes astray.

IAA is more than a phytohormone alone

Other than serving as a phytohormone, IAA seems to have other roles as well. In bacteria, too, IAA can operate as a signal molecule in surviving adverse conditions, microbial competition and resisting plant defences (Duca *et al.*, 2014). For example, *P. syringae* pv. *syringae* IAA positively affected syringomycin synthesis, which is mandatory for full virulence (Xu and Gross, 1988a,b; Mazzola and White, 1994), and knockout mutants of *P. savastanoi* pv. *nerii* Psn23, in which IAA synthesis was impaired, were more sensitive to 8-hydroxyquinoline, a suspected antimicrobial phytosiderophore (Tewari *et al.*, 2015), in comparison to the wild-type (Cerboneschi *et al.*, 2016). For fungi, information on IAA functioning as an internal signalling molecule is thus far absent.

In addition to being a signal molecule, IAA was shown to be toxic for the plant parasitic root-knot nematode *Meloidogyne incognita* (Bogner *et al.*, 2017). By dose-response analysis, the concentration causing 50% lethality (LC_{50}) of preparasitic second-stage juveniles was assessed at 117 µg/ml after 72 h. Although the cuticle of the dead nematodes remained turgid, the cuticle had released from nematode's body. Within the body, large vacuolar-like lipid bodies had accumulated. A similar phenotype was obtained when the juveniles were incubated with the commercial nematocide carbofuran (LC_{50}: 64 µg/ml at 72 h).

Abscisic acid (ABA)

After its initial discovery in the 1960s, the hormone abscisic acid (ABA) has been found in an ever-growing number of plants and mosses (Wasilewska *et al.*, 2008). It is therefore considered to be a common hormone, having multiple roles in the development and physiology of these organisms, realizing stomatal closure, bud break, seed maturation and dormancy, tolerance to desiccation, salinity, heat, cold, frost and heavy metal ions, as well as the onset of senescence (Finkelstein and Zeevaart, 1994; Wasilewska *et al.*, 2008). Due to these roles, ABA is often regarded as an abiotic stress-related hormone. At the same time, ABA may restrict ethylene production (Sharp and LeNoble, 2002; Benschop *et al.*, 2007), indicating a self-propelling negative feedback mechanism, because ethylene was reported to induce ABA synthesis (Grossmann and Hansen, 2001; Chiwocha *et al.*, 2005). ABA may nevertheless play a role in biotic interactions, as sedentary plant pathogenic nematode *M. graminicola* were suggested to abuse the ABA pathway to increase rice susceptibility to nematode infection (Kyndt *et al.*, 2014, 2017).

In plant cells, ABA is synthesized in the cytosol from xanthoxin, which is produced in the chloroplast in multiple steps from phytoene, through the carotenoid zeaxanthin and then transferred into the cytosol (Wasilewska *et al.*, 2008).

Based on the analysis on phytohormone production in, primarily, soil- and root-associated bacteria, including *Azospirillum brasiliense*, through applying highly sensitive immunological methods (Müller *et al.*, 1989), it was believed for a long time that bacteria in general could not synthesize ABA. However, an endophytic *A. brasiliense* strain, isolated from the common sunflower (*Helianthus annuus*), was shown to produce ABA (Forchetti *et al.*, 2007). Enrichment of the maize rhizosphere with this strain alleviated water stress in maize (*Zea mays*) (Cohen *et al.*, 2009), and it was suggested that the bacterially produced ABA contributed to the ABA content in the roots, similar to what was observed for *Arabidopsis* (Cohen *et al.*, 2008). ABA is synthesized by many species of ascomycetes, fungi imperfecti and basidiomycetes, such as isolates from *Cercospora* sp., *Botrytis cinerea*, *Ceratocystis coerulescens*, *Fusarium oxysporum*, *F. culmorum*, *Rhizoctonia solani*, *Alternaria* sp., *Rhizopus nigricans*, *Monilia* sp., *Pleurotus florida* (Tudzynski and Sharon, 2002) and an endophytic *Nigropsora* sp. (Clark *et al.*, 2013)

Additional roles of ABA

From the medicinal plant *Fragaria virginiana*, the endophytic *Nigropsora* sp. isolate TC2-054 was obtained. *In vitro* cultures of this fungus showed significant antimycobacterial activity against *Mycobacterium tuberculosis*

H37Ra, which was attributed to both linoleic acid derivatives and ABA (Clark *et al.*, 2013). ABA may thus play a role in limiting the colonization of plants by mycobacteria.

9.3.2 4-Hydroxybenzoic acid (4-HBA)

Although less well studied than IAA, the phenolic metabolite 4-hydroxybenzoic acid (4-HBA) is another compound that can be synthesized by both plants and microorganisms. 4-HBA accumulation could be identified in a variety of plant species, like *Arabidopsis thaliana* (Tan *et al.*, 2004), carrot (*Daucus carota*) (Bach *et al.*, 1993), grape (*Vitis vinifera*), oil palm (*Elaeis guineensis*), betel palm (*Areca catechu*), cuban royal palm (*Roystonea regia*), East African satinwood (*Fagara macrophylla*), yellow leaf tree (*Xanthophyllum rubescens*), peroba (*Paratecoma peroba*), pink trumpet tree (*Handroanthus impetiginosus*), red sandalwood (*Pterocarpus santalinus*), southern catalpa (*Catalpa bignonioides*), Chinese chest tree (*Vitex negundo*), medlar (*Mespilus germanica*) (Khadem and Marles, 2010), cucumber (*Cucumis sativa*) (Smith-Becker *et al.*, 1998) and bamboo (*Dendrocalamus asper*) (Zhang *et al.*, 2018). Suspension-cultured carrot cells could produce 4-HBA (Bach *et al.*, 1993), indicating that plants themselves can produce this compound, in which chorismate serves as the precursor in the synthesis (Viitanen *et al.*, 2004). The synthesis of 4-HBA also seems widespread in the microbial world. Both the endophytic *F. oxysporum* Fo162, which also synthesized IAA, and a marine bacterium, *Pseudoalteromonas haloplanktis* TAC125, could produce 4-HBA (Bogner *et al.*, 2017; Sannino *et al.*, 2018). Mycelial cultures of *Paecilomyces variotii* MTCC 6581, isolated from mesocarp tissue of tender coconut, and a white rot fungus *Schizophyllum commune* could transform p-coumaric acid, which is adequately present in the plant cell wall, into 4-HBA (Sachan *et al.*, 2006, 2010). Compared to its structural analogue, SA, the function of 4-HBA in plant physiology and ecology is significantly less well described. Its biological activity was initially disregarded because it was reported not to trigger defence responses (Chen *et al.*, 1993). 4-HBA could nevertheless activate the wounding responsive octopine synthase (ocs) element present in the soybean auxin-inducible promoter of the GH2/4 gene, which is also induced by other stress-related chemicals, including inactive auxin analogues, SA and inactive SA analogues, methyl jasmonate and heavy metals, and stress-related treatments, including heat shock and wounding (Ulmasov *et al.*, 1994). Cell wall extracts of *Arabidopsis thaliana* were found to contain 4-HBA, which increased in concentration after *Pythium sylvaticum* infection (Tan *et al.*, 2004), and in cucumber, 4-HBA accumulated in the phloem fluid after *Pseudomonas syringae* pv. *syringae* infection and, together with SA, was additionally *de novo* synthesized, parallel to the increase of phenylalanine ammonialyase (PAL) activity (Smith-Becker *et al.*, 1998).

Remarkably, 4-HBA showed reciprocal activities under abiotic stress conditions when compared to SA. 4-HBA increased the drought tolerance of winter wheat and the freezing tolerance of spring wheat, whereas SA reduced the freezing tolerance of winter wheat and the drought tolerance of spring wheat when administered exogenously at 0.5 mM (Horváth *et al.*, 2007).

Additional roles of 4-HBA

4-HBA exhibited only low toxicity to freshwater green alga, *Pseudokirchneriella subcapitata*, with a median inhibition concentration (IC_{50}) of 1.4 mg/ml after 72 h, whereas it stimulated the algal growth at lower concentrations, which ranged from 14 to 138 μg/ml (Kamaya *et al.*, 2006). At 2.5 and 1 mg/ml, 4-HBA showed antimicrobial activities against the causal agent of basal stem rot in oil palm, the basidiomycete *Ganoderma boninense* (Chong *et al.*, 2009), and several probiotic *Lactobacillus* spp. (Cueva *et al.*, 2010), respectively. Compared to this, 4-HBA was relatively toxic to the root-knot nematode *M. incognita*, with an LC_{50} of 104 μg/ml at 72 h, with a similar phenotype as observed with IAA and the commercial nematocide carbofuran (Bogner *et al.*, 2017; Fig. 9.2).

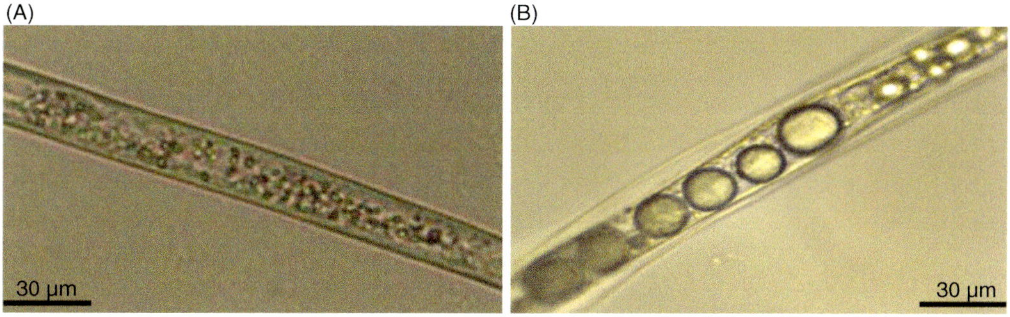

Fig. 9.2. Middle sections of root-knot nematodes, *Meloidigyne incognita*, untreated (panel A) and treated (panel B) with 4-hydroxybenzoic acid (4-HBA), observed with bright field microscopy. 4-HBA causes disintegration of inner structures, the accumulation of large vacuole-like lipid bodies and the releasing of the cuticle. Similar observations were made with indole-3-acetic acid (IAA) and the commercial nematicide carbofuran. (From Bogner *et al.*, 2017)

Conclusions and Outlook

Studies with *Arabidopsis* showed that the association between the endophytic *F. oxysporum* Fo162 significantly increased plant fresh weight, root length, average root diameter and lateral root formation (Martinuz *et al.*, 2015). This suggested an increase in hormonal activity congruent with that of IAA. Fewer root-knot juveniles infected the roots, thus producing fewer galls, an effect that was also observed in a split-root assay, in which the infecting nematode and root-associated endophyte were spatially separated (Martinuz *et al.*, 2015). Other studies with tomato showed that an association with *F. oxysporum* Fo162 resulted in a delayed or even attenuated nematode development and a reduced fecundity (Martinuz *et al.*, 2013). This attenuated or delayed development was also observed, together with an increase in the male–female ratio, for another root-knot nematode species, *M. graminicola*, in rice, when inoculated with the endophytic *Fusarium moniliforme* strain Fe14 (Le *et al.*, 2016). Sex determination among root-knot nematodes is primarily epigenetically driven in which sufficient access to nutrients leads to the development of more females (Papadopoulou and Triantaphyllou, 1982; Chan *et al.*, 2010).

Because fungal IAA gene expression was observed for other pathogenic *Fusarium* species during host colonization, it can be assumed that *F. oxysporum* Fo162 participated in the in planta accumulation of IAA. This may lead to a plethora of plant responses (Fig. 9.3), ranging from priming events to the immediate induction of defences through the systemic acquired resistance (SAR) or induced systemic resistance (ISR) pathways (Schouten, 2016; see also Chapter 2, this volume). Cucumber plants, inoculated with the nematode antagonistic root endophyte *T. asperellum* T203, showed an increase in the accumulation of lipoxygenase (Lox1), ethylene receptor 1 (ETR1) and constitutive triple response 1 gene B (CTR1) transcripts in roots and leaves, suggesting that the JA/ethylene-mediated induced defences were initiated (Shoresh *et al.*, 2005). Endophytic nematophagous *Arthrobotrys oligospora* and *Pochonia chlamydosporia* (syn. *Verticillium chlamydosporium*) provoked the formation of papillae and other cell wall appositions in barley and tomato, respectively (Escudero and Lopez-Llorca, 2012; Larriba *et al.*, 2015). Although associated with plant resistance, such cell wall structures can be initiated by both pathogenic and nonpathogenic fungi (Beswetherick and Bishop, 1993; Heitefuss, 1997; Bao and Lazarovits, 2001) and both endophytes could still proliferate inside the roots. Nevertheless, the activation of particular plant defence responses was suggested being an additional mechanism for nematode antagonism. A detailed transcriptome analysis of *P. chlamydosporia* root colonization in barley,

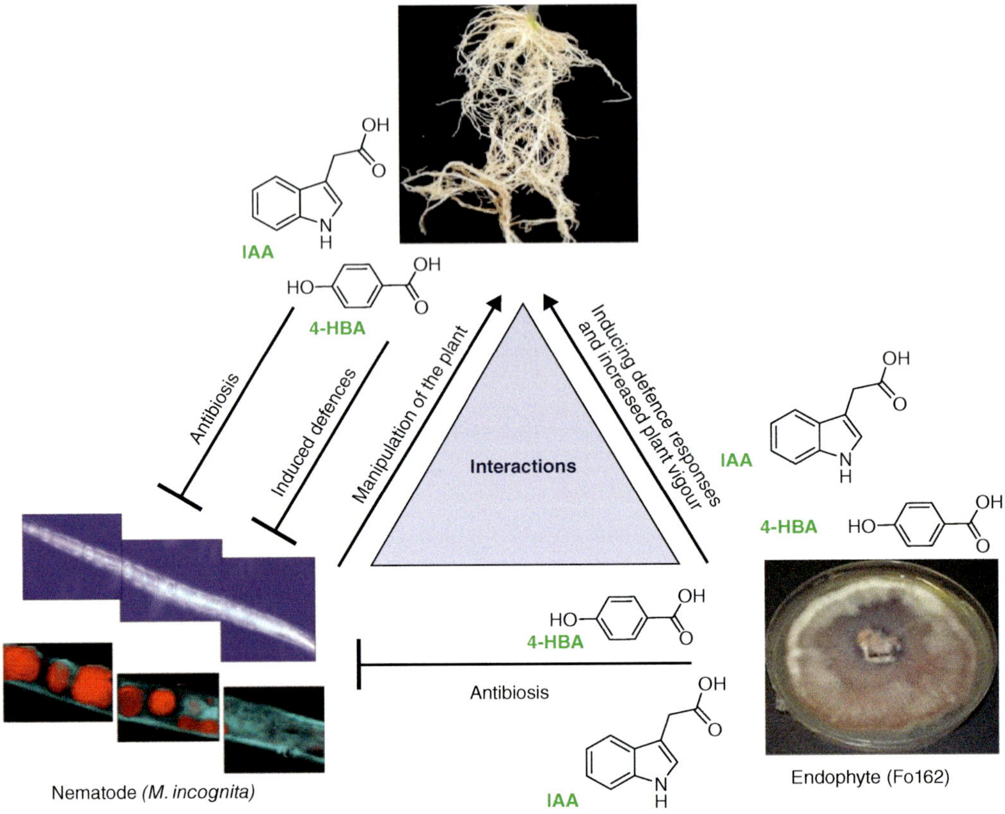

Fig. 9.3. Tripartite interactions between the endophytic *Fusarium oxysporum* Fo162, a plant and the root-knot nematode *Meloidigyne incognita*. The nematode manipulates the plant to facilitate the establishment and maintenance of a nurse cell. By producing indole-3-acetic acid (IAA) and 4-hydroxybenzoic acid (4-HBA), the endophyte induces defence responses in the plant against the nematode. At the same time IAA and 4-HBA, derived from both the endophyte and plant, are involved in antibiosis toward the nematode. (Modified from Bogner *et al*., 2017)

revealed an enrichment of genes associated with abiotic stress responses, predominantly those coding for heat shock proteins (Larriba *et al*., 2015). An increase of transcripts associated with biosynthesis of phytohormones, such as auxin, ethylene and jasmonic acid (JA), and of genes related to effector triggered immunity (ETI) and pattern-triggered immunity (PTI) was noted as well. Such responses may affect the performance of the infecting root-knot nematode and the proper functioning of the initiated feeding site, i.e. particular root cells that are converted into giant cells, an event which is crucial for sedentary nematodes for the duration of their development and reproduction (Schouten, 2016; see also Chapter 2, this volume). Other observations suggest that the SA-dependent systemic pathway, instead of the opposing (Caarls *et al*., 2015) JA-dependent defence pathway, is involved in the induced systemic defences. Upon challenging with the whitefly *Trialeurodes vaporariorum*, *F. oxysporum* Fo162-inoculated tomato plants accumulated PAL, PR1 and PR5 transcripts, whereas the LOX transcript accumulation did not change (Eschweiler *et al*., 2014). Another non-pathogenic root-associated F. oxysporum strain, Fo47, was also able to induce SA-dependent SAR responses (Fuchs *et al*., 1997; Duijff *et al*., 1998; Olivain *et al*., 2003). In *Radopholus similis*-resistant banana, PAL, the initial enzyme of the phenylpropanoid biosynthesis pathway is involved in the synthesis of a nematotoxic phytoalexin, anigorufone (Wuyts *et al.*, 2006; Hölscher *et al.*, 2014). Comparable to what was established for the pathogenic *F. oxysporum* f. sp. *cubense* (Luis *et al.*, 1994), *F. oxysporum* Fo162 may also elicit anigorufone accumulation in the banana roots (Schouten, 2016) and maybe other nematocidal phytoalexins in other plant species (Schouten, 2016; see also Chapter 2, this volume).

Striking is the observation that, next to the indirect effects, IAA can directly affect the health of the nematode (Figs 9.2 and 9.3), which may well affect its migration into the root, and probably even the initiation of the nurse cell (in case of sedentary nematodes) and further development as well (Bogner *et al.*, 2017). Even at sublethal doses such effects can be expected. 4-HBA can play a similar multi-functional role as IAA in this tripartite interaction (Figs 9.2 and 9.3), although the indirect effect, like induced defences and growth promotion through 4-HBA is still much less defined.

In all, both IAA and 4-HBA can be synthesized by the host plant and endophyte and both seem to act as multi-bladed swords, not only serving as phytohormones, facilitating the association with the host plant and stimulating defence against biotic and abiotic stress conditions, but also directly affecting plant parasitic nematodes, and possibly other microbial competitors, in a negative way (Fig. 9.3). The potential additive or synergistic effect on nematodes and induced plant resistance when IAA and 4-HBA are simultaneously present have thus far not been studied.

Beneficial associations with microorganisms seem to be ancient and crucial for sustaining plant life (Remy *et al.*, 1994; Taylor *et al.*, 1995, Redecker *et al.*, 2000; Taylor *et al.*, 2005; Taylor and Krings, 2005; Krings *et al.*, 2007; Rodriguez *et al.*, 2008; Rodriguez and Redman, 2008; Redman *et al.*, 2011; Labandeira and Prevec, 2014). Microbes that accumulate in the rhizosphere and endosphere may directly and indirectly support the plant to cope with biotic and abiotic stress situations. In this complex interaction, the chemical constituents produced by a beneficial microbe, the host plant and the two organisms combined play an important part, although this is in most cases still not well understood. Nevertheless, plants are well capable of mounting defensive measures towards pest, diseases and herbivory and seem to do this as economically as possible. Nutrients, like sugars and amino acids, are made temporarily inaccessible by sequestration to secondary metabolites. Different chemical constituents may be synthesized through one pathway and the constituents that are generated may have multiple activities. Plants may furthermore outsource the synthesis of constituents, thus even further economizing in their spending of valuable resources. Phytohormones can be synthesized as well by endophytes. Some of those not only affect the plant growth and development but also pathogens and pests, both directly and indirectly. The recent findings concerning IAA and 4-HBA and their effect on nematodes illustrate this, making it necessary to rethink the role of chemical constituents in plant life and ecology. Such insights will help to develop sustainable agricultural practices, particularly with the help of endophytes.

References

Aly, A.H., Debbab, A. and Proksch, P. (2011) Fungal endophytes: Unique plant inhabitants with great promises. *Applied Microbiology and Biotechnology* 90, 1829–1845. doi:10.1007/s00253-011-3270-y

Amna, T., Puri, S.C., Verma, V., Sharma, J.P., Khajuria, R.K., *et al.* (2006) Bioreactor studies on the endophytic fungus *Entrophospora infrequens* for the production of an anticancer alkaloid camptothecin. *Canadian Journal of Microbiology* 52, 189–196. doi:10.1139/w05-122

Bach, M., Schnitzler, J.P. and Seitz, H.U. (1993) Elicitor-induced changes in Ca^{2+} influx, K^+ efflux, and 4-hydroxybenzoic acid synthesis in protoplasts of *Daucus carota* L. *Plant Physiology* 103, 407–412. doi:10.1104/pp.103.2.407

Bacon, C.W., Hinton, D.M., Glenn, A.E., Macías, F.A. and Marin, D. (2007) Interactions of *Bacillus mojavensis* and *Fusarium verticillioides* with a benzoxazolinone (BOA) and its transformation product, APO. *Journal of Chemical Ecology* 33, 1885–1897. doi:10.1007/s10886-007-9347-5

Baker, E.A. and Smith, I.M. (1977) Antifungal compounds in winter-wheat resistant and susceptible to *Septoria nodorum*. *Annals of Applied Biology* 87, 67–73. doi:10.1111/j.1744-7348.1977.tb00660.x

Bao, J.R. and Lazarovits, G. (2001) Differential colonization of tomato roots by nonpathogenic and pathogenic *Fusarium oxysporum* strains may influence *Fusarium* wilt control. *Phytopathology* 91, 449–456.

Barker, S.J. and Tagu, D. (2000) The roles of auxins and cytokinins in mycorrhizal symbioses. *Journal of Plant Growth Regulation* 19, 144–154.

Barnes, J.P. and Putnam, A.R. (1987) Role of benzoxazinones in allelopathy by rye (*Secale cereale* L.). *Journal of Chemical Ecology* 13, 889–906. doi:10.1007/bf01020168

Benschop, J.J., Millenaar, F.F., Smeets, M.E., Van Zanten, M., Voesenek, L. and Peeters, A.J.M. (2007) Abscisic acid antagonizes ethylene-induced hyponastic growth in *Arabidopsis*. *Plant Physiology* 143, 1013–1023. doi:10.1104/pp.106.092700

Bertin, C., Yang, X. and Weston, L.A. (2003) The role of root exudates and allelochemicals in the rhizosphere. *Plant and Soil* 256, 67–83. doi:10.1023/A:1026290508166

Beswetherick, J.T. and Bishop, C.D. (1993) An ultrastructural-study of tomato roots inoculated with pathogenic and nonpathogenic necrotrophic fungi and a saprophytic fungus. *Plant Pathology* 42, 577–588.

Bogner, C.W., Kamdem, R.S.T., Sichtermann, G., Matthäus, C., Hölscher, D., *et al.* (2017) Bioactive secondary metabolites with multiple activities from a fungal endophyte. *Microbial Biotechnology* 10, 175–188. doi:10.1111/1751-7915.12467

Boutigny, A.L., Barreau, C., Atanasova-Penichon, V., Verdal-Bonnin, M.N., Pinson-Gadais, L. and Richard-Forget, F. (2009) Ferulic acid, an efficient inhibitor of type B trichothecene biosynthesis and *Tri* gene expression in *Fusarium* liquid cultures. *Mycological Research* 113, 746–753. doi:10.1016/j.mycres.2009.02.010

Buckingham, J. (1996) *Dictionary of Natural Products on CD-ROM*, 1st edn. Chapman & Hall, London.

Caarls, L., Pieterse, C.M.J. and Van Wees, S.C.M. (2015) How salicylic acid takes transcriptional control over jasmonic acid signaling. *Frontiers in Plant Science* 6, 170. doi:10.3389/fpls.2015.00170

Carpenter, S. and Kraus, G.A. (1991) Photosensitization is required for inactivation of equine infectious-anemia virus by hypericin. *Photochemistry and Photobiology* 53, 169–174. doi:10.1111/j.1751-1097.1991.tb03919.x

Cerboneschi, M., Decorosi, F., Biancalani, C., Ortenzi, M.V., Macconi, S., *et al.* (2016) Indole-3-acetic acid in plant-pathogen interactions: A key molecule for in planta bacterial virulence and fitness. *Research in Microbiology* 167, 774–787. doi:10.1016/j.resmic.2016.09.002

Chan, Y.L., Yang, A.H., Chen, J.T., Yeh, K.W. and Chan, M.T. (2010) Heterologous expression of taro cystatin protects transgenic tomato against *Meloidogyne incognita* infection by means of interfering sex determination and suppressing gall formation. *Plant Cell Reports* 29, 231–238. doi:10.1007/s00299-009-0815-y

Chanclud, E. and Morel, J.-B. (2016) Plant hormones: a fungal point of view. *Molecular Plant Pathology* 17, 1289–1297. doi:10.1111/mpp.12393

Chen, Z.X., Ricigliano, J.W. and Klessig, D.F. (1993) Purification and characterization of a soluble salicylic acid-binding protein from tobacco. *Proceedings of the National Academy of Sciences of the United States of America* 90, 9533–9537. doi:10.1073/pnas.90.20.9533

Chiwocha, S.D.S., Cutler, A.J., Abrams, S.R., Ambrose, S.J., Yang, J., Ross, A.R.S. and Kermode, A.R. (2005) The *etr1-2* mutation in *Arabidopsis thaliana* affects the abscisic acid, auxin, cytokinin and gibberellin metabolic pathways during maintenance of seed dormancy, moist-chilling and germination. *Plant Journal* 42, 35–48. doi:10.1111/j.1365-313X.2005.02359.x

Chong, K.P., Rossall, S. and Atong, M. (2009) In vitro synergy effect of syringic acid, caffeic acid and 4-hydroxybenzoic acid against *Ganoderma boninense*. *International Journal of Engineering and Technology* 1, 282–284.

Clark, T.N., Ellsworth, K., Li, H.X., Johnson, J.A. and Gray, C.A. (2013) Isolation of the plant hormone (+)-abscisic acid as an antimycobacterial constituent of the medicinal plant endophyte *Nigrospora* sp. *Natural Product Communications* 8, 1673–1674.

Cohen, A.C., Bottini, R. and Piccoli, P.N. (2008) *Azospirillum brasilense* Sp 245 produces ABA in chemically-defined culture medium and increases ABA content in *Arabidopsis* plants. *Plant Growth Regulation* 54, 97–103. doi:10.1007/s10725-007-9232-9

Cohen, A.C., Travaglia, C.N., Bottini, R. and Piccoli, P.N. (2009) Participation of abscisic acid and gibberellins produced by endophytic *Azospirillum* in the alleviation of drought effects in maize. *Botany* 87, 455–462. doi:10.1139/b09-023

Comai, L. and Kosuge, T. (1982) Cloning and characterization of *iaa*M, a virulence determinant of

Pseudomonas savastanoi. *Journal of Bacteriology* 149, 40–46.
Contreras-Cornejo, H.A., Macías-Rodríguez, L., Cortés-Penagos, C. and Lopéz-Bucio, J. (2009) *Trichoderma virens*, a plant beneficial fungus, enhances biomass production and promotes lateral root growth through an auxin-dependent mechanism in *Arabidopsis*. *Plant Physiology* 149, 1579–1592.
Corcuera, L.J., Woodward, M.D., Helgeson, J.P., Kelman, A. and Upper, C.D. (1978) 2,4-Dihydroxy-7-methoxy-2H-1,4-benzoxazin-3(4H)-one, an inhibitor from *Zea mays* with differential activity against soft rotting *Erwinia* species. *Plant Physiology* 61, 791–795. doi:10.1104/pp.61.5.791
Couture, R.M., Routley, D.G. and Dunn, G.M. (1971) Role of cyclic hydroxamic acids in monogenic resistance of maize to *Helminthosporium turcicum*. *Physiological Plant Pathology* 1, 515–521. doi:10.1016/0048-4059(71)90013-0
Cueva, C., Moreno-Arribas, M.V., Martín-Álvarez, P.J., Bills, G., Vicente, M.F., *et al.* (2010) Antimicrobial activity of phenolic acids against commensal, probiotic and pathogenic bacteria. *Research in Microbiology* 161, 372–382. doi:10.1016/j.resmic.2010.04.006
Dörffling, K. (1982) *Das Hormonsystem der Pflanzen*. Georg Thieme Verlag, Stuttgart, Germany/New York.
Duca, D., Lorv, J., Patten, C.L., Rose, D. and Glick, B.R. (2014) Indole-3-acetic acid in plant-microbe interactions. *Antonie Van Leeuwenhoek International Journal of General and Molecular Microbiology* 106, 85–125. doi:10.1007/s10482-013-0095-y
Du Fall, L.A. and Solomon, P.S. (20011) Role of cereal secondary metabolites involved in mediating the outcome of plant-pathogen interactions. *Metabolites* 1, 64–78. doi:10.3390/metabo1010064
Duijff, B.J., Pouhair, D., Olivain, C., Alabouvette, C. and Lemanceau, P. (1998) Implication of systemic induced resistance in the suppression of *Fusarium* wilt of tomato by *Pseudomonas fluorescens* WCS417r and by nonpathogenic *Fusarium oxysporum* Fo47. *European Journal of Plant Pathology* 104, 903–910.
Elmer, W.H., Mattina, M.J.I. and Maceachern, G.J. (1994) Sensitivity of plant-pathogenic fungi to taxane extracts from ornamental yews. *Phytopathology* 84, 1179–1185. doi:10.1094/Phyto-84-1179
Eschweiler, J., Messelink, G., Kasal, T., Main, D., Grundler, F. and Schouten, A. (2014) Integration of fungal endophytes in a greenhouse environment to control insect pests. In: *59. Deutsche Pflanzenschutztagung*, Freiburg, Germany. Julius Kühn Archiv.
Escudero, N. and Lopez-Llorca, L.V. (2012) Effects on plant growth and root-knot nematode infection of an endophytic GFP transformant of the nematophagous fungus *Pochonia chlamydosporia*. *Symbiosis* 57, 33–42. doi:10.1007/s13199-012-0173-3
Etemadi, M., Gutjahr, C., Couzigou, J.M., Zouine, M., Lauressergues, D., *et al.* (2014) Auxin perception is required for arbuscule development in arbuscular mycorrhizal symbiosis. *Plant Physiology* 166, 281–292. doi:10.1104/pp.114.246595
Etesami, H., Hosseini, H.M., Alikhani, H.A. and Mohammadi, L. (2014) Bacterial biosynthesis of 1-aminocyclopropane-1-carboxylate (ACC) deaminase and indole-3-acetic acid (IAA) as endophytic preferential selection traits by rice plant seedlings. *Journal of Plant Growth Regulation* 33, 654–670. doi:10.1007/s00344-014-9415-3
Eyberger, A.L., Dondapati, R. and Porter, J.R. (2006) Endophyte fungal isolates from *Podophyllum peltatum* produce podophyllotoxin. *Journal of Natural Products* 69, 1121–1124. doi:10.1021/np060174f
Feeny, P. (1992) The evolution of chemical ecology: Contributions from the study of herbivorous insects. In: Rosenthal, G.A. and Berenbaum, M.R. (eds) *Herbivores – Their Interactions with Secondary Plant Metabolites*, 2nd edn. Academic Press, Inc., San Diego, California.
Finkelstein, R.R. and Zeevaart, J.A.D. (1994) Gibbereilin and abscisic acid response. In: Meyerowitz, E.M. and Somerville, C.R. (eds) *Arabidopsis*. Cold Spring Harbor Laboratory Press, Cold Spring Harbor, New York.
Foo, E. and Reid, J.B. (2013) Strigolactones: New physiological roles for an ancient signal. *Journal of Plant Growth Regulation* 32, 429–442. doi:10.1007/s00344-012-9304-6
Forchetti, G., Masciarelli, O., Alemano, S., Alvarez, D. and Abdala, G. (2007) Endophytic bacteria in sunflower (*Helianthus annuus* L.): Isolation, characterization, and production of jasmonates and abscisic acid in culture medium. *Applied Microbiology and Biotechnology* 76, 1145–1152. doi:10.1007/s00253-007-1077-7
Fraenkel, G.S. (1959) Raison d'être of secondary plant substances. *Science* 129, 1466–1470. doi:10.1126/science.129.3361.1466
Fuchs, J.-G., Moënne-Loccoz, Y. and Défago, G. (1997) Nonpathogenic *Fusarium oxysporum* strain Fo47 induces resistance to *Fusarium* wilt in tomato. *Plant Disease* 81, 492–496.
Fuentes-Ramirez, L.E., Jimenez-Salgado, T., Abarca-Ocampo, I.R. and Caballero-Mellado, J. (1993) *Acetobacter diazotrophicus*, an indole

aceticacid producing bacterium isolated from sugarcane cultivars of México. *Plant and Soil* 154, 145–150. doi:10.1007/bf00012519

Furukawa, T., Koga, J., Adachi, T., Kishi, K. and Syono, K. (1996) Efficient conversion of L-tryptophan to indole-3-acetic acid and/or tryptophol by some species of *Rhizoctonia*. *Plant and Cell Physiology* 37, 899–905. doi:10.1093/oxfordjournals.pcp.a029037

Gao, R., Gao, C.F., Tian, X., Yu, X.Y., Di, X.D., Xiao, H. and Zhang, X. (2004) Insecticidal activity of deoxypodophyllotoxin, isolated from *Juniperus sabina* L, and related lignans against larvae of *Pieris rapae* L. *Pest Management Science* 60, 1131–1136. doi:10.1002/ps.922

Gay, G., Normand, L., Marmeisse, R., Sotta, B. and Debaud, J.C. (1994) Auxin overproducer mutants of *Hebeloma cylindrosporum* Romagnesi have increased mycorrhizal activity. *New Phytologist* 128, 645–657. doi:10.1111/j.1469-8137.1994.tb04029.x

Gianoli, E. and Niemeyer, H.M. (1998) DIBOA in wild Poaceae: Sources of resistance to the Russian wheat aphid (*Diuraphis noxia*) and the greenbug (*Schizaphis graminum*). *Euphytica* 102, 317–321. doi:10.1023/a:1018323121213

Glass, N.L. and Kosuge, T. (1986) Cloning of the gene for indoleacetic acid-lysine synthetase from *Pseudomonas syringae* subsp *savastanoi*. *Journal of Bacteriology* 166, 598–603. doi:10.1128/jb.166.2.598-603.1986

Glenn, A.E., Meredith, F.I., Morrison, W.H. and Bacon, C.W. (2003) Identification of intermediate and branch metabolites resulting from biotransformation of 2-benzoxazolinone by *Fusarium verticillioides*. *Applied and Environmental Microbiology* 69, 3165–3169. doi:10.1128/aem.69.6.3165-3169.2003

Glickmann, E., Gardan, L., Jacquet, S., Hussain, S., Elasri, M., Petit, A. and Dessaux, Y. (1998) Auxin production is a common feature of most pathovars of *Pseudomonas syringae*. *Molecular Plant-Microbe Interactions* 11, 156–162. doi:10.1094/mpmi.1998.11.2.156

Grossmann, K. and Hansen, H. (2001) Ethylene-triggered abscisic acid: A principle in plant growth regulation? *Physiologia Plantarum* 113, 9–14. doi:10.1034/j.1399-3054.2001.1130102.x

Gruen, H.E. (1959) Auxins and fungi. *Annual Review of Plant Physiology and Plant Molecular Biology* 10, 405–440. doi:10.1146/annurev.pp.10.060159.002201

Gulmon, S.L. and Mooney, H.A. (1986) *Costs of Defense and Their Effects on Plant Productivity*. Cambridge University Press, Cambridge, UK.

Guthrie, W.D., Wilson, R.L., Coats, J.R., Robbins, J.C., Tseng, C.T., Jarvis, J.L. and Russell, W.A. (1986) European corn-borer (Lepidoptera, Pyralidae) leaf-feeding resistance and DIMBOA content in inbred lines of dent maize grown under field versus greenhouse conditions. *Journal of Economic Entomology* 79, 1492–1496.

Hamberger, B., Ohnishi, T., Hamberger, B., Séguin, A. and Bohlmann, J. (2011) Evolution of diterpene metabolism: Sitka spruce CYP720B4 catalyzes multiple oxidations in resin acid biosynthesis of conifer defense against insects. *Plant Physiology* 157, 1677–1695. doi:10.1104/pp.111.185843

Hanlon, M.T. and Coenen, C. (2011) Genetic evidence for auxin involvement in arbuscular mycorrhiza initiation. *New Phytologist* 189, 701–709. doi:10.1111/j.1469-8137.2010.03567.x

Harborne, J.B. (1997) Plant secondary metabolism. In: Crawley, M.J. (ed.) *Plant Ecology*. Blackwell Science Ltd, Oxford, UK.

Hartley, S.E. and Jones, C.G. (1997) Plant chemistry and herbivory, or why the world is so green. In: Crawley, M.J. (ed.) *Plant Ecology*, 2nd edn. Blackwell Science Ltd, Oxford, UK.

Hartman, J.R., Kelman, A. and Upper, C.D. (1975) Differential inhibitory activity of a corn extract to *Erwinia* spp. causing soft rot. *Phytopathology* 65, 1082–1088.

Hartmann, T. (2007) From waste products to ecochemicals: Fifty years research of plant secondary metabolism. *Phytochemistry* 68, 2831–2846. doi:10.1016/j.phytochem.2007.09.017

Heitefuss, R. (1997) Cell wall modifications in relation to resistance. In: Hartleb, H., Heitefuss, R. and Hoppe, H.H. (eds) *Resistance of Crop Plants against Fungi*. Gustav Fischer Verlag, Stuttgart, Germany.

Herms, D.A. and Mattson, W.J. (1992) The dilemma of plants: To grow or defend. *Quarterly Review of Biology* 67, 283–335. doi:10.1086/417659

Hinsch, J., Vrabka, J., Oeser, B., Novák, O., Galuszka, P. and Tudzynski, P. (2015) De novo biosynthesis of cytokinins in the biotrophic fungus *Claviceps purpurea*. *Environmental Microbiology* 17, 2935–2951. doi:10.1111/1462-2920.12838

Hirsch, A.M. and Fang, Y.W. (1994) Plant hormones and nodulation: What's the connection? *Plant Molecular Biology* 26, 5–9. doi:10.1007/bf00039514

Hölscher, D., Dhakshinamoorthy, S., Alexandrov, T., Becker, M., Bretschneider, T., *et al.* (2014) Phenalenone-type phytoalexins mediate resistance of banana plants (*Musa* spp.) to the burrowing nematode *Radopholus similis*. *Proceedings of the National Academy of Sciences of the United States of America* 111, 105–110. doi:10.1073/pnas.1314168110

Horváth, E., Pál, M., Szalai, G., Páldi, E. and Janda, T. (2007) Exogenous 4-hydroxybenzoic acid and salicylic acid modulate the effect of short-term drought and freezing stress on wheat plants. *Biologia Plantarum* 51, 480–487. doi:10.1007/s10535-007-0101-1

Ioannou, Y.M., Dauterman, W.C. and Tucker, W.P. (1980) Degradation of diazinon by 2,4-dihydroxy-7-methoxy-2H-1,4-benzoxazin-3(4H)-one in maize. *Phytochemistry* 19, 1607–1611. doi:10.1016/s0031-9422(00)83779-6

Izhaki, I. (2002) Emodin – A secondary metabolite with multiple ecological functions in higher plants. *New Phytologist* 155, 205–217. doi:10.1046/j.1469-8137.2002.00459.x

Jørgensen, K., Rasmussen, A.V., Morant, M., Nielsen, A.H., Bjarnholt, N., *et al.* (2005) Metabolon formation and metabolic channeling in the biosynthesis of plant natural products. *Current Opinion in Plant Biology* 8, 280–291. doi:10.1016/j.pbi.2005.03.014

Kamaya, Y., Tsuboi, S., Takada, T. and Suzuki, K. (2006) Growth stimulation and inhibition effects of 4-hydroxybenzoic acid and some related compounds on the freshwater green alga *Pseudokirchneriella subcapitata*. *Archives of Environmental Contamination and Toxicology* 51, 537–541. doi:10.1007/s00244-005-0320-4

Khadem, S. and Marles, R.J. (2010) Monocyclic phenolic acids; hydroxy- and polyhydroxybenzoic acids: Occurrence and recent bioactivity studies. *Molecules* 15, 7985–8005. doi:10.3390/molecules15117985

King, D.J., Gleadow, R.M. and Woodrow, I.E. (2006) The accumulation of terpenoid oils does not incur a growth cost in *Eucalyptus polybractea* seedlings. *Functional Plant Biology* 33, 497–505. doi:10.1071/fp05304

Klun, J.A. and Brindley, T.A. (1966) Role of 6-methoxybenzoxazolinone in inbred resistance of host plant (maize) to first-brood larvae of European corn borer. *Journal of Economic Entomology* 59, 711–718.

Klun, J.A. and Robinson, J.F. (1969) Concentration of 2 1,4-benzoxazinones in dent corn at various stages of development of plant and its relation to resistance of host plant to European corn borer. *Journal of Economic Entomology* 62, 214–220.

Kremer, R.J. and Ben-Hammouda, M. (2009) Allelopathic plants. 19. Barley (*Hordeum vulgare* L.). *Allelopathy Journal* 24, 225–242.

Krings, M., Taylor, T.N., Hass, H., Kerp, H., Dotzler, N. and Hermsen, E.J. (2007) Fungal endophytes in a 400-million-yr-old land plant: Infection pathways, spatial distribution, and host responses. *New Phytologist* 174, 648–657. doi:10.1111/j.1469-8137.2007.02008.x

Kucht, S., Gross, J., Hussein, Y., Grothe, T., Keller, U., *et al.* (2004) Elimination of ergoline alkaloids following treatment of *Ipomoea asarifolia* (Convolvulaceae) with fungicides. *Planta* 219, 619–625. doi:10.1007/s00425-004-1261-2

Kusari, S., Lamshöft, M., Zühlke, S. and Spiteller, M. (2008) An endophytic fungus from *Hypericum perforatum* that produces hypericin. *Journal of Natural Products* 71, 159–162. doi:10.1021/np070669k

Kusari, S., Lamshöft, M. and Spiteller, M. (2009a) *Aspergillus fumigatus* Fresenius, an endophytic fungus from *Juniperus communis* L. Horstmann as a novel source of the anticancer pro-drug deoxypodophyllotoxin. *Journal of Applied Microbiology* 107, 1019–1030. doi:10.1111/j.1365-2672.2009.04285.x

Kusari, S., Zühlke, S., Košuth, J., Cellárová, E. and Spiteller, M. (2009b) Light-independent metabolomics of endophytic *Thielavia subthermophila* provides insight into microbial hypericin biosynthesis. *Journal of Natural Products* 72, 1825–1835. doi:10.1021/np9002977

Kusari, S., Zühlke, S. and Spiteller, M. (2011) Effect of artificial reconstitution of the interaction between the plant *Camptotheca acuminata* and the fungal endophyte *Fusarium solani* on camptothecin biosynthesis. *Journal of Natural Products* 74, 764–775. doi:10.1021/np1008398

Kusari, S., Verma, V.C., Lamshoeft, M. and Spiteller, M. (2012) An endophytic fungus from *Azadirachta indica* A. Juss. that produces azadirachtin. *World Journal of Microbiology and Biotechnology* 28, 1287–1294. doi:10.1007/s11274-011-0876-2

Kusari, S., Pandey, S.P. and Spiteller, M. (2013) Untapped mutualistic paradigms linking host plant and endophytic fungal production of similar bioactive secondary metabolites. *Phytochemistry* 91, 81–87. doi:10.1016/j.phytochem.2012.07.021

Kyndt, T., Fernandez, D. and Gheysen, G. (2014) Plant-parasitic nematode infections in rice: Molecular and cellular insights. *Annual Review of Phytopathology* 52, 135–153. doi:10.1146/annurev-phyto-102313-050111

Kyndt, T., Nahar, K., Haeck, A., Verbeek, R., Demeestere, K. and Gheysen, G. (2017) Interplay between carotenoids, abscisic acid and jasmonate guides the compatible rice-*Meloidogyne graminicola* interaction. *Frontiers in Plant Science* 8, 951. doi:10.3389/fpls.2017.00951

Labandeira, C.C. and Prevec, R. (2014) Plant paleopathology and the roles of pathogens and insects. *International Journal of Paleopathology* 4, 1–16. doi:10.1016/j.ijpp.2013.10.002

Larriba, E., Jaime, M.D., Nislow, C., Martin-Nieto, J. and Lopez-Llorca, L.V. (2015) Endophytic colonization of barley (*Hordeum vulgare*) roots by the nematophagous fungus *Pochonia chlamydosporia* reveals plant growth promotion and a general defense and stress transcriptomic response. *Journal of Plant Research* 128, 665–678. doi:10.1007/s10265-015-0731-x

Lattanzio, V., De Cicco, V., Di Venere, D., Lima, G. and Salerno, M. (1994) Antifungal activity of phenolics against fungi commonly encountered during storage. *Italian Journal of Food Science* 1, 23–30.

Laurans, F., Pepin, R. and Gay, G. (2001) Fungal auxin overproduction affects the anatomy of *Hebeloma cylindrosporum-Pinus pinaster* ectomycorrhizas. *Tree Physiology* 21, 533–540. doi:10.1093/treephys/21.8.533

Le, H.T.T., Padgham, J.L., Hagemann, M.H., Sikora, R. and Schouten, A. (2016) Developmental and behavioral effects of the endophytic *Fusarium moniliforme* Fe14 toward *Meloidogyne graminicola* in rice. *Annals of Applied Biology* 169, 134–143.

Leinhos, V. and Vacek, O. (1994) Biosynthesis of auxins by phosphate-solubilizing rhizobacteria from wheat (*Triticum aestivum*) and rye (*Secale cereale*). *Microbiological Research* 149, 31–35.

Long, B.J., Dunn, G.M. and Routley, D.G. (1978) Relationship of hydroxamate concentration in maize and field reaction to *Helminthosporium turcicum*. *Crop Science* 18, 573–575.

Loper, J.E. and Schroth, M.N. (1986) Influence of bacterial sources of indole-3-acetic-acid on root elongation of sugar-beet. *Phytopathology* 76, 386–389. doi:10.1094/Phyto-76-386

Ludwig-Müller, J. (2015) Plants and endophytes: Equal partners in secondary metabolite production? *Biotechnology Letters* 37, 1325–1334. doi:10.1007/s10529-015-1814-4

Luis, J.G., Fletcher, W.Q., Echeverri, F. and Grillo, T.A. (1994) Phenalenone-type phytoalexins from *Musa acuminata* synthesis of 4-phenyl-phenalenones. *Tetrahedron* 50, 10963–10970. doi:10.1016/S0040-4020(01)85707-0

Makowska, B., Bakera, B. and Rakoczy-Trojanowska, M. (2015) The genetic background of benzoxazinoid biosynthesis in cereals. *Acta Physiologiae Plantarum* 37, 1–12. doi:10.1007/s11738-015-1927-3

Maor, R., Haskin, S., Levi-Kedmi, H. and Sharon, A. (2004) In planta production of indole-3-acetic acid by *Colletotrichum gloeosporioides* f. sp *aeschynomene*. *Applied and Environmental Microbiology* 70, 1852–1854. doi:10.1128/aem.70.3.1852-1854.2004

Martin, C., Schnee, S., Lintz, J., Vogelgsang, S., Gindro, K., Mauch-Mani, B. and Mascher, F. (2017) Role of ferulic acid in *Fusarium* head blight resistance of wheat spikes. In: *Future IPM 3.0 Towards a Sustainable Agriculture*, 15–20 October 2017, Riva del Garda, Italy. IOBC-WPRS.

Martinuz, A., Schouten, A. and Sikora, R.A. (2013) Post-infection development of *Meloidogyne incognita* on tomato treated with the endophytes *Fusarium oxysporum* strain Fo162 and *Rhizobium etli* strain G12. *BioControl* 58, 95–104. doi:10.1007/s10526-012-9471-1

Martinuz, A., Zewdu, G., Ludwig, N., Grundler, F., Sikora, R.A. and Schouten, A. (2015) The application of *Arabidopsis thaliana* in studying tripartite interactions among plants, beneficial fungal endophytes and biotrophic plant-parasitic nematodes. *Planta* 241, 1015–25. doi:10.1007/s00425-014-2237-5

Mazzola, M. and White, F.F. (1994) A mutation in the indole-3-acetic-acid biosynthesis pathway of *Pseudomonas syringae* pv *syringae* affects growth in *Phaseolus vulgaris* and syringomycin production. *Journal of Bacteriology* 176, 1374–1382. doi:10.1128/jb.176.5.1374-1382.1994

McKeehen, J.D., Busch, R.H. and Fulcher, R.G. (1999) Evaluation of wheat (*Triticum aestivum* L.) phenolic acids during grain development and their contribution to *Fusarium* resistance. *Journal of Agricultural and Food Chemistry* 47, 1476–1482. doi:10.1021/jf980896f

McSteen, P. (2010) Auxin and monocot development. *Cold Spring Harbor Perspectives in Biology* 2, a001479. doi:10.1101/cshperspect.a001479

Meixner, C., Ludwig-Müller, J., Miersch, O., Gresshoff, P., Staehelin, C. and Vierheilig, H. (2005) Lack of mycorrhizal autoregulation and phytohormonal changes in the supernodulating soybean mutant nts1007. *Planta* 222, 709–715. doi:10.1007/s00425-005-0003-4

Mu, J.H., Bollon, A.P. and Sidhu, R.S. (1999) Analysis of β-tubulin cDNAs from taxol-resistant *Pestalotiopsis microspora* and taxol-sensitive *Pythium ultimum* and comparison of the taxol-binding properties of their products. *Molecular and General Genetics* 262, 857–868. doi:10.1007/s004380051151

Müller, M., Deigele, C. and Ziegler, H. (1989) Hormonal interactions in the rhizosphere of maize (*Zea mays* L.) and their effects on plant development. *Zeitschrift für Pflanzenernährung und Bodenkunde* 152, 247–254. doi:10.1002/jpln.19891520217

Mutka, A.M., Fawley, S., Tsao, T. and Kunkel, B.N. (2013) Auxin promotes susceptibility to *Pseudomonas syringae* via a mechanism independent of suppression of salicylic acid-mediated defenses. *Plant Journal* 74, 746–754. doi:10.1111/tpj.12157

Neilson, E.H., Goodger, J.Q.D., Woodrow, I.E. and Møller, B.L. (2013) Plant chemical defense: At what cost? *Trends in Plant Science* 18, 250–258. doi:10.1016/j.tplants.2013.01.001

Niemeyer, H.M. (1988) Hydroxamic acids (4-hydroxy-1,4-benzoxazin-3-ones), defense chemicals in the Gramineae. *Phytochemistry* 27, 3349–3358. doi:10.1016/0031-9422(88)80731-3

Niemeyer, H.M. (2009) Hydroxamic acids derived from 2-hydroxy-2H-1,4-benzoxazin-3(4H)-one: Key defense chemicals of cereals. *Journal of Agricultural and Food Chemistry* 57, 1677–1696. doi:10.1021/jf8034034

Olivain, C., Trouvelot, S., Binet, M.N., Cordier, C., Pugin, A. and Alabouvette, C. (2003) Colonization of flax roots and early physiological responses of flax cells inoculated with pathogenic and non-pathogenic strains of *Fusarium oxysporum*. *Applied and Environmental Microbiology* 69, 5453–5462.

Ou, S.Y. and Kwok, K.C. (2004) Ferulic acid: Pharmaceutical functions, preparation and applications in foods. *Journal of the Science of Food and Agriculture* 84, 1261–1269. doi:10.1002/jsfa.1873

Papadopoulou, J. and Triantaphyllou, A.C. (1982) Sex-differentiation in *Meloidogyne incognita* and anatomical evidence of sex reversal. *Journal of Nematology* 14, 549–566.

Parniske, M. (2008) Arbuscular mycorrhiza: The mother of plant root endosymbioses. *Nature Reviews Microbiology* 6, 763–775.

Peach, K. (1950) *Biochemie und Physiologie der Sekundären Pflanzenstoffe*. Springer, Berlin/Heidelberg.

Pérez, F.J. and Ormeño-Núñez, J. (1993) Weed growth interference from temperate cereals: The effect of a hydroxamic-acids-exuding rye (*Secale cereale* L.) cultivar. *Weed Research* 33, 115–119. doi:10.1111/j.1365-3180.1993.tb01924.x

Přikryl, Z., Vančura, V. and Wurst, M. (1985) Auxin formation by rhizosphere bacteria as a factor of root-growth. *Biologia Plantarum* 27, 159–163. doi:10.1007/bf02902155

Puri, S.C., Verma, V., Amna, T., Qazi, G.N. and Spiteller, M. (2005) An endophytic fungus from *Nothapodytes foetida* that produces camptothecin. *Journal of Natural Products* 68, 1717–1719. doi:10.1021/np0502802

Puri, S.C., Nazir, A., Chawla, R., Arora, R., Riyaz-ul-Hasan, S., *et al.* (2006) The endophytic fungus *Trametes hirsuta* as a novel alternative source of podophyllotoxin and related aryl tetralin lignans. *Journal of Biotechnology* 122, 494–510. doi:10.1016/j.jbiotec.2005.10.015

Radulović, N., Stankov-Jovanović, V., Stojanović, G., Šmelcerović, A., Spiteller, M. and Asakawa, Y. (2007) Screening of *in vitro* antimicrobial and antioxidant activity of nine *Hypericum* species from the Balkans. *Food Chemistry* 103, 15–21. doi:10.1016/j.foodchem.2006.05.062

Reddy, M.V.B., Arul, J., Angers, P. and Couture, L. (1999) Chitosan treatment of wheat seeds induces resistance to *Fusarium graminearum* and improves seed quality. *Journal of Agricultural and Food Chemistry* 47, 1208–1216.

Redecker, D., Kodner, R. and Graham, L.E. (2000) Glomalean fungi from the Ordovician. *Science* 289, 1920–1921.

Redman, R.S., Kim, Y.O., Woodward, C.J., Greer, C., Espino, L., Doty, S.L. and Rodriguez, R.J. (2011) Increased fitness of rice plants to abiotic stress via habitat adapted symbiosis: A strategy for mitigating impacts of climate change. *PLOS ONE* 6, e14823. doi:10.1371/journal.pone.0014823

Rehman, S., Shawl, A.S., Kour, A., Andrabi, R., Sudan, P., *et al.* (2008) An endophytic *Neurospora* sp. from *Nothapodytes foetida* producing camptothecin. *Applied Biochemistry and Microbiology* 44, 203–209. doi:10.1134/s0003683808020130

Reineke, G., Heinze, B., Schirawski, J., Buettner, H., Kahmann, R. and Basse, C.W. (2008) Indole-3-acetic acid (IAA) biosynthesis in the smut fungus *Ustilago maydis* and its relevance for increased IAA levels in infected tissue and host tumour formation. *Molecular Plant Pathology* 9, 339–355. doi:10.1111/j.1364-3703.2008.00470.x

Remy, W., Taylor, T.N., Hass, H. and Kerp, H. (1994) Four hundred-million-year-old vesicular-arbuscular mycorrhizae. *Proceedings of the National Academy of Sciences of the United States of America* 91, 11841–11843. doi:10.1073/pnas.91.25.11841

Reznik, H. (1960) Vergleichende Biochemie der Phenylpropane. In: Autrum, H. (ed.) *Ergebnisse der Biologie*. Springer, Berlin/Heidelberg.

Robinson, M., Riov, J. and Sharon, A. (1998) Indole-3-acetic acid biosynthesis in *Colletotrichum gloeosporioides* f. sp. *aeschynomene*. *Applied and Environmental Microbiology* 64, 5030–5032.

Rodriguez, R. and Redman, R. (2008) More than 400 million years of evolution and some plants still can't make it on their own: Plant stress tolerance via fungal symbiosis. *Journal of Experimental Botany* 59, 1109–1114.

Rodriguez, R.J., Henson, J., Van Volkenburgh, E., Hoy, M., Wright, L., *et al.* (2008) Stress tolerance in plants via habitat-adapted symbiosis. *ISME Journal* 2, 404–416.

Rowinsky, E.K., Cazenave, L.A. and Donehower, R.C. (1990) Taxol: A novel investigational antimicrotubule agent. *Journal of the National Cancer Institute* 82, 1247–1259. doi:10.1093/jnci/82.15.1247

Sachan, A., Ghosh, S. and Mitra, A. (2006) Biotransformation of p-coumaric acid by *Paecilomyces variotii*. *Letters in Applied Microbiology* 42, 35–41. doi:10.1111/j.1472-765X.2005.01801.x

Sachan, A., Ghosh, S. and Mitra, A. (2010) Transforming p-coumaric acid into p-hydroxybenzoic acid by the mycelial culture of a white rot fungus *Schizophyllum commune*. *African Journal of Microbiology Research* 4, 267–273.

Sannino, F., Sansone, C., Galasso, C., Kildgaard, S., Tedesco, P., *et al.* (2018) *Pseudoalteromonas haloplanktis* TAC125 produces 4-hydroxybenzoic acid that induces pyroptosis in human A459 lung adenocarcinoma cells. *Scientific Reports* 8, 1190. doi:10.1038/s41598-018-19536-2

Schiff, P.B., Fant, J. and Horwitz, S.B. (1979) Promotion of microtubule assembly *in vitro* by taxol. *Nature* 277, 665–667. doi:10.1038/277665a0

Schouten, A. (2016) Mechanisms involved in nematode control by endophytic fungi. *Annual Review of Phytopathology* 54, 121–142.

Schouten, A. (2018) Natural antifungal compounds for the control of diseases in wheat and other cereals. In: Oliver, R. (ed.) *Integrated Disease Management of Wheat and Barley*. Burleigh Dodds Science Publishing Ltd, Cambridge, UK. doi:10.19103/AS.2018.0039.22.

Schulz, M., Marocco, A., Tabaglio, V., Macias, F.A. and Molinillo, J.M.G. (2013) Benzoxazinoids in rye allelopathy: From discovery to application in sustainable weed control and organic farming. *Journal of Chemical Ecology* 39, 154–174. doi:10.1007/s10886-013-0235-x

Seigler, D. and Price, P.W. (1976) Secondary compounds in plants: Primary functions. *American Naturalist* 110, 101–105. doi:10.1086/283050

Seigler, D.S. (1977) Primary roles for secondary compounds. *Biochemical Systematics and Ecology* 5, 195–199. doi:10.1016/0305-1978(77)90004-7

Sharp, R.E. and LeNoble, M.E. (2002) ABA, ethylene and the control of shoot and root growth under water stress. *Journal of Experimental Botany* 53, 33–37. doi:10.1093/jexbot/53.366.33

Shimabukuro, R.H. (1967) Atrazine metabolism and herbicidal selectivity. *Plant Physiology* 42, 1269–1276. doi:10.1104/pp.42.9.1269

Shoresh, M., Yedidia, I. and Chet, I. (2005) Involvement of jasmonic acid/ethylene signaling pathway in the systemic resistance induced in cucumber by *Trichoderma asperellum* T203. *Phytopathology* 95, 76–84. doi:10.1094/PHYTO-95-0076

Shrestha, K., Strobel, G.A., Shrivastava, S.P. and Gewali, M.B. (2001) Evidence for paclitaxel from three new endophytic fungi of Himalayan yew of Nepal. *Planta Medica* 67, 374–376. doi:10.1055/s-2001-14307

Shweta, S., Zuehlke, S., Ramesha, B.T., Priti, V., Kumar, P.M., *et al.* (2010) Endophytic fungal strains of *Fusarium solani*, from *Apodytes dimidiata* E. Mey. ex Arn (Icacinaceae) produce camptothecin, 10-hydroxycamptothecin and 9-methoxycamptothecin. *Phytochemistry* 71, 117–122.doi:10.1016/j.phytochem.2009.09.030

Smidt, M. and Kosuge, T. (1978) The role of indole-3-acetic acid accumulation by alpha methyl tryptophan-resistant mutants of *Pseudomonas savastanoi* in gall formation on oleanders. *Physiological Plant Pathology* 13, 203–210. doi:10.1016/0048-059(78)90035-8

Smith-Becker, J., Marois, E., Huguet, E.J., Midland, S.L., Sims, J.J. and Keen, N.T. (1998) Accumulation of salicylic acid and 4-hydroxybenzoic acid in phloem fluids of cucumber during systemic acquired resistance is preceded by a transient increase in phenylalanine ammonia-lyase activity in petioles and stems. *Plant Physiology* 116, 231–238. doi:10.1104/pp.116.1.231

Soliman, S.S.M., Tsao, R. and Raizada, M.N. (2011) Chemical inhibitors suggest endophytic fungal paclitaxel is derived from both mevalonate and non-mevalonate-like pathways. *Journal of Natural Products* 74, 2497–2504. doi:10.1021/np200303v

Soliman, S.S.M., Greenwood, J.S., Bombarely, A., Mueller, L.A., Tsao, R., Mosser, D.D. and Raizada, M.N. (2015) An endophyte constructs fungicide-containing extracellular barriers for its host plant. *Current Biology* 25, 2570–2576. doi:10.1016/j.cub.2015.08.027

Søltoft, M., Jorgensen, L.N., Svensmark, B. and Fomsgaard, I.S. (2008) Benzoxazinoid concentrations show correlation with Fusarium Head Blight resistance in Danish wheat varieties. *Biochemical Systematics and Ecology* 36, 245–259. doi:10.1016/j.bse.2007.10.008

Spaepen, S., Vanderleyden, J. and Remans, R. (2007) Indole-3-acetic acid in microbial and microorganism-plant signaling. *FEMS Microbiology Reviews* 31, 425–448. doi:10.1111/j.1574-6976.2007.00072.x

Stahl, E. (1888) Pflanzen und Schnecken: Biologische Studien über Schutzmittel der Pflanzen gegen Schneckenfrass. *Jenaer Zeitschrift für Medizin und Naturwissenschaften* 22, 557–684.

Staniek, A., Woerdenbag, H.J. and Kayser, O. (2009) *Taxomyces andreanae*: A presumed paclitaxel producer demystified? *Planta Medica* 75, 1561–1566. doi:10.1055/s-0029-1186181

Steiner, U., Ahimsa-Müller, M.A., Markert, A., Kucht, S., Gross, J., *et al.* (2006) Molecular characterization of a seed transmitted clavicipitaceous fungus occurring on dicotyledoneous plants

(Convolvulaceae). *Planta* 224, 533–544. doi:10.1007/s00425-006-0241-0

Stierle, A., Strobel, G. and Stierle, D. (1993) Taxol and taxane production by *Taxomyces andreanae*, an endophytic fungus of Pacific yew. *Science* 260, 214–216. doi:10.1126/science.8097061

Tan, J.W., Bednarek, P., Liu, H.K., Schneider, B., Svatoš, A. and Hahlbrock, K. (2004) Universally occurring phenylpropanoid and species-specific indolic metabolites in infected and uninfected *Arabidopsis thaliana* roots and leaves. *Phytochemistry* 65, 691–699. doi:10.1016/j.phytochem.2003.12.009

Tan, R.X. and Zou, W.X. (2001) Endophytes: A rich source of functional metabolites. *Natural Product Reports* 18, 448–459. doi:10.1039/b100918o

Taylor, T.N. and Krings, M. (2005) Fossil microorganisms and land plants: Associations and interactions. *Symbiosis* 40, 119–135.

Taylor, T.N., Remy, W., Hass, H. and Kerp, H. (1995) Fossil arbuscular mycorrhizae from the early Devonian. *Mycologia* 87, 560–573. doi:10.2307/3760776

Taylor, T.N., Hass, H., Kerp, H., Krings, M. and Hanlin, R.T. (2005) Perithecial ascomycetes from the 400 million year old Rhynie chert: An example of ancestral polymorphism. *Mycologia* 97, 269–285.

Tewari, R.K., Bachmann, G. and Hadacek, F. (2015) Iron in complex with the alleged phytosiderophore 8-hydroxyquinoline induces functional iron deficiency and non-autolytic programmed cell death in rapeseed plants. *Environmental and Experimental Botany* 109, 151–160. doi:10.1016/j.envexpbot.2014.07.016

Thackray, D.J., Wratten, S.D., Edwards, P.J. and Niemeyer, H.M. (1990) Resistance to the aphids *Sitobion avenae* and *Rhopalosiphum padi* in Gramineae in relation to hydroxamic acid levels. *Annals of Applied Biology* 116, 573–582. doi:10.1111/j.1744-7348.1990.tb06640.x

Tsavkelova, E.A., Klimova, S.Y., Cherdyntseva, T.A. and Netrusov, A.I. (2006) Microbial producers of plant growth stimulators and their practical use: A review. *Applied Biochemistry and Microbiology* 42, 117–126. doi:10.1134/s0003683806020013

Tsavkelova, E.A., Oeser, B., Oren-Young, L., Israeli, M., Sasson, Y., Tudzynski, B. and Sharon, A. (2012) Identification and functional characterization of indole-3-acetamide-mediated IAA biosynthesis in plant-associated *Fusarium* species. *Fungal Genetics and Biology* 49, 48–57. doi:10.1016/j.fgb.2011.10.005

Tudzynski, B. and Sharon, A. (2002) Biosynthesis, biological role and application of fungal phytohormones. In: Osiewacz, H.D. (ed.) *The Mycota (A Comprehensive Treatise on Fungi as Experimental Systems for Basic and Applied Research): Industrial Applications*. Springer, Berlin/Heidelberg. doi:10.1007/978-3-662-10378-4_9

Ulmasov, T., Hagen, G. and Guilfoyle, T. (1994) The *ocs* element in the soybean *GH2/4* promoter is activated by both active and inactive auxin and salicylic-acid analogs. *Plant Molecular Biology* 26, 1055–1064. doi:10.1007/bf00040688

Viitanen, P.V., Devine, A.L., Khan, M.S., Deuel, D.L., Van Dyk, D.E. and Daniell, H. (2004) Metabolic engineering of the chloroplast genome using the *Echerichia coli ubi*C gene reveals that chorismate is a readily abundant plant precursor for p-hydroxybenzoic acid biosynthesis. *Plant Physiology* 136, 4048–4060. doi:10.1104/pp.104.050054

Virtanen, A.I. and Hietala, P.K. (1955) 2(3)-Benzoxazolinone, an anti-Fusarium factor in rye seedlings. *Acta Chemica Scandinavica* 9, 1543–1544. doi:10.3891/acta.chem.scand.09-1543b

Wahlroos, O. and Virtanen, A.I. (1959) The precursors of 6-methoxy-benzoxazolinone in maize and wheat plants, their isolation and some of their properties. *Acta Chemica Scandinavica* 13, 1906–1908. doi:10.3891/acta.chem.scand.13-1906

Wang, D., Pajerowska-Mukhtar, K., Culler, A.H. and Dong, X.N. (2007) Salicylic acid inhibits pathogen growth in plants through repression of the auxin signaling pathway. *Current Biology* 17, 1784–1790. doi:10.1016/j.cub.2007.09.025

Wang, J.F., Li, G.L., Lu, H.Y., Zheng, Z.H., Huang, X.J. and Su, W.J. (2000) Taxol from *Tubercularia* sp strain TF5, an endophytic fungus of *Taxus mairei*. *FEMS Microbiology Letters* 193, 249–253. doi:10.1111/j.1574-6968.2000.tb09432.x

Wasilewska, A., Vlad, F., Sirichandra, C., Redko, Y., Jammes, F., *et al.* (2008) An update on abscisic acid signaling in plants and more.... *Molecular Plant* 1, 198–217. doi:10.1093/mp/ssm022

Whittaker, R.H. and Feeny, P.P. (1971) Allelochemics: Chemical interactions between species. *Science* 171, 757–770. doi:10.1126/science.171.3973.757

Wink, M. (1998) Chemical ecology of alkaloids. In: Roberts, M.F. and Wink, M. (eds) *Alkaloids: Biochemistry, Ecology, and Medical Applications*. Plenum Press, New York.

Wink, M. (1999) Introduction. In: Wink, M. (ed.) *Functions of Plant Secondary Metabolites and Their Exploitation in Biotechnology*. Sheffield Academic Press Ltd/CRC Press LLC, Sheffield, UK/Boca Raton, Florida.

Wink, M. and Schimmer, O. (1999) Modes of action of defensive secondary metabolites. In: Wink, M. (ed.) *Functions of Plant Secondary Metabolites and Their Exploitation in Biotechnology*. Sheffield Academic Press Ltd/CRC Press LLC, Sheffield, UK/Boca Raton, Florida.

Wink, M., Schmeller, T. and Latz-Bruning, B. (1998) Modes of action of allelochemical alkaloids: Interaction with neuroreceptors, DNA, and other molecular targets. *Journal of Chemical Ecology* 24, 1881–1937. doi:10.1023/a:1022315802264

Wouters, F.C., Gershenzon, J. and Vassão, D.G. (2016) Benzoxazinoids: Reactivity and modes of action of a versatile class of plant chemical defenses. *Journal of the Brazilian Chemical Society* 27,1379–1397.doi:10.5935/0103-5053.20160177

Wright, G.A., Baker, D.D., Palmer, M.J., Stabler, D., Mustard, J.A., *et al.* (2013) Caffeine in floral nectar enhances a pollinator's memory of reward. *Science* 339, 1202–1204. doi:10.1126/science.1228806

Wu, H.-S., Luo, J., Raza, W., Liu, Y.-X., Gu, M.-A., *et al.* (2010) Effect of exogenously added ferulic acid on in vitro *Fusarium oxysporum* f. sp *niveum*. *Scientia Horticulturae* 124, 448–453. doi:10.1016/j.scienta.2010.02.007

Wuyts, N., Swennen, R. and De Waele, D. (2006) Effects of plant phenylpropanoid pathway products and selected terpenoids and alkaloids on the behaviour of the plant-parasitic nematodes *Radopholus similis, Pratylenchus penetrans* and *Meloidogyne incognita*. *Nematology* 8, 89–101. doi:10.1163/156854106776179953

Xu, G.W. and Gross, D.C. (1988a) Evaluation of the role of syringomycin in plant pathogenesis by using Tn5-mutants of *Pseudomonas syringae* pv. *syringae* defective in syringomycin production. *Applied and Environmental Microbiology* 54, 1345–1353.

Xu, G.W. and Gross, D.C. (1988b) Physical and functional analyses of the syrA-gene and syrB-genes involved in syringomycin production by *Pseudomonas syringae* pv *syringae*. *Journal of Bacteriology* 170, 5680–5688. doi:10.1128/jb.170.12.5680-5688.1988

Young, D.H., Michelotti, E.L., Swindell, C.S. and Krauss, N.E. (1992) Antifungal properties of taxol and various analogs. *Experientia* 48, 882–885. doi:10.1007/bf02118425

Yu, X., Che, Z.P. and Xu, H. (2017) Recent advances in the chemistry and biology of podophyllotoxins. *Chemistry – A European Journal* 23, 4467–4526. doi:10.1002/chem.201602472

Zhang, H.W., Song, Y.C. and Tan, R.X. (2006) Biology and chemistry of endophytes. *Natural Product Reports* 23, 753–771. doi:10.1039/b609472b

Zhang, J.L., Mohamad, F.H., Wong, J.H., Mohamad, H., Ismail, A.H., *et al.* (2018) The effects of 4-hydroxybenzoic acid identified from bamboo (*Dendrocalamus asper*) shoots on Kv1.4 channel. *Malaysian Journal of Medical Sciences* 25, 101–113. doi:10.21315/mjms2018.25.1.12

10 Bioprospecting of Endophytes

Weaam Ebrahim,[1,2a*] Sherif S. Ebada,[1,3,4a*] and Peter Proksch[1]

[1]*Institute of Pharmaceutical Biology and Biotechnology, Heinrich-Heine-University Düsseldorf, Düsseldorf, Germany;* [2]*Department of Pharmacognosy, Faculty of Pharmacy, Mansoura University, Mansoura, Egypt;* [3]*Department of Pharmacognosy, Faculty of Pharmacy, Ain Shams University, Cairo, Egypt;* [4]*Department of Pharmaceutical Chemistry, Faculty of Pharmacy, Mu'tah University, Al-Karak, Jordan*

Abstract

Endophytic microorganisms constitute a prolific source of bioactive compounds mainly in the field of antineoplastic compounds and anti-infective agents. This chapter provides an overview on bioactive secondary metabolites from endophytes that were reported within the last five years (2013–2018), highlighting their often-unique chemical structures and/or their mechanism(s) of action.

10.1 Introduction

Endophytes are microorganisms, often fungi and bacteria (actinomycetes or mycoplasma), that thrive within living plants for at least part of their life and do not cause any apparent symptoms of diseases to their hosts (Stone *et al.*, 2000). Endophytes colonize inter- and/or intracellular spaces of plants (Pimentel *et al.*, 2011). This definition includes a wide range of complex interactions in which endophytes and their host plants participate: mutualism, commensalism and parasitism (Stone *et al.*, 2000). There are approximately 300,000 species of higher plants on earth; each one very likely constitutes a host to one or more endophytic microorganisms (Strobel, 2003). Endophytes are known to provide protection by improving the plant's ability to combat numerous abiotic and biotic stress conditions. In addition, they ameliorate plant growth and development through the production of phytohormones and other interesting bioactive secondary metabolites (Joseph and Priya, 2011).

Recent studies have reported hundreds of novel natural products isolated from endophytes, including alkaloids, flavonoids, benzopyranones, quinones, tetralones, phenolic acids, saponins, tannins, steroids, terpenoids, xanthones and many other compounds featuring new or unique skeletons. A vast array of biological activities have been reported for these fungal natural products such as antibacterial, antifungal, anticancer, antiviral or anti-inflammatory activities. This notion has inspired medicinal chemists to work on the development of synthetic or semisynthetic derivatives (Lam, 2007; Gouda *et al.*, 2016).

The diterpenoid 'Taxol', one of the most important anticancer drugs approved by the FDA for the treatment of advanced breast cancer, lung cancer and refractory ovarian cancer (Cremasco *et al.*, 2009), was initially

[a] Authors contributed equally to this book chapter.

* Corresponding authors e-mails: weaamnabil@mans.edu.eg; ss_ebada@mutah.edu.jo; sherif_elsayed@pharma.asu.edu.eg

 Endophyte Biotechnology: Potential for Agriculture and Pharmacology (ed. A. Schouten)

obtained from the bark of the yew tree *Taxus brevifolia*. The natural supply of the drug from yew trees is severely limited. However, the detection of taxol in the endophytic fungus *Taxomyces andreanae* isolated from *T. brevifolia* as well as in several other fungal endophytes harboured in *Taxus* species or even in non-*Taxus* plants may provide an alternative, cheaper and faster way to obtain this anticancer drug in significant quantities in the future (Stierle *et al.*, 1993; Li *et al.*, 1996; Pandi *et al.*, 2010).

Other pharmacologically important endophyte-derived compounds that had first been reported from plants include the cytotoxic agent camptothecin from *Nothapodytes foetida* (Puri *et al.*, 2005; Kusari *et al.*, 2009a), the cholinesterase inhibitor huperzine A from *Huperzia serrata* (Nair and Padmavathy, 2014), the triterpenoid helvolic acid obtained from the endophytic fungus *Cytonaema* sp., which shows potent antibacterial activity (Kumar *et al.*, 2014), and the aryl tetralin lignan podophyllotoxin isolated from many endophytic fungi such as *Aspergillus fumigatus*, *Phialocephala fortinii* and *Fusarium oxysporum*. The latter compound is clinically relevant as the precursor of the anticancer drugs teniposide, etoposide and etopophos (Kour *et al.*, 2008; Kusari *et al.*, 2009b). Further examples of important endophyte-derived compounds include the antibiotic kakadumycin A isolated from an endophytic *Streptomyces* sp. that exhibits an impressive activity against Gram-positive bacteria and against the malaria parasite *Plasmodium falciparum* (Castillo *et al.*, 2003). Noteworthy in this regard are also the antifungal compounds chaetomugilin A and D isolated from *Chaetomium globosum* (Qin *et al.*, 2009) and cytosporone B and C obtained from the mangrove endophytic fungus *Phomopsis* sp. that inhibit *Candida albicans* and *F. oxysporum* (Huang *et al.*, 2008). Furthermore, numerous natural antioxidants have been reported from endophytic microorganisms. Examples include pestacin and isopestacin isolated from *Pestalotiopsis microspora*. Pestacin exhibited an 11 times stronger antioxidant activity than vitamin E (Harper *et al.*, 2003), while graphislactone A, a phenolic compound obtained from *Cephalosporium* sp. isolated from *Trachelospermum jasminoides*, displayed free radical-scavenging and antioxidant activities stronger than ascorbic acid (Song *et al.*, 2005). Moreover, exopolysaccharides produced by the endophytic bacterium *Paenibacillus polymyxa* also showed strong scavenging activities towards superoxide and hydroxyl radicals (Liu *et al.*, 2009).

The use of advanced analytical techniques, microbial genomics and metagenomics, as well as triggering silent gene clusters in endophytes, has yielded intriguing opportunities for the discovery of novel natural products with application in medicine and industry (Gao *et al.*, 2018). These findings emphasize the importance of endophytes as irreplaceable resources of novel biologically active natural products for biopharmaceutical development. This chapter presents an overview of the literature dealing with secondary metabolites originating from fungal or bacterial endophytes between 2013 and 2018. The data were compiled with emphasis on structure–activity relationships as well as on mechanisms of action of these compounds whenever reported.

10.2 Cytotoxic Metabolites from Endophytes

Endophytic fungi have been shown to afford numerous cytotoxic compounds of diverse chemical classes, thereby proving their powerful role in the discovery of new anticancer leads (Aly *et al.*, 2010). Investigation of the endophytic fungus *C. globosum* TY1, isolated from the bark of *Ginkgo biloba* growing in China, led to the discovery of six *N*-containing azaphilones, including three novel congeners, chaetomugilides A–C. Chaetomugilide A (**1**) showed significant cytotoxic activity against human hepatoblastoma cells (HepG-2) with an IC_{50} value of 1.7 μM (Li *et al.*, 2013).

The Hsp90 chaperone machine has become an interesting target for cancer treatment due to its proteostatic maintenance of oncoprotein stability (such as protein kinases and transcription factors) in addition to its buffering action on the cellular stresses of the tumour environment (Whitesell and Lindquist, 2005). Chemical analysis of the endophytic fungus *Chaetomium aureum*, isolated from stem tissue of the medicinal plant *Thymelaea lythroides* collected in Morocco, yielded (+)-sclerotiorin (**2**). Interestingly, this compound efficiently inhibited the Hsp90 chaperoning of the progesterone receptor (PR) *in vitro*. This inhibitory activity may be due to its ability to modify primary amines, such as in lysine residues or in *N*-terminal amino acids in cholesterol ester transfer protein (CETP), thereby leading to inactivation of Hsp90, Hsp70, Hsp40, Hop and p23. The oxygen atom of the heterocycle of (+)-sclerotiorin (**2**) was observed to be essential for this inhibitory activity. (+)-Sclerotiorin (**2**) showed no cytotoxic activity against the breast cancer cell lines Hs578T, MCF7, MDA-MB-231 and MDA-MB-453; the prostate cancer cell line LNCaP; or the cervical cancer cell line HeLa. However, chemical modification of (+)-sclerotiorin (**2**) by deacetylation rendered the compound cytotoxic to Hs578T, MDA-MB-231 and LNCaP cell lines, although this derivative was less efficient in inhibiting the Hsp90 *in vitro* (Kabbaj *et al.*, 2015).

Activation of the nuclear factor kappa B (NF-κB) is involved in cell adhesion as well as cell proliferation and is therefore an important target for treatment of progression of carcinogenesis (Chaturvedi *et al.*, 2011). Altersolanol A (**3**), a polyketide obtained from several *Alternaria* species, is also produced by other fungi such as the endophytic fungus *Stemphylium globuliferum* isolated from stem tissue of *Mentha pulegium* (Lamiaceae) growing in Morocco (Debbab *et al.*, 2009). This compound was reported to be cytotoxic against L5178 cells presumably mediated through inhibition of protein kinase activity (Debbab *et al.*, 2009). In a study evaluating its mechanism of action and its structure–activity relationships, altersolanol A (**3**) exhibited cytotoxic, cytostatic, anti-inflammatory and antimigratory activity against human chronic myeloid K562 leukemia and A549 lung cancer cells at low micromolar concentrations without affecting the tumour micro-environment. It induced cell death by apoptosis through activation of specific executors of apoptosis (caspases 3, 8 and 9) and decrease of the expression of the anti-apoptotic proteins Bcl-xL, XIAP and Mcl-1. Moreover, it decreased the TNFα-activated NF-κB transcriptional activity in a dose-dependent manner, suggesting its effect on NF-κB signalling pathway. The biological activity of altersolanol A was suggested to be mediated by the presence of the p-quinone moiety. These results warrant future research on the clinical application of the compound in treatments involving apoptotic cell death processes (Teiten *et al.*, 2013).

The two new alkaloids, embellicines A and B (**4** and **5**), obtained from the endophytic fungus *Embellisia eureka*, isolated from the Moroccan plant *Cladanthus arabicus* (Asteraceae), exhibited cytotoxic and cytostatic activities against human chronic myelogenous leukaemia (K562) cells in a dose- and time-dependent manner. Both compounds were able to induce cell death with IC_{50} values lower than 10 μM. Embellicine B (**5**), which lacks the $\Delta^{17,18}$ double bond, exhibited five to ten times stronger activity than embellicine A (**4**), which possesses a reactive α,β-unsaturated carbonyl group in the pyrrolidinone ring. This result highlighted the importance of the hydroxylation pattern of the pyrrolidinone ring of **5** for the cytotoxicity in this cellular system rather than the presence of the $\Delta^{17,18}$ double bond, which may act as a Michael acceptor. The mechanism of action of both compounds may be due to their pronounced inhibition of tumour necrosis factor alpha (TNF-α)-induced NF-κB transcriptional activity. These results suggested that both metabolites could be important leads for anticancer chemotherapy and for the design of semisynthetic congeners (Ebrahim *et al.*, 2013).

Phomoxanthone A (PXA) (**6**), a dimeric tetrahydroxanthone derivative, was obtained from the endophytic fungus *Phomopsis longicolla*, isolated from leaves of the mangrove plant *Sonneratia caseolaris* (Lythraceae) growing in South China. This natural product displayed strong inhibition of proliferation of the lymphoma cell line L5178Y, as well as of several human cancer cell lines including cisplatin-resistant cancer cells. Moreover, phomoxanthone was up to 100-fold less toxic to healthy blood cells. Its cytostatic activity is linked to its pro-apoptotic potential. It exhibited also a noticeable upregulation of murine $CD69^+$ T-lymphocytes, NK cells and macrophages, revealing a dual effect of this potent antitumour compound in combating cancer cells through apoptosis and immunostimulation, which could effectively help avoiding resistance development of tumour cells during chemotherapy. This study also indicated that the positions of the biaryl axis and of the acetyl groups are essential for the cytotoxic activity of this natural product (Rönsberg *et al.*, 2013).

In a recent study, the mycotoxin phomoxanthone A (PXA, **6**), derived from the same endophytic fungus *P. longicolla*, was identified as a mitochondrial toxin (Böhler *et al.*, 2018). PXA (**6**) disturbs the mitochondrial form and function in several ways. Mitochondria produce ATP through oxidative phosphorylation, which depends on the electron transport chain (ETC) embedded in the inner mitochondrial membrane (IMM). The ETC pumps protons from the mitochondrial matrix into the mitochondrial intermembrane space. This creates a proton gradient and, as a result, a membrane potential across the IMM that drives the mitochondrial ATP synthase. This study demonstrated that PXA (**6**) produces a strong release of Ca^{2+} from the mitochondria. In addition, it depolarizes the mitochondria membrane similar to protonophoric uncouplers. However, it does not increase but rather inhibits cellular respiration and electron transport chain activity. Mitochondria are rapidly fragmented upon PXA (**6**) treatment as obvious from EM studies. Mechanistically, PXA perform stress-induced OPA1-gene cleavage that is dependent on the metalloendopeptidase OMA1, whereas expression of ATP-dependent metalloprotease YME1L1 did not have any noticeable effect on PXA-induced OPA1 cleavage. The mitochondrial fragmentation is independent from the canonical mitochondrial fission and fusion mediators OPA1 and DRP1 and only affects the inner mitochondrial membrane, leading to cristae disruption as well as the release of pro-apoptotic proteins, and apoptosis. These results suggest that PXA (**6**) is a mitochondrial toxin possessing a novel mode of action and might provide a useful tool for the study of mitochondrial ion homoeostasis and membrane dynamics (Böhler *et al.*, 2018).

PM181110 (**7**) is a novel depsipeptide that was obtained from the endophytic fungus *Phomopsis glabrae* isolated from leaves of *Pongamia pinnata* (family Fabaceae) collected in Maharashtra, India. The antitumour activity of PM181110 (**7**) was tested *in vitro* against 40 human tumour cell lines derived from bladder, colon, gastric, head and neck, liver, lung (non-small-cell lung carcinoma), mammary, ovarian, pancreatic, prostate, renal and uterus cancer, as well as against melanoma, pleuramesothelioma and sarcoma. The compound showed concentration-dependent activity in all tested cell lines. It exhibited pronounced activity against the pancreatic cancer cell line PAXF 546L (IC_{50} = 0.016 μM) and the lung cancer cell line LXFA 526L (IC_{50} = 0.021 μM). Moreover, PM181110 (**7**) exerted *ex vivo* antitumour efficacy in a panel of 24 human tumour xenografts with IC_{50} values ranging from 0.03 μM to 0.422 μM. The most sensitive tumour model found was the bladder cancer cell line BXF 1218 with an IC_{50} value for **7** of 0.03 μM (Verekar *et al.*, 2014).

Cytochalasans are a class of metabolites isolated from several fungal genera (including *Phomopsis*, *Aspergillus* and *Penicillium*) that revealed interesting biological activities (Jiao *et al.*, 2004). Structurally, these compounds are composed of a tricyclic core, consisting of a macrocyclic ring fused to an isoindolone moiety formed from a reduced polyketide backbone and an amino acid (e.g. leucine or phenylalanine). Periconiasins A and B (**8** and **9**) are two new cytochalasans, obtained from the endophytic fungus *Periconia* sp. F-31, isolated from the medicinal plant *Annona muricata* from Hainan Province, China. Both compounds were tested for their cytotoxicity against five human cancer cell lines (A2780, A549, Bel-7402, BGC-823 and HCT-8). Periconiasin A (**6**) exhibited selective and significant cytotoxicity against the HCT-8 and BGC-823 cell lines with IC_{50} values of 0.9 and 2.1 μM, respectively, whereas periconiasin B (**9**) exhibited selective cytotoxic activity against HCT-8, Bel-7402 and BGC-823 cell lines with IC_{50} values of 0.8, 5.1 and 9.4 μM, respectively. These results emphasize the impact of periconiasins A and B (**6** and **7**) as potential lead compounds against HCT-8 cancer cells (Zhang *et al.*, 2013).

A new bicyclic lactone with a rare ring-fusion system, myrotheciumone A (**10**) was obtained from the endophytic fungus *Myrothecium roridum*, isolated from stems of the medicinal herb *Ajuga decumbens*, naturally occurring in Japan and China. The cytotoxic activity of this compound was tested against numerous cell lines including HepG2 (hepatocellular carcinoma), SMMC-7721 (human hepatocellular carcinoma), A549 (human lung adenocarcinoma), MCF-7 (human breast adenocarcinoma), QSG-7701 (human hepatocyte cells) and HL-7702 (human hepatocyte cells). Myrotheciumone A (**10**) strongly inhibited growth of HepG2 (IC_{50} = 5.36 ± 0.26 μM), SMMC-7721 (IC_{50} = 6.56 ± 0.58 μM), A549 (IC_{50} = 5.88 ± 0.68 μM), MCF-7 cells (IC_{50} = 7.56 ± 0.76 μM), QSG-7701 (IC_{50} = 16.30 ± 0.31 μM) and HL-7702 cells (IC_{50} = 20.69 ± 4.69 μM). Myrotheciumone A (**10**) showed stronger cytotoxicity towards cancer cells than towards normal cells. It acts by induction of apoptosis and promotes cytochrome C release from mitochondria and induces poly(ADP-ribose) polymerase (PARP) cleavage in a time- and dose-dependent manner. These data highlight the importance of further studies in order to exploit myrotheciumone A (**10**) for cancer chemotherapy (Lin *et al.*, 2014a; Fig. 10.1).

Numerous macrocyclic trichothecenes have been isolated from different fungi such as *Cylindrocarpon* sp., *Myrothecium* sp., *Verticimonosporium* sp., *Phomopsis* sp. and *Stachybotrys* sp. They are classified according to the macrolide ring, which is a typical structural feature of these metabolites. Trichothecenes display significant biological activities, especially antitumour activity. Minor variations in the structures of macrocyclic trichothecenes strongly affect their activity, for instance: breakage of the macrocyclic ring greatly reduces the activity, while the presence of a tetrahydropyranyl ring considerably increases the potency (Lin *et al.*, 2014b). Therefore, structure–activity relationships of trichothecenes with regard to the development of new anticancer derivatives were extensively studied. Shen *et al.* isolated the new cytotoxic trichothecene macrolide, dihydromyrothecine C (**11**), from *M. roridum* IFB-E012, a fungal endophyte obtained from the traditional Chinese medicinal plant *Artemisia annua* (Asteraceae). The compound was isolated as a mixture of epimers due to the presence of an unstable cyclic hemiacetal structure. When evaluating its *in vitro* cytotoxicity against the human nasopharyngeal carcinoma cell line KB, the compound showed moderate cytotoxic activity with an IC_{50} value of 44.48 μM (Shen *et al.*, 2016).

Breast cancer is among the most common and serious global malignancies. It is the leading cause of death of young women in developed countries (Siegel *et al.*, 2015). Elsayed *et al.* investigated the antiproliferative, antimigratory and anti-invasive activities of the well-known indole alkaloid, meleagrin (**12**), isolated from the endophytic fungus *Penicillium chrysogenum* obtained from leaves of the olive tree *Olea europea* towards breast cancer. The compound inhibited the growth of human breast cancer cell lines MDA-MB-231, MDA-468, BT-474, SK BR-3, MCF7 and MCF7-dox, with no effect on the growth and viability of the non-tumourigenic human mammary epithelial cells MCF10A. Its activity was correlated to its inhibition of tyrosine-protein kinase Met (c-Met). c-Met triggers tumour growth and new blood vessel formation (angiogenesis) that supply the tumour with oxygen and nutrients, and as a consequence facilitate metastasis. Meleagrin (**12**) showed no activity against c-Met-independent breast cancer cells. Meleagrin exhibited also a dose-dependent inhibition of hepatocyte growth factor (HGF)-induced cell migration and invasion of breast cancer cell lines. *In vivo*, meleagrin (**12**) potently decreased the invasive triple negative breast tumour cell growth in a clinically relevant orthotopic mouse xenograft model. These results suggest this indole alkaloid to be a novel lead compound for the control of c-Met-dependent metastatic and invasive breast malignancies (Mady *et al.*, 2016).

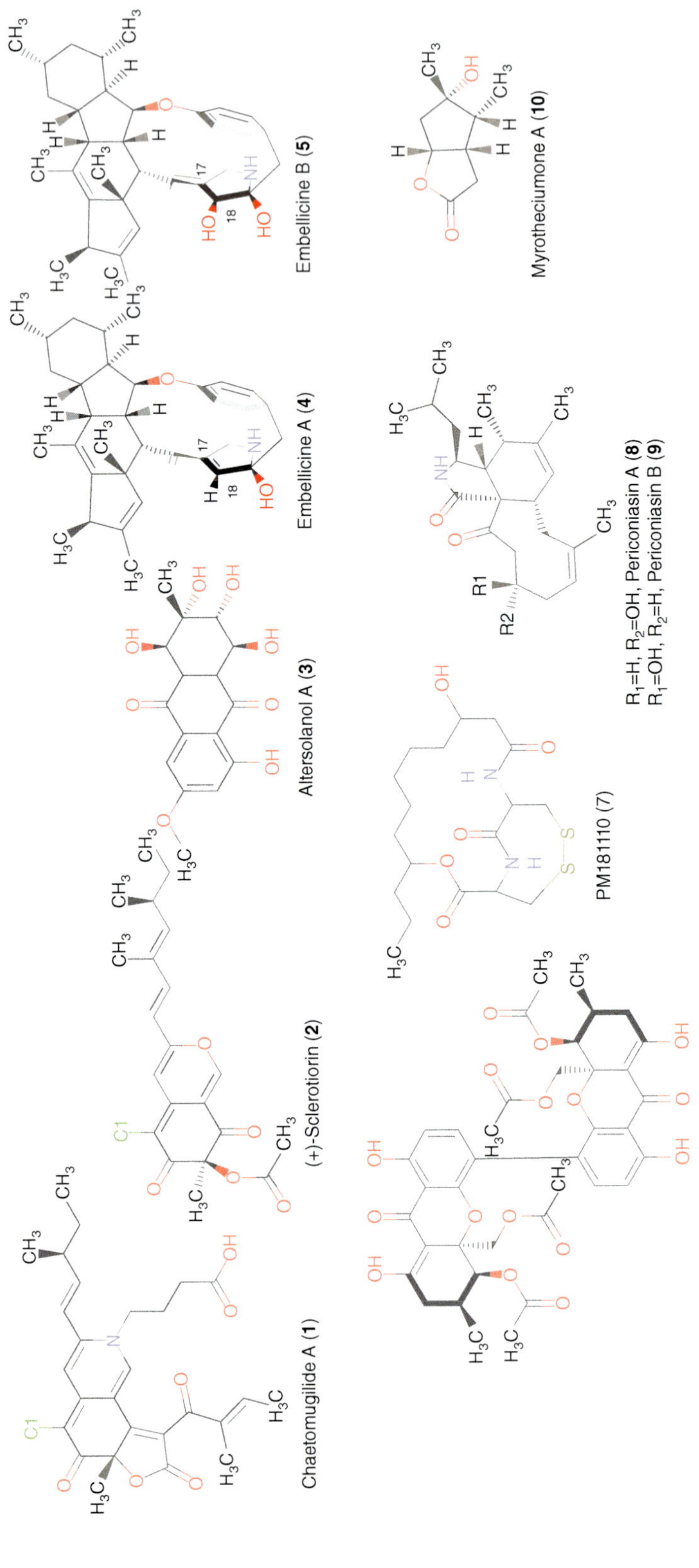

Fig. 10.1. Cytotoxic metabolites from endophytes, compounds (**1–10**).

Liverworts are one of the phyla belonging to *Bryophytes* (*Hepaticeae*), the latter being taxonomically placed between algae and *Pteridophytes* (ferns). Liverworts are characterized by growing in microbial-rich environments such as soil, wet rocks and rotten logs. Therefore, liverworts host numerous epiphytes and endophytes, which contribute effectively to the production of bioactive secondary metabolites (Asakawa *et al*., 2009). Gentisyl alcohol (**13**) and epoxydon (**14**) were isolated from cultures of the endophytic fungus *Penicillium concentricum* strain 4E-4, isolated from the liverwort *Trichocolea tomentella* (Trichocoleaceae) collected in Newport, Virginia. Gentisyl alcohol (**13**) displayed selective cytotoxicity against the HT-29 colon cancer cell line with an IC_{50} value of 6.4 μM, while epoxydon (**14**) was active against MCF-7 breast cancer cell line with IC_{50} of 5.7 μM (Ali *et al*., 2017).

The endophytic fungus *Peyronellaea coffeae-arabicae* FT238, isolated from the Hawaiian plant *Pritchardia lowreyana*, yielded the new epoxyphomalin analogue 11-dehydroxy epoxyphomalin A (**15**). The compound showed antiproliferative activity with an IC_{50} value of 0.5 μM against OVCAR3 cells, and it further inhibited A2780 CisR (cisplatin-resistant human ovarian cancer) cells with an IC_{50} value of 0.6 μM. 11-Dehydroxy epoxyphomalin A (**15**) inhibited intracellular phosphortyrosine Stat3 (pY705Stat3) (a member of DNA-binding factors that function to induce expression of responsive genes), suggesting potential inhibition of aberrant Stat3 activity also in tumour cells. Further studies are currently carried out to evaluate the role of this compound in p53 and Stat3 pathways (Li *et al*., 2016).

Prenylated isoindolone alkaloids are a class of meroterpenoids consisting of an isoindolone unit and a terpene moiety. These metabolites are widely distributed in fungi belonging to the genera *Aspergillus*, *Alternaria*, *Hericium*, *Emericella* and *Stachybotrys*. These compounds exhibit a wide range of bioactivities including cytotoxicity, antiviral and antihyperlipidemic activities. Emeriphenolicin E (**16**) is an isoindolone-derived meroterpenoid that features two farnesyl groups. It was obtained from the endophytic fungus *Emericella nidulans* (anamorph *Aspergillus nidulans*) HDN12-249, isolated from leaves of *Tamarix chinensis* collected from Laizhou Bay, China. The compound displayed selective cytotoxicity against three human cancer cell lines, including HeLa, A549 and HCT-116 with IC_{50} values of 4.77, 12.04 and 33.05 μM, respectively (Zhou *et al*., 2016).

Fungi of the genus *Fusarium* are prolific producers of therapeutically promising secondary metabolites with antiproliferative activity such as the cyclic depsipeptide sansalvamide A with strong *in vitro* cytotoxicity against pancreatic, colon and breast cancer cell lines, and enniatins, which are also currently being investigated as potential anticancer agents. Moreover, *Fusarium* species produce several antineoplastic agents previously isolated from plants such as paclitaxel (taxol), camptothecin and L-asparaginase, in addition to podophyllotoxin, which is the lead compound for the clinically used semisynthetic drugs etoposide and teniposide. Chemical investigation of the ethyl acetate extract of the endophytic fungus *Fusarium solani* strain AURE-4, isolated from roots of *Aponogeton undulatus*, growing in water in Rajshahi, Bangladesh, led to two new azaanthraquinone derivatives 7-desmethylscorpinone (**17**) and 7-desmethyl-6-methylbostrycoidin (**18**). When evaluated against four human tumour cell lines including MIA PaCa2 pancreatic cancer, NCI H1975 non-small-cell lung cancer, HeLa cervical carcinoma and MDA MB 231 breast cancer cells, both compounds showed cytotoxic activity with low micromolar to submicromolar IC_{50} values, with 7-desmethyl-6-methylbostrycoidin (**18**) being the most potent compound. 7-Desmethylscorpinone (**17**) and 7-desmethyl-6-methylbostrycoidin (**18**) act by intercalating DNA base pairs and forming stable complexes (Chowdhury *et al*., 2017).

Chemical investigation of *Fusarium tricinctum*, a fungal endophyte isolated from rhizomes of *Aristolochia paucinervis* collected in Beni-Mellal (Morocco), led to the isolation of fusarielin J (**19**). Fusarielin-type polyketides (17 derivatives identified so far) display antifungal, antibacterial and cytotoxic activities. Compound **19** showed cytotoxicity against the human ovarian cancer cell line A2780, with an IC_{50} value of 12.5 µM. The presence of a phenyl acetic acid residue seems to play an important role in the cytotoxic activity of fusarielin J (**19**) (Hemphill *et al.*, 2017).

Fungi belonging to the genus *Preussia* (Sporormiaceae) can be found in animal dung, plant debris, soil and wood, or as endophytes. They are rich sources of bioactive polyketides (Rangel-Grimaldo *et al.*, 2017). Preussilides A and C (**20** and **21**) are two novel bicyclic polyketides obtained from the endophytic fungus *Preussia similis* DSM 104666, isolated from roots of the medicinal plant *Globularia alypum* (Plantaginaceae) collected in Batna, Algeria. Both compounds were evaluated for their cytotoxic activity against different mammalian cancer cell lines. They exhibited moderate activity with IC_{50} values below 10 µM for murine cells (mouse fibroblasts L929), human cancer cell lines (HeLa KB.3.1 and human osteosarcoma U2OS), in addition to human breast adenocarcinoma MCF-7 (for preussilide C (**21**) only). Interestingly, compounds **20** and **21** affect the cell cycle provoking multipolar spindles and unequal cell divisions, resulting in cells that cannot multiply further and therefore die. However, apoptotic cell death appears not to be the predominant mechanism of action; these compounds might target an enzyme involved in cell division cycle. Hence, they might affect timing or spindle assembly mechanisms, leading to defects in chromosome segregation and/or spindle geometry (Noumeur *et al.*, 2017).

Species of the ascomycete genus *Teratosphaeria* are widespread as foliar pathogens in plants of the *Proteaceae* and are known to inhabit leaves as endophytes in other plant families (Ganley *et al.*, 2004). Members of the genus *Teratosphaeria* represent an understudied genus. Padumadasa *et al.* investigated the biology and chemistry of a cytotoxic fraction from the extract of the endophytic fungal strain *Teratosphaeria* sp. FL2137 isolated from senescent leaves (needles) of *Pinus clausa* obtained from a pine-dominated scrub forest at Archbold Biological Station in Florida (Fig. 10.2).

Teratosphaerone A (**22**), a new naphthoquinone, was isolated and tested for its cytotoxic activity against five tumour cell lines including SF-268 (human CNS cancer; glioma), NCI-H460 (human non-small-cell lung cancer), MCF-7 (human breast cancer), MDA-MB-231 (human breast adenocarcinoma) and PC-3 M (metastatic human prostate adenocarcinoma) cells. Compound **22** displayed selective cytotoxic activity against MDA-MB-231 cells with an IC_{50} value of 1.2 µM and moderate activity against MCF-7 and PC-3M cells but was inactive against NCI-H460 and SF-268 cell lines at the range of concentrations tested (Padumadasa *et al.*, 2018).

10.3 Anti-infective Metabolites from Endophytes

Endophytes continue to be an important source of leads for new anti-infective agents. The well-known antibiotic penicillin served as a prototype of a fungal anti-infective agent and was a milestone leading to the discovery of numerous further fungal compounds exhibiting anti-infective activity (Demain and Elander, 1999). The increasing resistance of microorganisms to currently used antibiotics is alarming and at least partly due to the misuse of antibiotics as animal food supplements aimed at faster animal growth, accelerating weight gain and disease prevention, thus increasing the profitability for animal breeders. Several studies have discussed the role of antibiotics as animal food supplements. It is estimated that about 80% of all produced antibiotics are fed to animals and about two-thirds of them are not fully metabolized, being passed on to the environment, which increases the spread of resistance against antibiotics (Chee-Sanford *et al.*, 2009).

The widespread and well-documented microbial resistance towards clinically used antibiotics strengthens the need for new antibiotics that feature novel structural types and/or address new targets (Signoretto *et al.*, 2012). Several fungal secondary metabolites revealed *in vitro* antimicrobial activity against Gram-positive and Gram-negative bacteria, including several food-borne pathogens (Venturini *et al.*, 2008), whereas others are

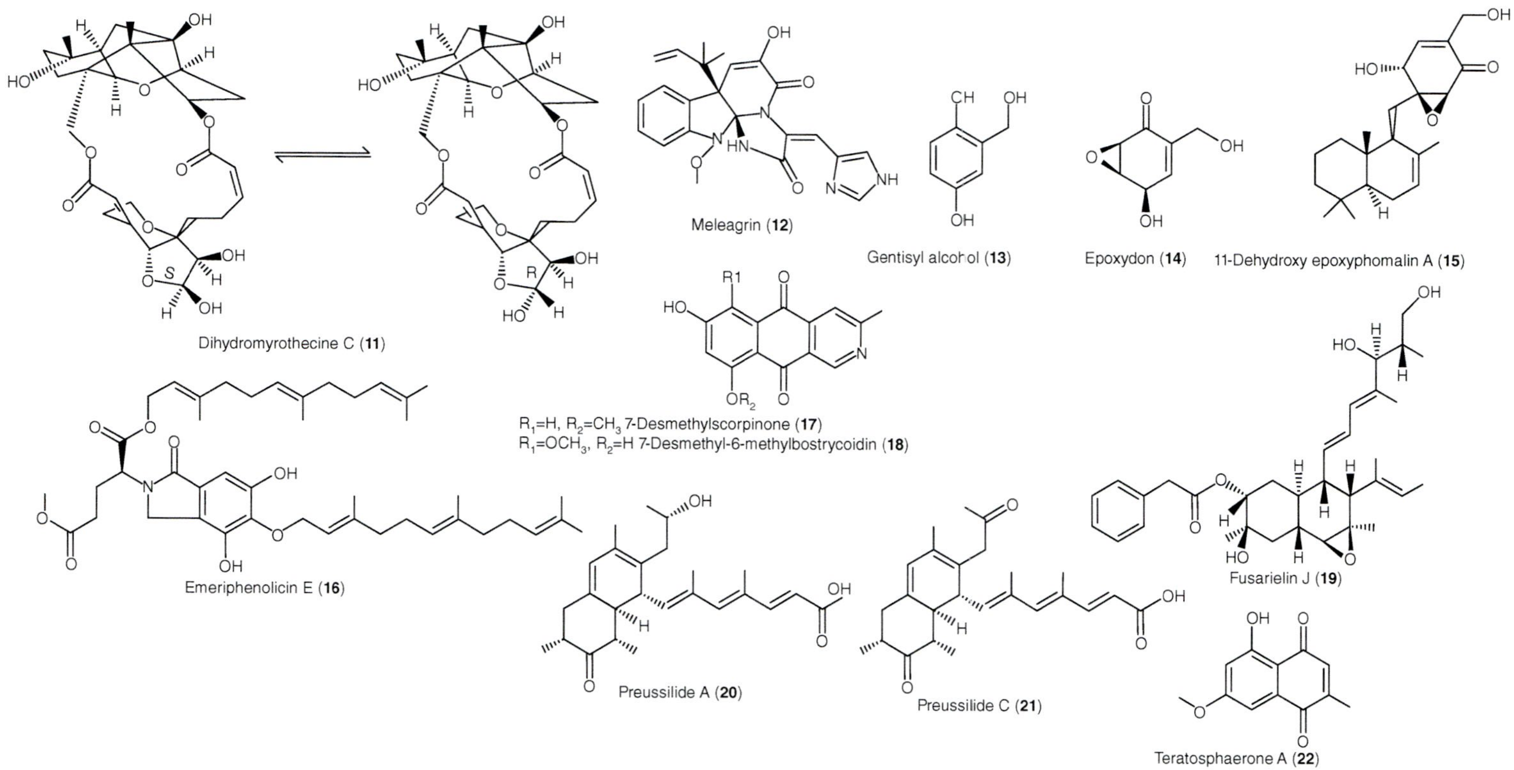

Fig. 10.2. Cytotoxic metabolites from endophytes, compounds (**11–22**).

active against fungi (Hearst *et al.*, 2009) or display antiviral activity (Wasser and Weis, 1999). In this part, we will survey fungal secondary metabolites reported during the past five years (2013–2018) showing anti-infective properties with a focus on their mechanism of action and on structure–activity relationships.

Asperterpenoid A (**23**) was reported as a novel sesterterpenoidal fungal metabolite isolated from the endophytic fungus *Aspergillus* sp. 16-5c (Huang *et al.*, 2013). Compound **23** features a unique structure with five fused rings and revealed a potent inhibitory activity against *Mycobacterium tuberculosis* protein tyrosine phosphatase B (*m*PTPB) with an IC_{50} value of 2.2 μM (Huang *et al.*, 2013). Based on the structure and the potential *in vitro* antitubercular activity, asperterpenoid A (**23**) has been suggested as a potential lead compound for constructing a library of antitubercular derivatives.

Talaromyces wortmannii, an endophytic fungus derived from *Aloe vera* leaves, yielded two bisdihydroanthracenone atropodiastereomers, including the homodimers flavomannins A and B (**24** and **25**) along with the heterodimers flavomannins C and D (**26** and **27**) (Bara *et al.*, 2013a). In addition, the same extract afforded two new mixed dihydroanthracenone/anthraquinone dimers, namely, talaromannins A and B (**28** and **29**) (Bara *et al.*, 2013a). The isolated compounds were assessed for their antibacterial activity against a panel of Gram-positive and Gram-negative bacteria comprising antibiotic-susceptible strains and multi-resistant clinical isolates. Dihydroanthracenone dimers with either homo- or heterodimeric skeletons possess antibacterial activity against staphylococci with IC_{50} values ranging from 4 to 8 μg/ml. Interestingly, these compounds exhibited even more potent antibacterial activity against methicillin-resistant *Staphylococcus aureus* (MRSA). Mechanistically, some of the tested compounds proved to induce the *yor*B promoter, a component of the universal transcriptional framework mediating the so-called SOS response in bacteria (Lazarevic *et al.*, 1999; Au *et al.*, 2005). The *yor*B promoter plays an important role in the cellular response against DNA synthesis, damage and metabolism. Therefore, *yor*B induction was found to be related to error-prone DNA repair, which eventually leads to cell arrest and death (Urban *et al.*, 2007). Among the isolated compounds, flavomannin B (**25**) revealed the most potent induction of the *yor*B promoter (Bara *et al.*, 2013a). None of the tested compounds displayed cytotoxic activity, thus making them particularly interesting for antibacterial drug development (Bara *et al.*, 2013a).

Unexpectedly, cultivating the same strain of *T. wortmannii* for four instead of two weeks afforded the new atropisomer dianthrone, biemodin (**30**), in addition to three wortmannin derivatives and several further known metabolites. All isolated compounds were assessed for their antibacterial activity against Gram-positive and Gram-negative bacteria. Skyrin (**31**) and rugulosin A (**32**) displayed potent antimicrobial activity against Gram-positive pathogenic isolates with MIC values between 4 and 16 μg/ml. Biemodin (**30**) was more potent against Gram-positive bacteria, in particular MRSA, but was less active compared to skyrin and rugulosin A (Bara *et al.*, 2013b). Hence, these compounds may serve as a basis for developing new antibacterial agents especially against drug-resistant bacterial pathogens.

An unprecedented 3*H*-oxepine-containing diketopiperazine-type alkaloid, varioxepine A (**33**), featuring a 3,6,8-trioxabicyclo[3.2.1] octane core structure, was obtained from the marine red algal-derived endophytic fungus *Paecilomyces variotii* EN-291 (Zhang *et al.*, 2014). Varioxepine A (**33**) was subjected to antimicrobial assays against a panel of several human- and aqua-pathogenic strains including both Gram-positive and Gram-negative bacteria. The compound possesses moderate to pronounced antibacterial activities against the tested strains with MICs of 16–64 μg/ml. However, the most potent antibacterial activity of **33** was exhibited against the plant-pathogenic fungus *Fusarium graminearum* (MIC = 4 μg/ml), which may draw attention towards its potential antifungal activity. The compound may be considered as a starting structure for developing new antifungal agents for both agricultural and human applications (Zhang *et al.*, 2014; Fig. 10.3).

Screening of a small panel of plant endophyte extracts for anti-HIV activity implementing a cellular assay using infected T-cells cells revealed the endophytic fungus *Alternaria tenuissima* QUE1Se, derived from *Quercus emoryi* stem tissue, as a potential candidate for further investigation. The fungus yielded five altertoxin derivatives including the two new congeners, altertoxin V (**37**) and VI, in addition to three known metabolites, altertoxin I–III (**34**–**36**) (Bashyal *et al.*, 2014). In a cell-based anti-HIV assay, all isolated altertoxins with the exception of altertoxin VI completely inhibited viral replication of the HIV-1 virus at concentrations ranging between 0.30 and 2.20 μM (Bashyal *et al.*, 2014). Remarkably, the HIV inhibitory activity exhibited by altertoxins was equal or even superior to that caused by azidothymidine (AZT, Zidovudine®) at a concentration of 20 μM, which was used as a positive control (Bashyal *et al.*, 2014).

Altertoxins revealed anti-HIV activity at concentrations below their cytotoxic and/or mutagenic levels, thus provoking a significant interest for those compounds as potential candidates for the development of novel effective anti-HIV agents. Based on the altertoxin structural features, the presence of epoxyperylene and [1,1′-biphenyl]-4,4′-diol moieties seemed to be essential for the anti-HIV activity (Bashyal *et al.*, 2014).

The endophytic fungus *Periconia* sp. F-31, derived from the medicinal plant *A. muricata*, was found to produce the four new polyketide synthase-nonribosomal peptides pericoannosins A (**38**), D, E and F (**39**) (Zhang *et al.*, 2015), which showed low *in vitro* anti-HIV activity, with IC_{50} values ranging from 30 to 70 nM compared to efavirenz [standard antiretroviral pharmaceutical (tenofovir, Sustiva®, BMS)], which was used as a positive control (IC_{50} = 1.4 nM) (Zhang *et al.*, 2015).

The EtOAc-soluble fraction of an extract from solid rice cultures of *Aspergillus* sp. CPCC 400735, an endophytic fungus isolated from the medicinal plant *Kadsura longipedunculata*, revealed significant anti-HIV activity and hence encouraged its chemical exploration. This afforded 33 metabolites including five new phenalenones, one new cytochalasin derivative and one new butenolide together with known phenyl derivatives (Pang *et al.*, 2017). Among the isolated compounds, asperphenalenones A–E were identified as new natural products together with aspochalasin R and aspulvinone R (Pang *et al.*, 2017). Unlike other known phenalenone-type metabolites, asperphenalenones A–E feature a linear diterpene linked to a phenalenone via a C–C bond. When assessed for their anti-HIV activity, several metabolites belonging to different chemical classes exhibited moderate to potent anti-HIV activity (IC_{50} values ranging from 2.4 to 22.1 μM) with asperphenalenones A (**40**), D (**41**), cytochalasin Z_8 (**42**) and epicocconigrone A (**43**) identified as the most active compounds having IC_{50} values of 4.5, 2.4, 9.2 and 6.6 μM, respectively, when compared to lamivudine (IC_{50} = 0.1 μM) and efavirenz (IC_{50} = 0.4 × 10^{-3} μM) (Pang *et al.*, 2017).

Two lanostane-type triterpenes, sclerodols A (**44**) and B (**45**) together with three known metabolites were obtained from the endophytic fungus *Scleroderma* UFSMSc1 (Persoon) Fries isolated from *Eucalyptus grandis* (Morandini *et al.*, 2016). Interestingly, sclerodol B (**45**) rather than **44** revealed stronger anticandidal activity with the best activity against *Candida krusei*, followed by *Candida parapsilosis*, and with lower potency against *C. albicans* and *C. tropicalis* with IC_{50} values between 6.25 and 25 μg/ml (Morandini *et al.*, 2016). Sclerodol B (**45**) was found to be fungistatic and fungicidal at these concentrations. Based on the structural similarity between sclerodols and lanosterol, the anticandidal activity of sclerodols is assumed to be caused by inhibition of the activity of fungal methyltransferases (SMT) (Kanagasabai *et al.*, 2004).

Neosartoryadins A (**46**) and B (**47**), two new fumiquinazoline alkaloids, were first reported from the endophytic fungus *Neosartorya udagawae* HDN13-313 derived from roots of the mangrove plant *Avicennia marina* (Yu *et al.*, 2015). Chemically, neosartoryadins are characterized by an unprecedented condensation between a unique 6/6/6/5 quinazoline ring system and a 6/5/5 imidazoindolone ring. At non-cytotoxic levels, neosartoryadins A (**46**) and B (**47**)

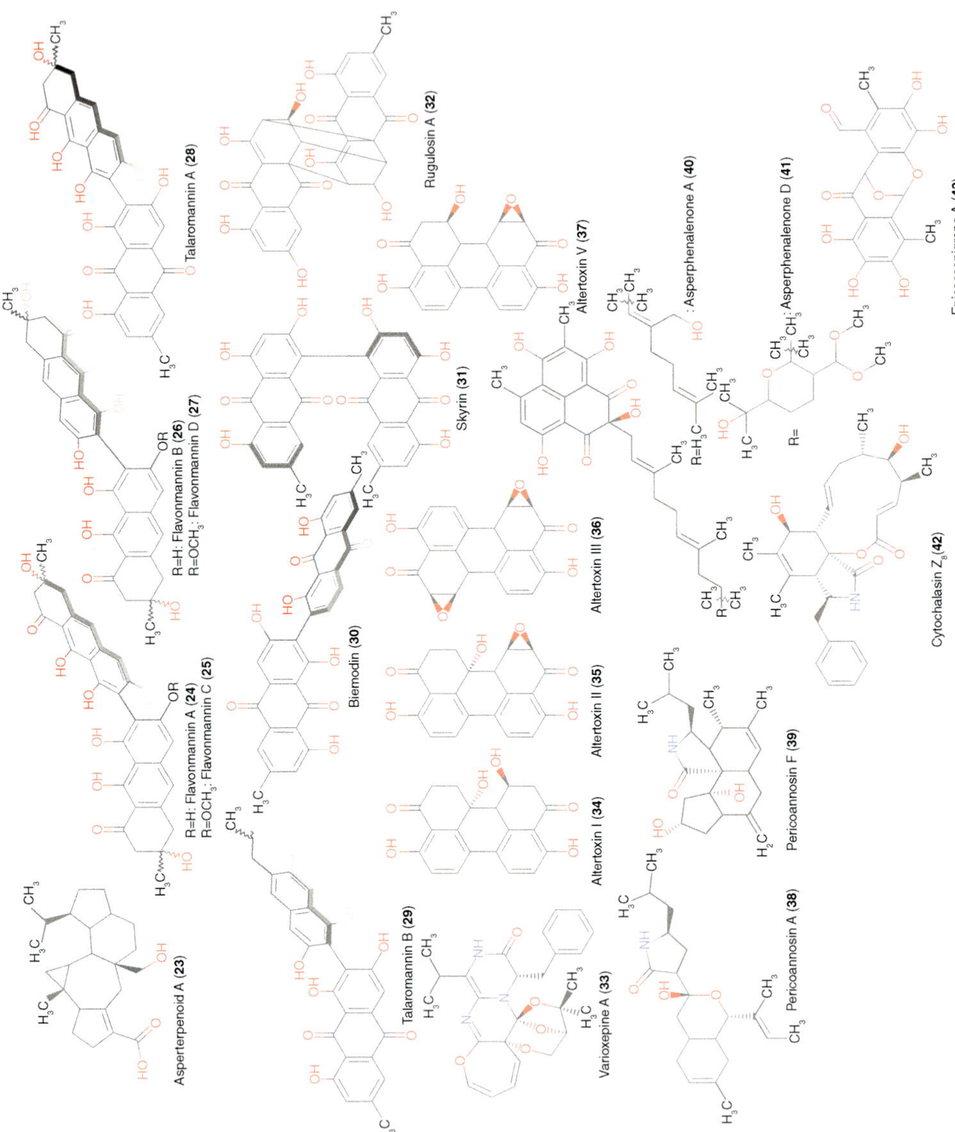

Fig. 10.3. Anti-infective metabolites from endophytes, compounds (**23–43**).

exhibited antiviral activity against influenza A (H1N1) virus with IC_{50} values of 66 and 58 μM, respectively, compared to ribavirin as a standard antiviral agent (IC_{50} = 94 μM) (Yu *et al.*, 2015).

Three new isoaigialones A–C (**48**–**50**) together with the parent compound aigialone (**51**) were reported from the endophytic fungus *Phaeoacremonium* sp. associated with the plant *Senna spectabilis* (Silva *et al.*, 2017). The furo[3,4-b]pyran scaffold is an uncommon ring system in natural products and it was previously only reported in one fungal metabolite, massarilactone B, isolated from the freshwater aquatic fungus *Massarina tunicata* (Oh *et al.*, 2001; Fig. 10.4).

Aigialone (**51**) and its three isomers (**48**–**50**) were tested for their antimicrobial activity against *Cladosporium cladosporioides* and against *Cladosporium sphaerospermum* using direct bioautography (Rahalison *et al.*, 1991), in which the microorganisms grow directly on the TLC plate of the tested samples and/or extracts at predetermined concentrations. Only isoaigialone B (**49**) and aigialone (**51**) showed antifungal activity at 5 μg detection limit compared to nystatin used as a standard antifungal drug (detection limit of 1 μg) (Silva *et al.*, 2017). The antifungal activity of **49** and **51** was far stronger than their cytotoxicity against the human cervical tumour (HeLa) cell line where both compounds exhibited IC_{50} values higher than 50 μM (Silva *et al.*, 2017).

Three new meroterpenoids, emervaridones A (**52**), B and C, together with two new polyketides, varioxiranediols A (**53**) and B, were isolated together with known derivatives from a culture of *Emericella* sp. TJ29, an endophytic fungus obtained from *Hypericum perforatum* roots (He *et al.*, 2017). The isolated compounds were assessed for their antibacterial activity against five drug-resistant microbial pathogens, namely MRSA, *Enterococcus faecalis*, *Klebsiella pneumoniae*, *Pseudomonas aeruginosa* and an extended-spectrum β-lactamase-producing *Escherichia coli* (ESBL-producing *E. coli*). Mervaridone A (**52**) and varioxiranediol A (**53**) were active against the five tested drug-resistant microorganisms, with the highest activity found against ESBL-producing *E. coli* and *P. aeruginosa* with MIC values of 2.0 and 4.0 μg/ml, respectively, compared to the positive control amikacin (MIC = 2.0 μg/ml) (He *et al.*, 2017). The time-kill assays conducted for ESBL-producing *E. coli* indicated that **52** was bacteriostatic while **53** was bactericidal. Based on these findings, both emervaridone A and varioxiranediol A can be considered as new scaffolds for the discovery of antibacterial agents for targeting drug-resistant microbial pathogens.

Another major microbial pathogen is *M. tuberculosis*, which can be fatal if not controlled and eradicated. During the course of ongoing exploration for bioactive fungal metabolites, three isoprenylisoindole alkaloid derivatives, diaporisoindoles A (**54**), B and C, together with their biosynthetic precursor, tenellone C (**55**), were obtained from culture media of the endophytic fungus *Diaporthe* sp. SYSU-HQ3 derived from the mangrove plant *Excoecaria agallocha* (Cui *et al.*, 2017). All isolated compounds were assessed for their inhibitory activity against *M. tuberculosis* protein tyrosine phosphatase B. Compounds **54** and **55** were shown to possess potent inhibitory activities with IC_{50} values of 4.2 and 5.2 μM, respectively, which was even more potent than the positive control oleanolic acid (IC_{50} = 22.1 μM) (Cui *et al.*, 2017). Therefore, diaporisoindole A (**54**) and tenellone C (**55**) can be considered as a basis for the development of new pharmaceuticals for the treatment of tuberculosis.

The flavonoid derivative chlorflavonin (**56**) was isolated from the endophytic fungus *Mucor irregularis* obtained from the Cameroonian medicinal plant *Moringa stenopetala* (Rehberg *et al.*, 2018). Chlorflavonin (**56**) exhibited potent *in vitro* antitubercular activity (MIC_{90} = 1.56 μM) but no cytotoxic effects against the human cell lines MRC-5 and THP-1 up to concentrations of 100 μM (Rehberg *et al.*, 2018). Detailed mechanistic investigation including mapping of resistance-mediating mutations, employing whole-genome sequencing, chemical supplementation assays and molecular docking studies, together with enzymatic characterization, disclosed that chlorflavonin (**56**) specifically inhibits acetohydroxyacid synthase catalytic subunit IlvB1, causing auxotrophies to

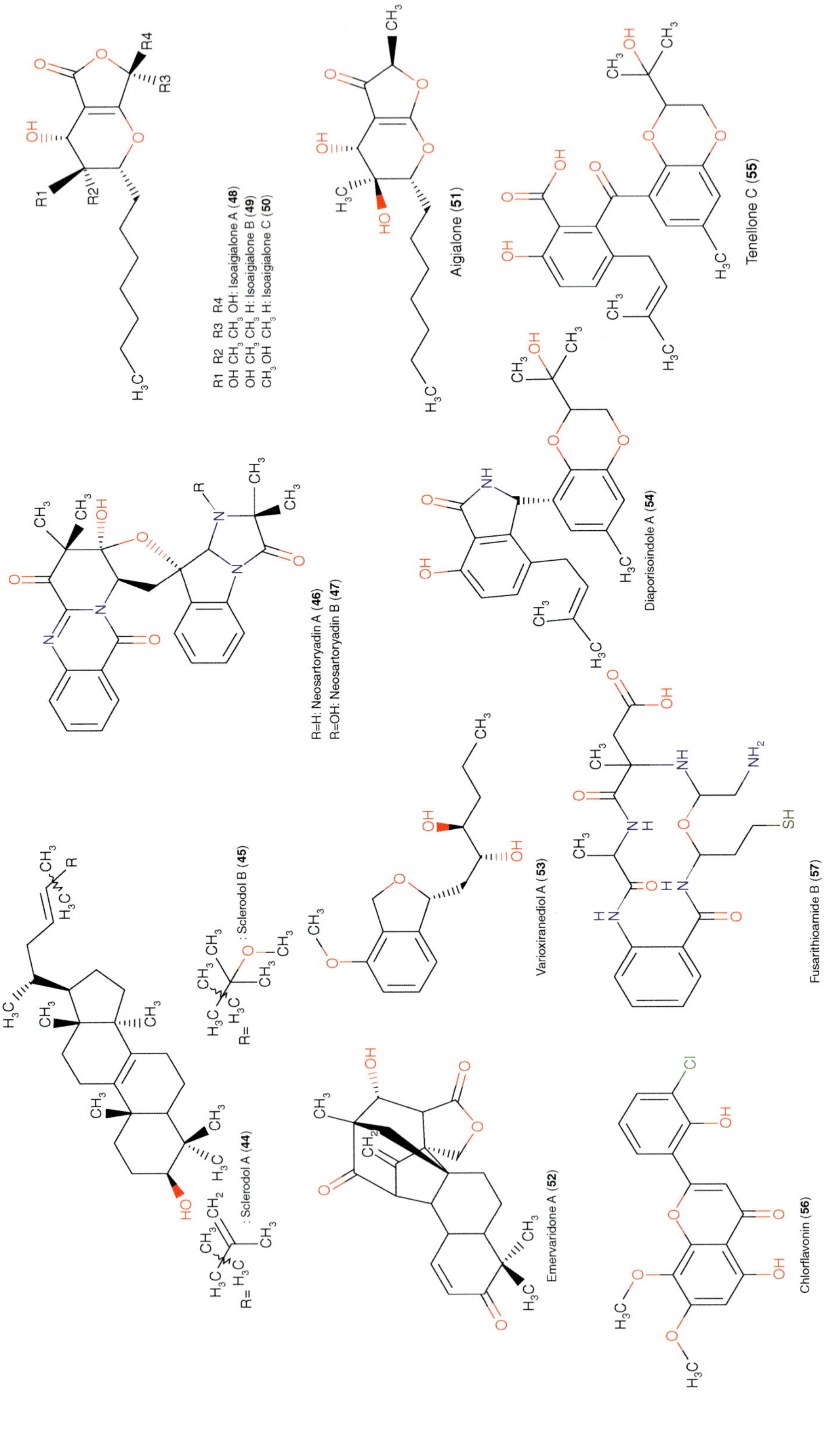

Fig. 10.4. Anti-infective metabolites from endophytes, compounds (**44–57**).

leucine, isoleucine and valine as branched-chain amino acids in addition to pantothenic acid. These amino acids and pantothenic acid cannot be acquired from the host during infection, and hence inhibition of their biosynthesis will attenuate *M. tuberculosis* and cause an increased vulnerability to antibiotic treatment. Interestingly, chlorflavonin (**56**) was found to impart a bacteriostatic effect when used in monotreatment; however, when combined with each of the first-line antibiotics isoniazid and delamanid, it displayed potent synergistic effects leading to a complete sterilization in liquid cultures (Rehberg *et al.*, 2018). Intracellulary, chlorflavonin revealed superior activity compared to streptomycin against *M. tuberculosis* within infected macrophages as assessed using a fluorescent reporter strain of *M. tuberculosis* (Rehberg *et al.*, 2018).

Fusarithioamide B (**57**), a new aminobenzamide derivative with an unprecedented sulfur-containing carbon skeleton, was isolated together with other metabolites from the EtOAc extract of the endophytic fungus *Fusarium chlamydosporium* obtained from *Anvillea garcinii* leaves (Ibrahim *et al.*, 2018). When assessed for its antimicrobial activity against different bacteria and fungi, fusarithioamide B (**57**) revealed potent activity as antifungal agent against *C. albicans* (MIC = 4.0 μM, IZD = 14.5 mm) and was even more potent than the standard antifungal drug clotrimazole (MIC = 8.0 μM, IZD = 17.9 mm) (Ibrahim *et al.*, 2018).

In addition, **57** exhibited potent antibacterial activities against *E. coli*, *Bacillus cereus* and *S. aureus* compared to ciprofloxacin. However, the importance of these results diminished when the compound was subjected to a cytotoxicity (MTT) assay against a panel of six different cell lines revealing potent antiproliferative effects against four of them, namely, BT-549 (breast), MCF-7 (breast), SKOV-3 (ovary) and HCT-116 (colon). In these cell lines the compound showed IC_{50} values between 0.09 and 1.23 μM, which was comparable to the standard anticancer drug doxorubicin (IC_{50} values of 0.05–0.32 μM) (Ibrahim *et al.*, 2018). This may indicate that the antimicrobial activity of fusarithioamide B (**57**) can be due to its cytotoxic effects. In addition, being such a potent cytotoxic compound can be a major limitation in developing antimicrobials based on the core structure of **57**.

Conclusion

Cancer and infectious diseases continue to be leading causes of death worldwide due to resistance evolving against well-established treatment protocols. Therefore, the need for discovering new entities for the treatment of antibiotic-resistant infectious microbial strains and malignant cancer is an urgent demand.

Endophytes have been shown to be a renewable and highly promising source of secondary metabolites featuring novel and/or new scaffolds that possess significant activities.

Therefore, chemical exploration of endophyte cultures for discovering unprecedented molecules with potent anticancer and anti-infective activities against a vast array of viral, fungal or bacterial infections appears to be a promising strategy for bioprospecting.

In this chapter, we surveyed the literature published during the last five years (2013–2018) with emphasis on endophyte-derived compounds showing promising antiproliferative or antimicrobial activities. Among the reported compounds, periconiasins and cytochalasan derivatives revealed selective *in vitro* antiproliferative activity against human ileocecal adenocarcinoma (HCT-8) cells with IC_{50} values in the nanomolar range.

Meleagrin, an indole alkaloidal fungal metabolite, was found to possess potent *in vitro* and *in vivo* antitumour activity against human breast cancer by inhibiting c-Met-dependent metastatic and invasive breast malignancies.

Several further reported metabolites highlighted in this survey may also have the potential to be developed as lead compounds such as altertoxins, which revealed anti-HIV activity equal or even superior to that for azidothymidine (AZT, Zidovudine®), which highlights altertoxins as potential candidates for developing new anti-HIV pharmaceuticals.

Biemodin, an endophytic dianthrone, revealed potent antimicrobial activity against MRSA and can be considered as a potential lead compound for developing new antibiotics combating antibiotic-resistant strains.

Chlorflavonin, identified as an endophytic flavonoidal metabolite, revealed potent *in vitro* antitubercular activity at non-toxic levels through inhibition of acetohydroxy acid synthase catalytic subunit IlvB1. The compound was superior to the well-known antitubercular drug streptomycin within infected macrophages. These findings support chlorflavonin as a potential lead compound for developing new antitubercular drugs.

References

Ali, T., Inagaki, M., Chai, H.-B., Wieboldt, T., Rapplye, C. and Rakotondraibe, L.H. (2017) Halogenated compounds from directed fermentation of *Penicillium concentricum*, an endophytic fungus of the liverwort *Trichocolea tomentella*. *Journal of Natural Products* 80(5), 1397–1403.

Aly, A.H., Debbab, A., Kjer, J. and Proksch, P. (2010) Fungal endophytes from higher plants: A prolific source of phytochemicals and other bioactive natural products. *Fungal Diversity* 41(1), 1–16.

Asakawa, Y., Ludwiczuk, A. and Nagashima, F. (2009) Bryophytes: Bio-and chemical diversity, bioactivity and chemosystematics. Heterocycles 77(1), 99–150.

Au, N., Kuester-Schoeck, E., Mandava, V., Bothwell, L.E., Canny, S.P., *et al.* (2005) Genetic composition of the *Bacillus subtilis* SOS system. *Journal of Bacteriology* 187(22), 7655–7666.

Bara, R., Aly, A.H., Pretsch, A., Wray, V., Wang, B., Proksch, P. and Debbab, A. (2013a) Antibiotically active metabolites from *Talaromyces wortmannii*, an endophyte of *Aloe vera*. *The Journal of Antibiotics* 66(8), 491.

Bara, R., Zerfass, I., Aly, A.H., Goldbach-Gecke, H., Raghavan, V., *et al.* (2013b) Atropisomeric dihydroanthracenones as inhibitors of multiresistant *Staphylococcus aureus*. *Journal of Medicinal Chemistry* 56(8), 3257–3272.

Bashyal, B.P., Wellensiek, B.P., Ramakrishnan, R., Faeth, S.H., Ahmad, N. and Gunatilaka, A.L. (2014) Altertoxins with potent anti-HIV activity from *Alternaria tenuissima* QUE1Se, a fungal endophyte of *Quercus emoryi*. *Bioorganic & Medicinal Chemistry* 22(21), 6112–6116.

Böhler, P., Stuhldreier, F., Anand, R., Kondadi, A.K., Schlütermann, D., *et al.* (2018) The mycotoxin phomoxanthone A disturbs the form and function of the inner mitochondrial membrane. *Cell Death & Disease* 9(3), 286.

Castillo, U., Harper, J.K., Strobel, G.A., Sears, J., Alesi, K., *et al.* (2003) Kakadumycins, novel antibiotics from *Streptomyces* sp. NRRL 30566, an endophyte of *Grevillea pteridifolia*. *FEMS Microbiology Letters* 224(2), 183–190.

Chaturvedi, M., Sung, B., Yadav, V., Kannappan, R. and Aggarwal, B. (2011) NF-κB addiction and its role in cancer: 'One size does not fit all'. *Oncogene* 30(14), 1615.

Chee-Sanford, J.C., Mackie, R.I., Koike, S., Krapac, I.G., Lin, Y.-F., *et al.* (2009) Fate and transport of antibiotic residues and antibiotic resistance genes following land application of manure waste. *Journal of Environmental Quality* 38(3), 1086–1108.

Chowdhury, N.S., Sohrab, M.H., Rana, M.S., Hasan, C.M., Jamshidi, S. and Rahman, K.M. (2017) Cytotoxic naphthoquinone and azaanthraquinone derivatives from an endophytic *Fusarium solani*. *Journal of Natural Products* 80(4), 1173–1177.

Cremasco, M., Hritzko, B. and Linda Wang, N.-H. (2009) Experimental purification of paclitaxel from a complex mixture of taxanes using a simulated moving bed. *Brazilian Journal of Chemical Engineering* 26(1), 207–218.

Cui, H., Lin, Y., Luo, M., Lu, Y., Huang, X. and She, Z. (2017) Diaporisoindoles A–C: Three isoprenylisoindole alkaloid derivatives from the mangrove endophytic fungus *Diaporthe* sp. SYSU-HQ3. *Organic Letters* 19(20), 5621–5624.

Debbab, A., Aly, A.H., Edrada-Ebel, R., Wray, V., Müller, W.E.G., *et al.* (2009) Bioactive metabolites from the endophytic fungus *Stemphylium globuliferum* isolated from *Mentha pulegium*. *Journal of Natural Products* 72(4), 626–631.

Demain, A.L. and Elander, R.P. (1999) The β-lactam antibiotics: Past, present and future. *Antonie van Leeuvenhoek* 75, 5–19.

Ebrahim, W., Aly, A.H., Wray, V., Mándi, A., Teiten, M.-H., *et al.* (2013) Embellicines A and B: Absolute configuration and NF-κb transcriptional inhibitory activity. *Journal of Medicinal Chemistry* 56(7), 2991–2999.

Ganley, R.J., Brunsfeld, S.J. and Newcombe, G. (2004) A community of unknown, endophytic fungi in western white pine. *Proceedings of the National Academy of Sciences of the United States of America* 101(27), 10107–10112.

Gao, H., Li, G. and Lou, H.-X. (2018) Structural diversity and biological activities of novel secondary metabolites from endophytes. *Molecules* 23(3), 646.

Gouda, S., Das, G., Sen, S.K., Shin, H.-S. and Patra, J.K. (2016) Endophytes: A treasure house of bioactive compounds of medicinal importance. *Frontiers in Microbiology* 7, 1538.

Harper, J.K., Arif, A.M., Ford, E.J., Strobel, G.A., Porco, J.A., *et al.* (2003) Pestacin: A 1,3-dihydro isobenzofuran from *Pestalotiopsis microspora* possessing antioxidant and antimycotic activities. *Tetrahedron* 59(14), 2471–2476.

He, Y., Hu, Z., Li, Q., Huang, J., Li, X.-N., *et al.* (2017) Bioassay-guided isolation of antibacterial metabolites from *Emericella* sp. TJ29. *Journal of Natural Products* 80(9), 2399–2405.

Hearst, R., Nelson, D., McCollum, G., Millar, B.C., Maeda, Y., *et al.* (2009) An examination of antibacterial and antifungal properties of constituents of Shiitake (*Lentinula edodes*) and Oyster (*Pleurotus ostreatus*) mushrooms. *Complementary Therapies in Clinical Practice* 15(1), 5–7.

Hemphill, C.F.P., Sureechatchaiyan, P., Kassack, M., Orfali, R., Lin, W., Daletos, G. and Proksch, P. (2017) OSMAC approach leads to new fusarielin metabolites from *Fusarium tricinctum*. *The Journal of Antibiotics* 70(6), 726–732.

Huang, X., Huang, H., Li, H., Sun, X., Huang, H., *et al.* (2013) Asperterpenoid A, a new sesterterpenoid as an inhibitor of *Mycobacterium tuberculosi* s protein tyrosine phosphatase B from the culture of *Aspergillus* sp. 16-5c. *Organic Letters* 15(4), 721–723.

Huang, Z., Cai, X., Shao, C., She, Z., Xia, X., *et al.* (2008) Chemistry and weak antimicrobial activities of phomopsins produced by mangrove endophytic fungus *Phomopsis* sp. ZSU-H76. *Phytochemistry* 69(7), 1604–1608.

Ibrahim, S.R., Mohamed, G.A., Al Haidari, R.A., Zayed, M.F., El-Kholy, A.A., Elkhayat, E.S. and Ross, S.A. (2018) Fusarithioamide B, a new benzamide derivative from the endophytic fungus *Fusarium chlamydosporium* with potent cytotoxic and antimicrobial activities. *Bioorganic & Medicinal Chemistry* 26(3), 786–790.

Jiao, W., Feng, Y., Blunt, J.W., Cole, A.L.J. and Munro, M.H.G. (2004) Chaetoglobosins Q, R and T, three further new metabolites from *Chaetomium globosum*. *Journal of Natural Products* 67(10), 1722–1725.

Joseph, B. and Priya, R.M. (2011) Bioactive compounds from endophytes and their potential in. *American Journal of Biochemistry and Molecular Biology* 1(3), 291–309.

Kabbaj, F.Z., Lu, S., Faouzi, M.E.A., Meddah, B., Proksch, P., *et al.* (2015) Bioactive metabolites from *Chaetomium aureum*: Structure elucidation and inhibition of the Hsp90 machine chaperoning activity. *Bioorganic & Medicinal Chemistry* 23(1), 126–131.

Kanagasabai, R., Zhou, W., Liu, J., Minh Nguyen, T.T., Veeramachaneni, P. and Nes, W.D. (2004) Disruption of ergosterol biosynthesis, growth, and the morphological transition in *Candida albicans* by sterol methyltransferase inhibitors containing sulfur at C-25 in the sterol side chain. *Lipids* 39(8), 737–746.

Kour, A., Shawl, A.S., Rehman, S., Sultan, P., Qazi, P.H., *et al.* (2008) Isolation and identification of an endophytic strain of *Fusarium oxysporum* producing podophyllotoxin from *Juniperus recurva*. *World Journal of Microbiology and Biotechnology* 24(7), 1115–1121.

Kumar, S., Aharwal, R.P., Shukla, H., Rajak, R. and Sandhu, S.S. (2014) Endophytic fungi: As a source of antimicrobials bioactive compounds. *World Journal of Pharmacy and Pharmaceutical Sciences* 3, 1179–1197.

Kusari, S., Zühlke, S. and Spiteller, M. (2009a) An endophytic fungus from *Camptotheca acuminata* that produces camptothecin and analogues. *Journal of Natural Products* 72(1), 2–7.

Kusari, S., Lamshöft, M. and Spiteller, M. (2009b) *Aspergillus fumigatus* Fresenius, an endophytic fungus from *Juniperus communis* L. Horstmann as a novel source of the anticancer pro-drug deoxypodophyllotoxin. *Journal of Applied Microbiology* 107(3), 1019–1030.

Lam, K.S. (2007) New aspects of natural products in drug discovery. *Trends in Microbiology* 15(6), 279–289.

Lazarevic, V., Dusterhoft, A., Soldo, B., Hilbert, H., Manuel, C. and Karamata, D. (1999) Nucleotide sequence of the *Bacillus subtilis* temperate bacteriophage SPbetac2. *Microbiology* 145(5), 1055–1067.

Li, C.-S., Ren, G., Yang, B.-J., Miklossy, G., Turkson, J., *et al.* (2016) Meroterpenoids with antiproliferative activity from a Hawaiian-plant associated fungus *Peyronellaea coffeae-arabicae* FT238. *Organic Letters* 18(10), 2335–2338.

Li, J.-Y., Strobel, G., Sidhu, R., Hess, W. and Ford, E.J. (1996) Endophytic taxol-producing fungi from bald cypress, *Taxodium distichum*. *Microbiology* 142(8), 2223–2226.

Li, X., Tian, Y., Yang, S.-X., Zhang, Y.-M. and Qin, J.-C. (2013) Cytotoxic azaphilone alkaloids from *Chaetomium globosum* TY1. *Bioorganic & Medicinal Chemistry Letters* 23(10), 2945–2947.

Lin, T., Wang, G., Shan, W., Zeng, D., Ding, R., *et al.* (2014a) Myrotheciumones: Bicyclic cytotoxic lactones isolated from an endophytic fungus of *Ajuga decumbens*. *Bioorganic & Medicinal Chemistry Letters* 24(11), 2504–2507.

Lin, T., Wang, G., Zhou, Y., Zeng, D., Liu, X., *et al.* (2014b) Structure elucidation and biological activity of two new trichothecenes from an endophyte *Myrothecium roridum*. *Journal of Agricultural and Food Chemistry* 62(25), 5993–6000.

Liu, J., Luo, J., Ye, H., Sun, Y., Lu, Z. and Zeng, X. (2009) Production, characterization and antioxidant activities *in vitro* of exopolysaccharides from endophytic bacterium *Paenibacillus polymyxa* EJS-3. *Carbohydrate Polymers* 78(2), 275–281.

Mady, M.S., Mohyeldin, M.M., Ebrahim, H.Y., Elsayed, H.E., Houssen, W.E., *et al.* (2016) The indole alkaloid meleagrin, from the olive tree endophytic fungus *Penicillium chrysogenum*, as a novel lead for the control of c-Met-dependent breast cancer proliferation, migration and invasion. *Bioorganic & Medicinal Chemistry* 24(2), 113–122.

Morandini, L.M., Neto, A.T., Pedroso, M., Antoniolli, Z.I., Burrow, R.A., *et al.* (2016) Lanostane-type triterpenes from the fungal endophyte *Scleroderma* UFSMSc1 (Persoon) Fries. *Bioorganic & Medicinal Chemistry Letters* 26(4), 1173–1176.

Nair, D.N. and Padmavathy, S. (2014) Impact of endophytic microorganisms on plants, environment and humans. *The Scientific World Journal* 2014, 250693.

Noumeur, S.R., Helaly, S.E., Jansen, R., Gereke, M., Stradal, T.E.B., Harzallah, D. and Stadler, M. (2017) Preussilides A–F, bicyclic polyketides from the endophytic fungus *Preussia similis* with antiproliferative activity. *Journal of Natural Products* 80(5), 1531–1540.

Oh, H., Swenson, D.C., Gloer, J.B. and Shearer, C.A.(2001) Massarilactones A and B: Novel secondary metabolites from the freshwater aquatic fungus *Massarina tunicata*. *Tetrahedron Letters* 42(6), 975–977.

Padumadasa, C., Xu, Y.-M., Wijeratne, E.M.K., Espinosa-Artiles, P., U'Ren, J.M., Arnold, A.E. and Gunatilaka, A.A.L. (2018) Cytotoxic and noncytotoxic metabolites from *Teratosphaeria* sp. FL2137, a fungus associated with *Pinus clausa*. *Journal of Natural Products* 81(3), 616–624.

Pandi, M., Manikandan, R. and Muthumary, J. (2010) Anticancer activity of fungal taxol derived from *Botryodiplodia theobromae* Pat., an endophytic fungus, against 7,12 dimethyl benz(a)anthracene (DMBA)-induced mammary gland carcinogenesis in Sprague Dawley rats. *Biomedicine and Pharmacotherapy* 64(1), 48–53.

Pang, X., Zhao, J.-Y., Fang, X.-M., Zhang, T., Zhang, D.-W., *et al.* (2017) Metabolites from the plant endophytic fungus *Aspergillus* sp. CPCC 400735 and their anti-HIV activities. *Journal of Natural Products* 80(10), 2595–2601.

Pimentel, M.R., Molina, G., Dionísio, A.P., Maróstica, M.R. Jr and Pastore, G.M. (2011) The use of endophytes to obtain bioactive compounds and their application in biotransformation process. *Biotechnology Research International* 2011, 576286. doi:10.4061/2011/576286

Puri, S.C., Verma, V., Amna, T., Qazi, G.N. and Spiteller, M. (2005) An endophytic fungus from *Nothapodytes foetida* that produces camptothecin. *Journal of Natural Products* 68(12), 1717–1719.

Qin, J.-C., Zhang, Y.-M., Gao, J.-M., Bai, M.-S., Yang, S.-X., Laatsch, H. and Zhang, A.-L. (2009) Bioactive metabolites produced by *Chaetomium globosum*, an endophytic fungus isolated from *Ginkgo biloba*. *Bioorganic & Medicinal Chemistry Letters* 19(6), 1572–1574.

Rahalison, L., Hamburger, M., Hostettmann, K., Monod, M. and Frenk, E. (1991) A bioautographic agar overlay method for the detection of antifungal compounds from higher plants. *Phytochemical Analysis* 2(5), 199–203.

Rangel-Grimaldo, M., Rivero-Cruz, I., Madariaga-Mazón, A., Figueroa, M. and Mata, R. (2017) α-Glucosidase inhibitors from *Preussia minimoides*. *Journal of Natural Products* 80(3), 582–587.

Rehberg, N., Akone, H.S., Ioerger, T.R., Erlenkamp, G., Daletos, G., *et al.* (2018) Chlorflavonin targets acetohydroxyacid synthase catalytic subunit IlvB1 for synergistic killing of *Mycobacterium tuberculosis*. *ACS Infectious Diseases* 4(2), 123–134.

Rönsberg, D., Debbab, A., Mándi, A., Vasylyeva, V., Böhler, P., *et al.* (2013) Pro-apoptotic and immunostimulatory tetrahydroxanthone dimers from the endophytic fungus *Phomopsis longicolla*. *The Journal of Organic Chemistry* 78(24), 12409–12425.

Shen, L., Zhu, L., Tan, Q., Wan, D., Xie, J. and Peng, J. (2016) New cytotoxic trichothecene macrolide epimers from endophytic *Myrothecium roridum* IFB-E012. *The Journal of Antibiotics* 69, 652.

Siegel, R.L., Miller, K.D. and Jemal, A. (2015) Cancer statistics, 2015. *CA: A Cancer Journal for Clinicians* 65(1), 5–29.

Signoretto, C., Canepari, P., Stauder, M., Vezzulli, L. and Pruzzo, C. (2012) Functional foods and strategies contrasting bacterial adhesion. *Current Opinion in Biotechnology* 23(2), 160–167.

Silva, G.H., Zeraik, M.L., de Oliveira, C.M., Teles, H.L., Trevisan, H.C., *et al.* (2017) Lactone derivatives produced by a *Phaeoacremonium* sp., an endophytic fungus from *Senna spectabilis*. *Journal of Natural Products* 80(5), 1674–1678.

Song, Y.C., Huang, W.Y., Sun, C., Wang, F.W. and Tan, R.X. (2005) Characterization of graphislactone A as the antioxidant and free radical-scavenging substance from the culture of

Cephalosporium sp. IFB-E001, an endophytic fungus in *Trachelospermum jasminoides*. *Biological and Pharmaceutical Bulletin* 28(3), 506–509.

Stierle, A., Strobel, G. and Stierle, D. (1993) Taxol and taxane production by *Taxomyces andreanae*, an endophytic fungus of Pacific yew. *Science* 260(5105), 214–216.

Stone, J.K., Bacon, C.W. and White J.F. Jr (2000) An overview of endophytic microbes: Endophytism defined. In: Bacon, C.W. and White J.F. Jr (eds) *Microbial Endophytes*. Marcel Dekker, Inc., New York, pp. 3–30.

Strobel, G.A. (2003) Endophytes as sources of bioactive products. *Microbes and Infection* 5(6), 535–544.

Teiten, M.-H., Mack, F., Debbab, A., Aly, A.H., Dicato, M., Proksch, P. and Diederich, M. (2013) Anticancer effect of altersolanol A, a metabolite produced by the endophytic fungus *Stemphylium globuliferum*, mediated by its pro-apoptotic and anti-invasive potential via the inhibition of NF-κB activity. *Bioorganic & Medicinal Chemistry* 21(13), 3850–3858.

Urban, A., Eckermann, S., Fast, B., Metzger, S., Gehling, M., *et al.* (2007) Novel whole-cell antibiotic biosensors for compound discovery. *Applied and Environmental Microbiology* 73(20), 6436–6443.

Venturini, M., Rivera, C., Gonzalez, C. and Blanco, D. (2008) Antimicrobial activity of extracts of edible wild and cultivated mushrooms against foodborne bacterial strains. *Journal of Food Protection* 71(8), 1701–1706.

Verekar, S.A., Mishra, P.D., Sreekumar, E.S., Deshmukh, S.K., Fiebig, H.-H., Kelter, G. and Maier, A. (2014) Anticancer activity of new depsipeptide compound isolated from an endophytic fungus. *The Journal of Antibiotics* 67(10), 697–701.

Wasser, S.P. and Weis, A.L. (1999) Medicinal properties of substances occurring in higher basidiomycetes mushrooms: Current perspectives. *International Journal of Medicinal Mushrooms* 1(1), 47–50.

Whitesell, L. and Lindquist, S.L. (2005) HSP90 and the chaperoning of cancer. *Nature Reviews Cancer* 5, 761.

Yu, G., Zhou, G., Zhu, M., Wang, W., Zhu, T., Gu, Q. and Li, D. (2015) Neosartoryadins A and B, fumiquinazoline alkaloids from a mangrove-derived fungus *Neosartorya udagawae* HDN13-313. *Organic Letters* 18(2), 244–247.

Zhang, D., Ge, H., Xie, D., Chen, R., Zou, J.-H., Tao, X. and Dai, J. (2013) Periconiasins A-C, new Cytotoxic cytochalasans with an unprecedented 9/6/5 tricyclic ring system from endophytic fungus *Periconia* sp. *Organic Letters* 15(7), 1674–1677.

Zhang, D., Tao, X., Chen, R., Liu, J., Li, L., *et al.* (2015) Pericoannosin A, a polyketide synthase-nonribosomal peptide synthetase hybrid metabolite with new carbon skeleton from the endophytic fungus *Periconia* sp. *Organic Letters* 17(17), 4304–4307.

Zhang, P., Mandi, A., Li, X.-M., Du, F.-Y., Wang, J.-N., *et al.* (2014) Varioxepine A, a 3H-oxepine-containing alkaloid with a new oxa-cage from the marine algal-derived endophytic fungus *Paecilomyces variotii*. *Organic Letters* 16(18), 4834–4837.

Zhou, H., Sun, X., Li, N., Che, Q., Zhu, T., Gu, Q. and Li, D. (2016) Isoindolone-containing meroperpenoids from the endophytic fungus *Emericella nidulans* HDN12-249. *Organic Letters* 18(18), 4670–4673.

11 Prospects for Biotechnological Exploitation of Endophytes Using Functional Metagenomics

Melanie-Maria Obermeier[1,2] and Christina A. Müller Bogotá[1*]

[1]Institute of Environmental Biotechnology, Graz University of Technology, Graz, Austria; [2]Austrian Centre of Industrial Biotechnology, Graz, Austria

Abstract

The usage of natural products, especially in the treatment of diseases, has a long history. While natural products used to be administered directly, they today serve as lead compounds and structural scaffolds for the development of new drugs and other market products. The success of combinatorial approaches to develop new products strongly depends on natural product-likeness. This exemplifies the importance of natural products as structural leads during product development and demonstrates natural product discovery to be as important as ever. This chapter highlights endophytes as a rich bioresource for the identification of novel natural compounds and emphasizes functional metagenomics as a promising method to source the endophytic potential. With the majority of microorganisms not readily cultivable under laboratory conditions, a vast amount of natural products synthesized by endophytes remains inaccessible. Functional metagenomics circumvents current cultivation limitations by direct cloning of bacterial community DNA. This procedure is, however, rarely performed exclusively on endophytes. This chapter outlines the procedures underlying this methodology with focus on its application to endophytes.

11.1 Introduction

Plants, animals and microorganisms naturally produce innumerable metabolites and metabolic by-products of medical or industrial value. This value has been recognized long-since. The molecules produced by living organisms, generally referred to as natural products (NPs), find application in the medical, agricultural and industrial sectors as pharmaceuticals, agrochemicals and food additives (David *et al.*, 2015; Schmitt *et al.*, 2016; Sparks *et al.*, 2017; Lorenzo *et al.*, 2018; Ribes *et al.*, 2018). They either are commercialized directly or serve as lead compounds and structural templates for effective product development. Today NPs and their derivatives constitute a high share among market products (Newman and Cragg, 2016; Patridge *et al.*, 2016). Due to their great structural diversity and manifold and specific bioactivity, they continue to constitute a highly promising resource for the identification of novel drugs, pesticides or preservatives.

Novel therapeutics and pesticides are required in the wake of the ever-increasing emergence of resistances to treat infections and ensure food security. Rising numbers of cancer patients and those suffering from chronic diseases urges the development of new treatments. Industries have been attempting to meet this demand for new lead compounds and scaffolds, primarily by combinatorial chemistry, since the 1990s. By parallel synthesis utilizing several building

* Corresponding author e-mail: christina.mueller@tugraz.at

blocks simultaneously, this approach enables the synthesis of large compound collections consisting of diverse variants of the starting material. Combinatorial chemistry, hence, represents until today a rapid way to generate molecular diversity. With the simultaneously expanding high-throughput screening methodologies, which allowed for fast screening of large compound libraries, this approach was believed to lead to compound discovery much quicker as compared to identifying a bioactive compound of interest from the complex NP mixture present in biological samples. This caused, among other things, the decline of NP discovery programmes. Over the last decades various synthesis methods were established, and while success stories were reported. Combinatorial chemistry could not yet meet the high expectations that were placed on this approach (Ortholand and Ganesan, 2004; Lindell *et al.*, 2009; Liu *et al.*, 2017). The limited structural diversity of the generated molecules counts as one potential reason. Two studies conducted early on found significant differences in the chemical structures of NPs and those of combinatorial compounds. Both evaluations found NPs to span a wider range in molecular weight and to generally comprise a higher amount of oxygen atoms but fewer nitrogen, sulfur and halogen atoms. Furthermore, they showed that chiral centres, which add to steric complexity and account as an important determinant for selectivity, are prominent in NPs and mostly missing in synthetic compounds. NPs were generally determined to be of more complex steric structures (Henkel *et al.*, 1999; Feher and Schmidt, 2002). Especially, Henkel *et al.* took a strong stand emphasizing the importance of NPs with their vast chemical diversity as source for lead discovery. By comparing compounds from an NP and a synthetics database, they estimated that synthetic compounds only represent 60% of the NPs' structural diversity (Henkel *et al.*, 1999).

As the generation of compound libraries continued and evolved to increase the number of synthesized molecules, the concept of the 'chemical space' arose. Seen as the possible structural diversity of molecules, estimates range in several orders of magnitude. Depending on the chosen parameters, such as the number and types of atoms, or whether it is a chemical or a peptide, calculations determined the possible number of molecules between 10^8 and 10^{390}, whereby only a fraction might show bioactivity (Medina-Franco *et al.*, 2008). Chemoinformatic analyses to compare and evaluate the chemical space of NPs, combinatorial libraries, pharmaceuticals and other marketed molecules unroll 'NP-likeness' as an important criterion to be considered when expanding the current available chemical space by designing and generating new compound collections (Medina-Franco, 2012). This necessity is exemplified in the latest survey of Newman and Cragg, which states that the antitumour drugs sorafenib and vemurafenib and the pharmaceutical ataluren account as the only three de novo synthesized combinatorial molecules entering the pharmaceutical market (Newman and Cragg, 2016). However, NPs, their derivatives and NP-like molecules contribute to more than half of the market products and represent, especially as lead compounds, a valuable asset for product development (Newman and Cragg, 2016; Sparks *et al.*, 2017). Along with the trend of generating more NP-like compound libraries, NP discovery experiencing a revival (McChesney *et al.*, 2007; Shen, 2015).

Similarly, novel enzymes, which are generally not considered NPs, are in demand for the food and pharmaceutical industries (Coughlan *et al.*, 2015). As outlined in detail by Coughlan *et al.*, microbial enzymes find inter alia application in food processing, flavouring, dairy products, brewing and baking, and novel bio-catalysts for the various processes are required: lipases for milk fat hydrolysis, esterases for flavour production in the beverage industry, α-amylases for starch modification in the baking industry, among other things. Another class of enzymes which caught considerable attention over recent years is non-ribosomal peptide synthetases and polyketide synthases. As producers of secondary metabolites, many bioactive compounds synthesized by these enzymes have been identified (Nikolouli and Mossialos, 2012).

Only a fraction of all the natural resources are exploited (Henkel *et al.*, 1999; Bérdy,

2012). The diversity of microbial NPs is mostly not yet investigated. According to estimates, the diversity of bacterial compounds smaller than 1 kDa reaches 10^9 molecules (Davies, 2007) of which merely 60–80,000 metabolites are identified (Bérdy, 2012). Interestingly, Bérdy (2012) reported that almost half of the known microbial NPs showed bioactivity, exemplifying the great potential of microbial NPs in general. This goes well in hand with an evaluation by Patridge *et al.* showing that half of the Food and Drug Administration (FDA)-approved NPs originate from bacteria and fungi (Patridge *et al.*, 2016). While bacteria and fungi regain importance as bio-resource after a drop-down following the golden era in microbial drug discovery from 1960 to 1980, screenings focus primarily on microbial communities in soil, sludge and rhizosphere. The plant-associated and especially plant-inhabiting bacteria and fungi are widely unexplored (Strobel and Daisy, 2003; Müller *et al.*, 2016a).

Each plant is inhabited by a diverse and specific community of microorganisms which fulfil important functions for the host including nutrient supply, plant development, pathogen defence and stress resilience (Hardoim *et al.*, 2008; Brader *et al.*, 2014; Hardoim *et al.*, 2015). It was estimated that between 250,000 and 500,000 higher plant species populate the planet (McChesney *et al.*, 2007), of which only a fraction has been investigated for their endophytes (Strobel and Daisy, 2003). The species richness and diversity of endophytes is staggering. According to estimates, 1 million fungal endophytes can be found (Dreyfuss and Chapela, 1994). Drivers for the endophytic community composition are: (i) biotic factors such as plant genotype, plant physiology, the microorganisms present in the surrounding bulk soil; and (ii) abiotic factors like soil type, nutrient availability, temperature (Hardoim *et al.*, 2015; Kandel *et al.*, 2017). Interestingly, plants actively recruit certain endophytes from the bulk soil (Lemanceau *et al.*, 1995; Kloepper *et al.*, 2004; Rudrappa *et al.*, 2008) and even propagate them to their offspring (Berg *et al.*, 2014; Truyens *et al.*, 2015; Frank *et al.*, 2017). Similarly, mosses and other lower plants comprise endophytic communities, whereby the community structure underlies the same principles (Bragina *et al.*, 2012). These highly adapted microbial consortia represent widely unexplored bio-resources which comprise a unique set of metabolic pathways and a tremendous amount of NPs and enzymes (Strobel and Daisy, 2003; Gunatilaka, 2006; Brader *et al.*, 2014; Müller *et al.*, 2016a).

This chapter focuses on the discovery of NPs and enzymes of microbial origin, highlighting endophytes as a plenteous and yet less explored bio-resource. Thereby, we assess the potential of sourcing endophytes by functional metagenomics, a methodology which facilitates the access to unexplored NPs and enzymes of yet uncultivable microorganisms.

11.2 Functional Metagenomics of Endophytes

The biotechnological potential of endophytes has long been recognized and sourced for NPs, which led to the isolation and identification of many novel compounds as summarized in various reviews (Tan and Zou, 2001; Strobel and Daisy, 2003; Gunatilaka, 2006; Zhang *et al.*, 2006; Chen *et al.*, 2014; Deshmukh *et al.*, 2014; Newman and Cragg, 2015; Martinez-Klimova and Rodríguez-Peña, 2017; Nalini and Prakash, 2017; Gao *et al.*, 2018). The therein described work exploited the endophytic potential exclusively by cultivation-dependent techniques. Such screenings, however, allow only a glimpse into the endophytic treasure chest of metabolites, metabolic by-products and enzymes. The great plate count anomaly (Stewart, 2012), the fact that most microorganisms are not readily cultivable under laboratory conditions, applies to endophytes. Generally, about 99% of all microorganisms account as not yet cultivable by standard methods (Strobel and Daisy, 2003). Following this, NPs and enzymes of the majority of the species inhabiting a plant are not accessible by cultivation-dependent techniques.

Cloning of microbial community DNA from a sample directly into a surrogate host represents a promising methodology to circumvent cultivation dependency. The great potential lies therein, that this can be applied to any microbial community (Ravin *et al.*, 2015). As heterologous expression of such DNA facilitates access to hidden genes and biosynthetic gene clusters, this method allows identification of novel NPs and enzymes from yet uncultivable microorganisms (Coughlan *et al.*, 2015; Katz *et al.*, 2016), including endophytes. This methodology belongs to the omics approaches and generally investigates the collective genomes of all members of a microbial community, the metagenome (Handelsman *et al.*, 1998). Mainly grouped into sequence- and function-based approaches, metagenomics comprises different methods as outlined by different reviews (Handelsman, 2004; Simon and Daniel, 2011; Coughlan *et al.*, 2015; Ravin *et al.*, 2015). Metagenomics in its broadest interpretation also comprises amplicon sequencing of conserved gene regions, mostly the 16S rRNA marker gene for bacteria or the ITS region for fungi. Inferred from such species-specific marker genes, the phylogenetic community composition can be analysed. Based on knowledge about functional traits of single phyla, the functional potential of a microbiome can in part be evaluated. In comparison, shotgun sequencing of adapter-ligated metagenomic DNA fragments or metagenome clone libraries allows a detailed analysis of functional properties based on gene homologies. Annotated genes can then be cloned for heterologous expression. Neither approach, however, leads directly to the identification of new genes and consequently novel NPs or enzymes, which is where functional metagenomics comes into play. This methodology can lead to the identification of truly novel NPs and enzymes. Functional metagenomics is based on direct cloning of metagenomic DNA and, therefore, requires the isolation of high-quality metagenomic DNA, fragmentation and cloning of the metagenomic DNA into a vector and transformation into a host. The generated metagenomic library can then either be explored by polymerase chain reaction (PCR)-based screenings using degenerated primers which bind to conserved regions of the gene target of interest or by phenotypic screenings for a desired activity. As this chapter assesses the exploitation of the endophytic potential by phenotypic-based functional metagenomics, this approach will be elaborated on in the following.

Historically, the first milestone for functional metagenomics was laid by Pace *et al.* After proposing direct cloning of metagenomic DNA in 1986 (Pace *et al.*, 1986), they constructed the first metagenomic library five years later for 16S rRNA gene sequencing and phylogenetic analysis of picoplankton-associated microorganisms (Schmidt *et al.*, 1991). Almost another five years later, the first successful phenotypic screening of a metagenomic library resulted in the identification of four novel cellulases (Healy *et al.*, 1995). Functional metagenomics has since then led to the discovery of many novel enzymes and bioactives from uncultivable microorganisms (Coughlan *et al.*, 2015). The identification process generally involves two steps: the generation and subsequently the functional screening of metagenomic libraries. The process can be further broken down into several steps: plant selection and sampling, surface sterilization including the indispensable usage of a DNA-degrading agent such as sodium hypochlorite (as an additional step when metagenomic DNA of only endophytes is desired), microbial enrichment, DNA isolation, cloning, host transformation and screening (Fig. 11.1). Once a metagenomic clone has been identified, biosynthesis genes contained on the metagenomic DNA insert are more readily identified as compared to identifying the gene clusters within the whole genome of isolated microorganisms.

The selection of a plant species for constructing of a metagenomic clone library (Fig. 11.1A) certainly depends on the molecule or the activity that is targeted for screening. A very promising way of selecting a plant as source of endophytes, not only for isolation of microorganisms but for isolation of metagenomic DNA, is the so-called ethnobotanical approach (Cox and Balick, 1994). Thus, the ethnobotanical knowledge from native people or the traditional use of plants in herbal medicine for treatment of diseases is employed to find interesting candidates. Once the selected plant is collected,

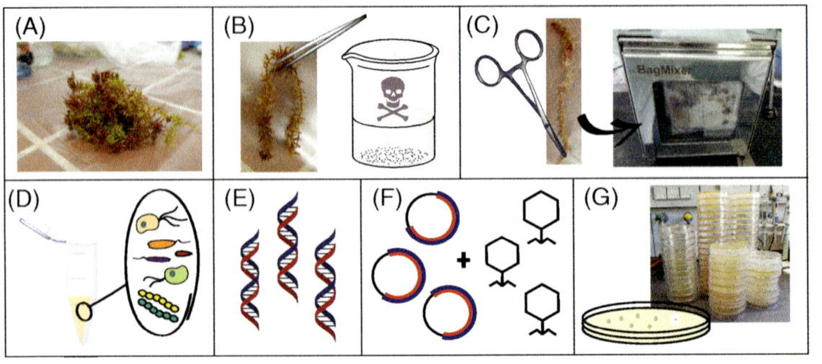

Fig. 11.1. Generating of a metagenomic clone library from endophytes. The following steps are necessary for the construction of a clone library, here exemplarily shown for moss: (A) Plant selection and sampling. (B) Surface sterilization of the fresh plant material including a DNA-degrading step using, for instance, sodium hypochlorite. (C) The plant material is treated by mechanical or enzymatic methods to enrich the microbial fraction (e.g. cutting, homogenization, treatment with salt or detergents, bag-mixing, centrifugation). (D and E) The enriched endophytic microbiome is used for isolation of metagenomic DNA. (F) The DNA is ligated into an appropriate vector system and cloned into the library host, e.g. using fosmids and packaging them into phages prior to transfection of the host. (G) The generated clone library undergoes an activity- or sequence-based screening, which is designed to identify a desired activity or specific genetic traits.

the treatment of the plant material for construction of the library should proceed without delay to avoid contamination or loss of microbial diversity.

Surface sterilization of the plant material is the most commonly employed technique to obtain solely endophytic microorganisms (Hallmann *et al.*, 2006; Fig. 11.1B). Furthermore, high-quality DNA must be obtained from the selected material. In general, different isolation protocols and commercial kits are available but mainly for soil and water samples (Leis *et al.*, 2013). When working with endophyte communities a big challenge resides in the extraction of endophyte DNA in enough quality and quantity for library construction. A study by Gabor *et al.* reports that direct DNA isolation, mechanically, enzymatically or using detergents, yields higher DNA amounts than indirect approaches for which the microbial cells are extracted and enriched prior to cell lysis. When evaluating the isolated DNA, they showed that lysates obtained by direct methods contained considerably higher amounts of eukaryotic DNA (>50%) as compared with indirect methods which yielded lysates containing more than 90% bacterial DNA (Gabor *et al.*, 2003). As metagenomic libraries contain several thousand clones of which only a small amount will exhibit the desired phenotype (Handelsman, 2004), it is desirable that metagenomic libraries contain as much endophytic DNA as possible. As mentioned above, extracting enough high-quality DNA from endophytes for library production is challenging. The high yield of plant DNA obtained by standard methods and the fact that the plant DNA interferes with further processing and analysis steps make usually an enrichment of the microbial fraction indispensable (Jiao *et al.*, 2006; Fig. 11.1C). Due to this restraint, only a few examples of library construction from endophyte DNA are available.

A few methods for enriching the plant microbiome, e.g. from stems, and simultaneously reducing or eliminating the plant DNA have been described (Wang *et al.*, 2008; Ikeda *et al.*, 2009). Wang *et al.* extracted and enriched microbial DNA from stem bark material using a combined treatment with salt and detergent (0.9% NaCl, 0.063% sodium dodecyl sulfate) for disrupting plastids and eliminating plant DNA (Wang *et al.*, 2008). A similar method was proposed by Ikeda *et al.* for enrichment of the bacterial fraction in soybean stems (Ikeda *et al.*, 2009). In this case Triton X-100 was used as a mild detergent for disrupting the membranes of chloroplasts, in combination with a density gradient centrifugation using the non-ionic medium Nycodenz. Other microbial enrichment methods, for example, for sugarcane stems (Dos-Santos *et al.*, 2017) or for leaves from the *Maytenus hookeri* tree (Jiao *et al.*, 2006), have been reported; the latter involves the enzymatic hydrolysis of plant cell walls and differential centrifugation. The method of choice will ultimately depend on the plant morphology and composition and has to be adapted and evaluated for each individual case.

For library establishment (Fig. 11.1D–F), several considerations have to be taken into account. Choice of the vector–host system is crucial, as successful expression is a necessity for the later screening. Heterologous gene expression depends on various factors. Different codon usage and the metabolic background of the surrogate host in comparison to that of the species from which the metagenomic DNA insert derives may differ greatly and can hamper gene expression (Liebl *et al.*, 2014). The most widely used host strain is *E. coli* for which many protocols and commercial kits for DNA extraction and library generation exist, aiding in the procedure and making it more efficient (Simon and Daniel, 2011). Yet the predicted potential of *E. coli* to heterologously express metagenomic DNA varies greatly ranging from as low as 7% and up to 73%, whereby expression of one-third of the genes depends on the vector promoter (Gabor *et al.*, 2004). This drives the interest towards different species such as *Pseudomonas* or *Streptomyces* strains for library generation (Liebl *et al.*, 2014). Yeast could be theoretically employed as an alternative host for the directed expression of eukaryotic metagenomic DNA. Cloning vectors containing a broad-host replicon such as the RK2-based plasmid have been developed (Aakvik *et al.*, 2009). This vector system can be, in principle, transferred to Gram-negative and Gram-positive bacteria and also eukaryotic hosts like yeast. Nevertheless, there are no reports available detailing the successful application of yeast as a metagenomic library host. Although being more efficient producers and more promising hosts for the identification of novel NPs, the establishment of metagenomic libraries in alternative hosts remains a laborious procedure. Another important question is that of the insert size to be used. With increasing insert size, the likelihood rises that complete biosynthesis pathways and gene clusters are cloned. This is interesting when the screenings target metabolites or big enzyme complexes such as non-ribosomal peptide synthetases. Cosmid or fosmid libraries holding high-molecular-weight DNA up to 40 kb or bacterial artificial chromosomes (>40 kb) are generated to this end (Simon and Daniel, 2017). Yet high-molecular-weight inserts affect the cloning efficiency, limiting the resulting number of library clones. Small DNA fragments are cloned more readily, which favours libraries with low-molecular-weight inserts that are generally used for screenings of different enzyme classes like hydrolases (Simon and Daniel, 2017).

As mentioned earlier, metagenomic libraries comprise several thousand clones, whereby only a small fraction will exert the activity of interest (Handelsman, 2004). Therefore, a high-throughput screening to process many clones simultaneously for the desired phenotype needs to be in place to make the screening of large clone collections feasible. Such screenings, however, come with the drawback of potentially missing clones of interest (Coughlan *et al.*, 2015). As reviewed by Leis *et al.* as well as by Simon and David, screenings may involve the detection of enzymes by supplementing the growth medium with indicator reagents specific for tracing the desired enzymatic activity (called phenotypic detection) or the use of reporter genes for which expression is triggered only once the compound of interest is present. A further approach represents heterologous complementation of the host by the gene of interest. Only with the target gene being present will growth be observed under selection pressure (Simon and Daniel, 2011; Leis *et al.*, 2013). For instance, phenotypic detection led to the discovery of six novel polyesterases from a moss metagenomic library (Müller *et al.*, 2016b). The underlying high-throughput screening procedure used tributyrin-containing agar plates where hydrolytic activity could easily be spotted by halo formation. Novel 4′-phosphopantetheinyl transferases (PPT) from a metagenomic library in *E. coli* and *Streptomyces albus* were identified by a coupled *bpsA* reporter gene PPTase complementation approach (Owen *et al.*, 2012; Bitok *et al.*, 2017). Thereby, the PPT was deleted and pigment production by the PPT-dependent *bpsA* gene only restored upon expression of a functional PPT. Furthermore, Fluorescence-Activated Cell Sorting (FACS)-based screening methods have been established. The two methods, substrate-induced gene expression screening (SIGEX) (Uchiyama *et al.*, 2005) and metabolite-regulated expression screening (METREX) (Williamson *et al.*, 2005), are based on the induction of *gfp* reporter gene expression. While the latter employs a quorum-sensing promoter upstream of the *gfp* gene, for SIGEX the reporter gene is promoter-less. Hence, for SIGEX reporter gene, expression depends on the presence of a promoter on the metagenomic DNA insert. Uchiyama *et al.* used this approach to identify clones that express enzymes which convert a substrate of interest – in their case hydrocarbons. By adding the substrate of interest, the desired promoters get activated and drive the expression of not only the downstream enzyme but also the reporter gene. METREX, on the contrary, requires the synthesis of the promoter-activating molecules and facilitates the identification of signal molecules. Interestingly, both methodologies employing FACS for cell sorting have the advantage of a very efficient high-throughput screening.

A metagenomic analysis targeting the community composition and functions of endophytes in rice roots was reported for the first time by Sessitsch *et al.* (2011). Using the metagenome data, the authors predicted main microbial adaption mechanisms supporting an endophytic lifestyle, for example, the availability of plant-polymer–degrading enzymes, iron acquisition and storage, protein secretion systems, among others. Later on, the endophyte community of other plant species and plant compartments including *Arabidopsis thaliana* roots (Bulgarelli *et al.*, 2012), grapevine branches (Campisano *et al.*, 2014), sugar beet (Shi *et al.*, 2014), *Aloe vera* root, stem and leaves (Akinsanya *et al.*, 2015), tomato roots (Tian *et al.*, 2015) or floating fern (*Azolla filliculoides*) (Dijkhuizen *et al.*, 2018) has been evaluated. However, functional metagenomics, i.e. the screening of clone libraries for identification of new NPs, has not been in the focus of research yet. To the best of our knowledge, there is only one published study by Nikolic *et al.* thus far, targeting exclusively the endophytic microbial community for clone library generation and subsequent screening (Nikolic *et al.*, 2011). Here, using a few selected examples, we will highlight the potential of endophytes for the discovery of new natural products and enzymes.

11.3 Natural Products from Endophytes

The production of functional secondary metabolites by endophytes is linked to the improvement of plant fitness (Tan and Zou, 2001). Besides plant-growth promotion, one major function of microbial metabolites is protecting the plant against biotic and abiotic stress, e.g. by inducing resistance against pathogens (Bailly and Weisskopf, 2012). The structural and chemical diversity of NPs is extensive, including alkaloids, steroids, terpenoids, peptides and aliphatic compounds, among others (Tan and Zou, 2001; Gao *et al.*, 2018). A main group of interesting secondary metabolites is composed of high-molecular compounds, such as peptides and polyketides, encoded by non-ribosomal peptide synthetases and polyketide synthases. This type of metabolites display complex structural diversity, along with a broad range of biological activities and functions, such as antibacterial, antifungal and cytotoxic activity, or acting as metal chelators (siderophores) (Cane and Walsh, 1999). In this way they also support the lifestyle of endophytes in association with the host. Promising sources for this type of NPs are actinobacteria and fungi. Especially endophytic actinobacteria are regarded as a nearly unexplored reservoir of bioactive secondary metabolites

(Qin *et al.*, 2010). Most clinically relevant antibiotics used today have, for instance, their origin in actinomycetes (Baltz, 2007). Comprehensive reviews on the discovery of antibiotics from endophytes were published by Deshmukh *et al.* (2014) and Martinez-Klimova and Rodríguez-Peña (2017).

The enhanced acquisition of iron has been generally hypothesized as a central aspect in the life cycle of endophytes (Reinhold-Hurek and Hurek, 2011), in particular for nitrogen-fixing bacteria, since this process is iron dependent. Siderophores, which are mostly non-ribosomal peptides (Crosa and Walsh, 2002), have high chelating affinity for iron, and they contribute to the nutritional requirement of both microorganism and host plant. A new class of siderophores, the so-called serobactins, were identified in the grass-endophyte *Herbaspirillum seropedicae* Z67, a bacterium of interest due to its nitrogen-fixation ability (Rosconi *et al.*, 2013). Another type of siderophore, epichloënin A, was discovered as a product of a fungal endophyte *Epichloë festucae*, which lives in perennial ryegrass (Koulman *et al.*, 2012). This type of fungal endosymbiosis in temperate grasses not only improves the herbivore resistance of the plant (Lane *et al.*, 2000) but it was also shown that production of epichloënin A is required for maintaining of a mutual beneficial interaction between the fungus and its host (Johnson *et al.*, 2013).

The cyclic depsipeptide FR900359 is a further example of a non-ribosomal peptide from an uncultivable endosymbiont, *Candidatus Burkholderia crenata*, living in the tropical plant *Ardisia crenata* (Crüsemann *et al.*, 2018). The depsipeptide produced mainly in the leaf nodules by the endosymbiotic partner functions probably as a protective defence chemical against plant herbivores like insects and nymphs. In medicinal applications, this peptide is an indispensable tool for pharmacological studies of cellular signalling processes, being a potent inhibitor of guanine nucleotide binding proteins. For more detailed examples, the review by Abdalla and Matasyoh gives a good overview of different peptide classes isolated from endophytes (Abdalla and Matasyoh, 2014).

Polyketides, which are synthetized by large and iterative multi-functional proteins, so-called polyketide synthases, are also interesting bioactive NP. Several new polyketides have been reported in fungal endophytes. For example, six novel bicyclic polyketides, the so-called preussilides (for details on structure and activity see Chapter 10, this volume), were isolated from the fungus *Preussia similis*, an endosymbiont of *Globularia alypum*, and showed antiproliferation activity on eukaryotic cell lines. Similarly, different types of compounds belonging to the family of *oblongolides* were isolated from *Phomopsis oblonga*, and endophyte, from wild banana. Some of the isolated *oblongolides* displayed cytotoxic activity (Bunyapaiboonsri *et al.*, 2010). Recently, another group of cytotoxic polyketides was found in a related fungus, *Phomopsis* sp. A818, isolated from mangrove (Zhang *et al.*, 2017). A dimeric anthraquinone called skyrin (for details on structure and activity see Chapter 10, this volume) was identified as a pigment of the fungal endophyte *Cyanodermella asteris* (Jahn *et al.*, 2017). *C. asteris* was isolated from the plant *Aster tataricus*, which has been employed in traditional Chinese medicine as expectorant and showing anti-inflammatory properties as well (Yu *et al.*, 2015). Through *in silico* analysis of the *C. asteris* genome, putative biosynthetic pathways for production of skyrin were elucidated, suggesting the involvement of a non-reducing polyketide synthase (Jahn *et al.*, 2017).

Another group of metabolites produced by the plant microbiota are volatile organic compounds (VOCs). VOCs enable chemical inter- and intra-species communication over longer distances than non-volatile compounds (Kanchiswamy *et al.*, 2015). Many microbial volatiles, or mixtures thereof, have been investigated for their bioactivity, and especially for the ability to antagonise plant pathogens (Berg, 2009). One of the most prominent and first discovered examples of VOC-producing microbes is the endophytic fungus *Muscodor albus*, isolated by Strobel *et al.* in the late 1990s from a cinnamon tree in a botanical garden in Honduras (Sears *et al.*, 2001). The mixture of VOCs produced by *M. albus* contained different classes of organic substances (esters, alcohols, lipids, ketones and acids), which showed antibiotic effect on several plant and human pathogenic bacteria and fungi. While the single volatiles only inhibited the growth of the test organisms, the mixture thereof showed a potent lethal activity, being the most active single volatile isoamyl acetate.

Since this first discovery, efforts have been undertaken to isolate new VOC producers and evaluate their bioactivity and biotechnological potential. For instance, volatiles and semi-volatiles from endophytic fungi like *Hypoxylon anthochroum*, *Gleosporium* sp. and *Geotrichum candidum* PF005 have been studied, showing growth inhibition of important plant pathogens like *Fusarium oxysporum*, *Phytophthora palmivora*, *Rhizoctonia solani* and other fungi (Schaible *et al.*, 2015; Ulloa-Benítez *et al.*, 2016; Medina-Romero *et al.*, 2017; Mookherjee *et al.*, 2018). Not only fungi but also bacteria are capable of producing VOCs, many of them having plant-modulating properties and disease-suppressing activities (Weisskopf, 2013). New volatiles identified from bacterial endophytes are scarcer than those reported for fungi. However, bacteria from the genera *Pseudomonas*, *Bacillus*, *Serratia* and *Stenotrophomonas* are well-known VOCs producers (Bailly and Weisskopf, 2012), many of them being capable of an endophytic lifestyle. This is the case for *Pseudomonas putida* BP25, an isolated endophyte from the root of black pepper (Sheoran *et al.*, 2015). This bacterium showed production of some well-known antimicrobial VOCs like 1-undecene and different types of pyrazines.

In general, sourcing endophytes for bioactive compounds attributed to plants represents a promising alternative to exploiting the plant itself. This is of interest when the commercial supply cannot be maintained, for instance due to the compound being isolated from slow-growing or rare plants and its molecular structure being highly complex so that chemical synthesis is not a suitable option (Strobel and Daisy, 2003; McChesney *et al.*, 2007). A well-known example of this dilemma is the anticancer drug paclitaxel (Taxol®), a diterpenoid originally isolated from the bark of *Taxus brevifolia*, Western Yew (Cragg, 1998). The search for alternative sources for paclitaxel extraction revealed not only other plant species to produce this compound, but also several plant endophytes, mainly fungi (Stierle *et al.*, 1993; Kharwar *et al.*, 2011). Other examples include the anticancer compound camptothecin, which was first isolated from a *Nyssaceae* (Wall *et al.*, 1966) but is also produced by endophytic fungus *Fusarium solani* (Kusari *et al.*, 2009); and the potentially antidepressant hypericin, first isolated from St. John's wort (Aly *et al.*, 2013) but later found to be produced by the fungal endophyte *Thielavia subthermophila* (Kusari *et al.*, 2008). These and several other examples are reviewed in detail by Aly *et al.* (2013).

11.4 Enzymes from Endophytes

Microorganisms living in close association with plants often produce a wealth of different (extracellular) enzymes for the degradation of plant polymers and oligomers. For example, the production of an endoglucanase, a cell wall-degrading enzyme, was reported as a key factor for the initial and active bacterial colonization of internal plant tissues (Reinhold-Hurek *et al.*, 2006).

A metagenomic investigation of the root gall-associated microbiome in tomato plants showed a high abundance of oligosaccharide-degrading genes; however, only a lower frequency of genes coding for enzymes acting on full-length polymers, like cellulases or hemicellulases, was detected (Tian *et al.*, 2015). Moreover, several endophytic fungi isolated from medicinal plants, mangrove or the shrub *Brucea javanica* were tested positive for extracellular enzyme activities, like cellulases, lipases, amylases, laccases, pectinases or proteases (Choi *et al.*, 2005; Maria *et al.*, 2005; Sunitha *et al.*, 2013).

Ligninolytic enzymes, in particular laccases, are a group of enzymes commonly found in wood-decomposing fungi (Singh Arora and Kumar Sharma, 2010). Compared to fungi, reports on laccases from bacterial origin are rare. Endophytes might represent a new source for this type of bacterial enzymes. A new bacterial laccase, showing lignin degradation, dye decoloration and acid-stable properties, was found in the rice endophyte *Pantoea ananatis* Sd-1 (Shi *et al.*, 2015).

Chitin is a major constituent of the fungal cell wall. Chitinases and chitin-modifying enzymes are of biotechnological interest. One of the first reported enzymes was an extracellular chitobiosidase from *Bacillus cereus*, an endophyte isolated from mustard. The presence of extracellularly produced proteins from this bacterial strain decreased the rate of germination of spores from the plant pathogen *Fusarium oxysporum* and supported the idea that the production of extracellular enzymes might protect the plant from fungal infection (Pleban *et al.*, 1997). The availability of chitin-modifying enzymes in fungal endophytes was also investigated by Govinda *et al.* A high genetic diversity of this class of enzymes was found for several fungal isolates (Govinda Rajulu *et al.*, 2011). In a further study, a new chitin deacetylase from the endophytic fungi *Pestalotiopsis* sp. was isolated and characterized. It was shown that the enzymatic deacetylation of the chitin oligomers could be part of the survival strategy of the fungus inside the plant, since the modified oligomers were no longer recognized by the plant's immune system (Cord-Landwehr *et al.*, 2016).

Studies have also focused on the investigation of beneficial enzymes for the host plant, like those derived from plant-growth-promoting bacteria. This is the case for 1-aminocyclopropane-1-carboxylic acid (ACC) deaminases. ACC deaminases are important catalysts for regulating the level of plant-produced ethylene. ACC is a precursor of ethylene and its cleavage reduces the level of this stress-induced hormone and increases the stress resistance of the plant (Glick, 2005). Nikolic *et al.* analysed the abundance and diversity of ACC deaminase genes (*acd*S) from bacterial endophytes colonizing field-grown potato plants. One complete *acd*S operon was identified and analysed, showing the presence of a transcriptional regulator (*acd*R), which may be exclusive for the phyla of *Alpha*- and *Betaproteobacteria* (Nikolic *et al.*, 2011).

Similarly, the population of endophytic bacteria from the nickel (Ni) hyperaccumulator plant *Thlaspi gosingense* was analysed by means of total DNA extraction from shoot-associated DNA. The division of *Proteobacteria* dominated the bacterial endophyte population, showing clear differences to the bacterial community from the plant's rhizosphere. The presence of genes or genetic treats responsible for a higher Ni resistance was analysed using bacterial isolates but not the uncultivable bacterial fraction. All endophytic isolates were positive for the production of siderophores, and some showed ACC deaminase activity (Idris *et al.*, 2004). This study underpins the potential of finding new types of siderophores and other heavy metal resistance determinants in endophytes from hyper-accumulating plants.

11.5 Conclusion

Our society is experiencing an urgent need for new bioactive compounds in medicine, agriculture and industry, like antibiotics, anticancer drugs or pesticides. The discovery of novel pharmaceuticals by combinatorial chemistry has, however, not delivered the expected results in regard to the amount and the bioactivity of new developed compounds. Growing interest in the search for new NPs, their corresponding biosynthetic pathways and enzymes as bio-catalysts has now awakened. In particular, plants have been historically used as a source of NPs. Likewise, endophytes, through a tight synergism with their host, have evolved a specialized metabolism for the production of bioactive substances of interest. Hence, endophytes represent an uttermost promising and yet less explored source for this type of molecules.

While cultivation-dependent techniques provide an enormous share of bioactive molecules, myriad other compounds from not-yet-cultivable microorganisms remain undiscovered. The application of functional metagenomics can aid in the search for those unexplored molecules from uncultivable microbes. In this chapter, we described the methodology and highlighted important considerations for the construction of metagenomic clone libraries from endophytes. As only a small fraction of a metagenomic library exhibits the desired bioactivity, it is of interest to generate endophytic metagenomic libraries containing no DNA other than from endophytes. This way the endophytic potential can be harnessed to the fullest. Several examples illustrate the potential for the discovery of new enzymes and NPs using this strategy. This is a promising new field of study since endeavours for using functional metagenomics from sole endophytic microbiomes are still limited. It is exciting to see how the field of endophyte biotechnology in combination with functional metagenomics develops, allowing for the discovery of novel bioactive molecules.

References

Aakvik, T., Degnes, K.F., Dahlsrud, R., Schmidt, F., Dam, R., *et al.* (2009) A plasmid RK2-based broad-host-range cloning vector useful for transfer of metagenomic libraries to a variety of bacterial species. *FEMS Microbiology Letters* 296(2), 149–158. doi:10.1111/j.1574-6968.2009.01639.x

Abdalla, M.A. and Matasyoh, J.C. (2014) Endophytes as producers of peptides: An overview about the recently discovered peptides from endophytic microbes. *Natural Products and Bioprospecting* 4(5), 257–270. doi:10.1007/s13659-014-0038-y

Akinsanya, M.A., Goh, J.K., Lim, S.P. and Ting, A.S.Y. (2015) Metagenomics study of endophytic bacteria in *Aloe vera* using next-generation technology. *Genomics Data* 6, 159–163. doi:10.1016/J.GDATA.2015.09.004

Aly, A.H., Debbab, A. and Proksch, P. (2013) Fungal endophytes – Secret producers of bioactive plant metabolites. *Die Pharmazie - An International Journal of Pharmaceutical Sciences* 68(7), 449–505. doi:10.1691/ph.2013.6517

Bailly, A. and Weisskopf, L. (2012) The modulating effect of bacterial volatiles on plant growth. *Plant Signaling & Behavior* 7(1), 79–85. doi:10.4161/psb.7.1.18418

Baltz, R. (2007) Antimicrobials from actinomycetes: Back to the future. *Microbe* 2(3), 125–131.

Bérdy, J. (2012) Thoughts and facts about antibiotics: Where we are now and where we are heading. *The Journal of Antibiotics* 65(8), 385–395. doi:10.1038/ja.2012.27

Berg, G. (2009) Plant-microbe interactions promoting plant growth and health: Perspectives for controlled use of microorganisms in agriculture. *Applied Microbiology and Biotechnology* 84, 11–18. doi:10.1007/s00253-009-2092-7

Berg, G., Grube, M., Schloter, M. and Smalla, K. (2014) Unraveling the plant microbiome: Looking back and future perspectives. *Frontiers in Microbiology* 5, 148. doi:10.3389/fmicb.2014.00148

Bitok, J.K., Lemetre, C., Ternei, M.A. and Brady, S.F. (2017) Identification of biosynthetic gene clusters from metagenomic libraries using PPTase complementation in a *Streptomyces* host. *FEMS Microbiology Letters* 364(16), 1–8. doi:10.1093/femsle/fnx155

Brader, G., Compant, S., Mitter, B., Trognitz, F. and Sessitsch, A. (2014) Metabolic potential of endophytic bacteria. *Current Opinion in Biotechnology* 27(100), 30–37. doi:10.1016/j.copbio.2013.09.012

Bragina, A., Berg, C., Cardinale, M., Shcherbakov, A., Chebotar, V. and Berg, G. (2012) *Sphagnum* mosses harbour highly specific bacterial diversity during their whole lifecycle. *The ISME Journal* 6(4), 802–813. doi:10.1038/ismej.2011.151

Bulgarelli, D., Rott, M., Schlaeppi, K., Ver Loren van Themaat, E., Ahmadinejad, N., *et al.* (2012) Revealing structure and assembly cues for *Arabidopsis* root-inhabiting bacterial microbiota. *Nature* 488(7409), 91–95. doi:10.1038/nature11336

Bunyapaiboonsri, T., Yoiprommarat, S., Srikitikulchai, P., Srichomthong, K. and Lumyong, S. (2010) Oblongolides from the endophytic fungus *Phomopsis* sp. BCC 9789. *Journal of Natural Products* 73(1), 55–59. doi:10.1021/np900650c

Campisano, A., Antonielli, L., Pancher, M., Yousaf, S., Pindo, M. and Pertot, I. (2014) Bacterial endophytic communities in the grapevine depend on pest management. *PLOS ONE* 9(11), e112763. doi:10.1371/journal.pone.0112763

Cane, D.E. and Walsh, C.T. (1999) The parallel and convergent universes of polyketide synthases and nonribosomal peptide synthetases. *Chemistry & Biology* 6(12), R319–R325. doi:10.1016/S1074-5521(00)80001-0

Chen, L., Zhang, Q.-Y., Jia, M., Ming, Q.-L., Yue, W., *et al.* (2014) Endophytic fungi with antitumor activities: Their occurrence and anticancer compounds. *Critical Reviews in Microbiology* 42(3), 1–20. doi:10.3109/1040841X.2014.959892

Choi, Y.W., Hodgkiss, I.J. and Hyde, K.D. (2005) Enzyme production by endophytes of *Brucea javanica*. *International Journal of Agricultural Technology* 1(1), 55–66.

Cord-Landwehr, S., Melcher, R.L.J., Kolkenbrock, S. and Moerschbacher, B.M. (2016) A chitin deacetylase from the endophytic fungus *Pestalotiopsis* sp. efficiently inactivates the elicitor activity of chitin oligomers in rice cells. *Scientific Reports* 6(1), 38018. doi:10.1038/srep38018

Coughlan, L.M., Cotter, P.D., Hill, C. and Alvarez-Ordóñez, A. (2015) Biotechnological applications of functional metagenomics in the food and pharmaceutical industries. *Frontiers in Microbiology* 6, 672. doi:10.3389/fmicb.2015.00672

Cox, P. A. and Balick, M.J. (1994) The ethnobotanical approach to drug discovery. *Scientific American* 270(6), 82–87. doi:10.2307/24942736

Cragg, G. M. (1998) Paclitaxel (Taxol®): A success story with valuable lessons for natural product drug discovery and development. *Medicinal Research Reviews* 18(5), 315–331. doi:10.1002/(SICI)1098-1128(199809)18:5<315::AID-MED3>3.0.CO;2-W

Crosa, J.H. and Walsh, C.T. (2002) Genetics and assembly line enzymology of siderophore biosynthesis in bacteria. *Microbiology and Molecular Biology Reviews* 66(2), 223–249. doi:10.1128/MMBR.66.2.223-249.2002

Crüsemann, M., Reher, R., Schamari, I., Brachmann, A. O., Ohbayashi, T., *et al.* (2018) Heterologous expression, biosynthetic studies, and ecological function of the selective Gq-signaling inhibitor FR900359. *Angewandte Chemie International Edition* 57(3), 836–840. doi:10.1002/anie.201707996

David, B., Wolfender, J.-L. and Dias, D.A. (2015) The pharmaceutical industry and natural products: Historical status and new trends. *Phytochemistry Reviews* 14(2), 299–315. doi:10.1007/s11101-014-9367-z

Davies, J. (2007) Small molecules: The lexicon of biodiversity. *Journal of Biotechnology* 129(1), 3–5. doi:10.1016/J.JBIOTEC.2006.11.023

Deshmukh, S.K., Verekar, S.A. and Bhave, S.V. (2014) Endophytic fungi: A reservoir of antibacterials. *Frontiers in Microbiology* 5, 715. doi:10.3389/fmicb.2014.00715

Dijkhuizen, L.W., Brouwer, P., Bolhuis, H., Reichart, G.-J., Koppers, N., *et al.* (2018) Is there foul play in the leaf pocket? The metagenome of floating fern *Azolla* reveals endophytes that do not fix N_2 but may denitrify. *New Phytologist* 217(1), 453–466. doi:10.1111/nph.14843

Dos-Santos, C.M., de Souza, D. G. Balsanelli, E., Cruz, L.M., de Souza, E.M. Baldani, J.I. and Schwab, S. (2017) A culture-independent approach to enrich endophytic bacterial cells from sugarcane stems for community characterization. *Microbial Ecology* 74(2), 453–465. doi:10.1007/s00248-017-0941-y

Dreyfuss, M.M. and Chapela, I.H. (1994) Potential of fungi in the discovery of novel, low-molecular weight pharmaceuticals. In: Gullo, V.P. (ed.) *Discovery of Novel Natural Products with Therapeutic Potential*. Newnes, Oxford, UK/Boston, Massachusetts, pp. 49–80. doi:10.1016/B978-0-7506-9003-4.50009-5

Feher, M. and Schmidt, J.M. (2002) Property distributions: Differences between drugs, natural products, and molecules from combinatorial chemistry. *Journal of Chemical Information and Computer Science* 43(1), 218–227. doi:10.1021/CI0200467

Frank, A.C., Saldierna Guzmán, J.P. and Shay, J.E. (2017) Transmission of bacterial endophytes. *Microorganisms* 5(4), 70. doi:10.3390/microorganisms5040070

Gabor, E.M., Vries, E.J. and Janssen, D.B. (2003) Efficient recovery of environmental DNA for expression cloning by indirect extraction methods. *FEMS Microbiology Ecology* 44(2), 153–163. doi:10.1016/S0168-6496(02)00462-2

Gabor, E.M., Alkema, W.B.L. and Janssen, D.B. (2004) Quantifying the accessibility of the metagenome by random expression cloning techniques. *Environmental Microbiology* 6(9), 879–886. doi:10.1111/j.1462-2920.2004.00640.x

Gao, H., Li, G. and Lou, H.-X. (2018) Structural diversity and biological activities of novel secondary metabolites from endophytes. *Molecules* 23(3), 646. doi:10.3390/molecules23030646

Glick, B.R. (2005) Modulation of plant ethylene levels by the bacterial enzyme ACC deaminase. *FEMS Microbiology Letters* 251(1), 1–7. doi:10.1016/j.femsle.2005.07.030

Govinda Rajulu, M.B., Thirunavukkarasu, N., Suryanarayanan, T.S., Ravishankar, J.P., El Gueddari, N.E. and Moerschbacher, B.M. (2011) Chitinolytic enzymes from endophytic fungi. *Fungal Diversity* 47(1), 43–53. doi:10.1007/s13225-010-0071-z

Gunatilaka, A.A.L. (2006) Natural products from plant-associated microorganisms: Distribution, structural diversity, bioactivity, and implications of their occurrence. *Journal of Natural Products* 69(3), 509–526. doi:10.1021/np058128n

Hallmann, J., Berg, G. and Schulz, B. (2006) Isolation procedures for endophytic microorganisms. In: Schulz, B.J.E., Boyle, C.J.C., Sieber, T.N. (eds) *Microbial Root Endophytes*. Springer, Berlin/Heidelberg, pp. 299–319. doi:10.1007/3-540-33526-9_17

Handelsman, J. (2004) Metagenomics: Application of genomics to uncultured microorganisms. *Microbiology and Molecular Biology Reviews* 68(4), 669–685. doi:10.1128/MMBR.68.4.669-685.2004

Handelsman, J., Rondon, M.R., Brady, S.F., Clardy, J. and Goodman, R.M. (1998) Molecular biological access to the chemistry of unknown soil microbes: A new frontier for natural products. *Chemistry & Biology* 5(10), R245–R249. doi:10.1016/S1074-5521(98)90108-9

Hardoim, P.R., van Overbeek, L.S. and Elsas, J.D. (2008) Properties of bacterial endophytes and their proposed role in plant growth. *Trends in*

Microbiology 16(10), 463–471. doi:10.1016/J.TIM.2008.07.008

Hardoim, P.R., van Overbeek, L.S., Berg, G., Pirttilä, A.M., Compant, S., *et al.* (2015) The hidden world within plants: Ecological and evolutionary considerations for defining functioning of microbial endophytes. *Microbiology and Molecular Biology Reviews* 79(3), 293–320. doi:10.1128/MMBR.00050-14

Healy, F.G., Ray, R.M., Aldrich, H.C., Wilkie, A.C., Ingram, L.O. and Shanmugam, K.T. (1995) Direct isolation of functional genes encoding cellulases from the microbial consortia in a thermophilic, anaerobic digester maintained on lignocellulose. *Applied Microbiology and Biotechnology* 43(4), 667–674. doi:10.1007/BF00164771

Henkel, T., Brunne, R.M., Müller, H. and Reichel, F. (1999) Statistical investigation into the structural complementarity of natural products and synthetic compounds. *Angewandte Chemie International Edition* 38(5), 643–647. doi:10.1002/(SICI)1521-3773(19990301)38:5<643::AID-ANIE643>3.0.CO;2-G

Idris, R., Trifonova, R., Puschenreiter, M., Wenzel, W.W. and Sessitsch, A. (2004) Bacterial communities associated with flowering plants of the Ni hyperaccumulator *Thlaspi goesingense*. *Applied and Environmental Microbiology* 70(5), 2667–2677. doi:10.1128/AEM.70.5.2667-2677.2004

Ikeda, S., Kaneko, T., Okubo, T., Rallos, L.E., Eda, S., *et al.* (2009). Development of a bacterial cell enrichment method and its application to the community analysis in soybean stems. *Microbial Ecology* 58(4), 703–714. doi:10.1007/s00248-009-9566-0

Jahn, L., Schafhauser, T., Wibberg, D., Rückert, C., Winkler, A., *et al.* (2017) Linking secondary metabolites to biosynthesis genes in the fungal endophyte *Cyanodermella asteris*: The anti-cancer bisanthraquinone skyrin. *Journal of Biotechnology* 257, 233–239. doi:10.1016/J.JBIOTEC.2017.06.410

Jiao, J.-Y., Wang, H.-X., Zeng, Y. and Shen, Y.-M. (2006) Enrichment for microbes living in association with plant tissues. *Journal of Applied Microbiology* 100(4), 830–837. doi:10.1111/j.1365-2672.2006.02830.x

Johnson, L.J., Koulman, A., Christensen, M., Lane, G.A., Fraser, K., *et al.* (2013) An extracellular siderophore is required to maintain the mutualistic interaction of *Epichloë festucae* with *Lolium perenne*. *PLOS Pathogens* 9(5), e1003332. doi:10.1371/journal.ppat.1003332

Kanchiswamy, C.N., Malnoy, M. and Maffei, M.E. (2015) Chemical diversity of microbial volatiles and their potential for plant growth and productivity. *Frontiers in Plant Science* 6, 151. doi:10.3389/fpls.2015.00151

Kandel, S.L., Joubert, P.M. and Doty, S.L. (2017) Bacterial endophyte colonization and distribution within plants. *Microorganisms* 5(4), 77. doi:10.3390/microorganisms5040077

Katz, M., Hover, B.M. and Brady, S.F. (2016) Culture-independent discovery of natural products from soil metagenomes. *Journal of Industrial Microbiology & Biotechnology* 43(2–3), 129–141. doi:10.1007/s10295-015-1706-6

Kharwar, R.N., Mishra, A., Gond, S.K., Stierle, A. and Stierle, D. (2011) Anticancer compounds derived from fungal endophytes: Their importance and future challenges. *Natural Product Reports* 28(7), 1208. doi:10.1039/c1np00008j

Kloepper, J.W., Ryu, C.-M. and Zhang, S. (2004) Induced systemic resistance and promotion of plant growth by *Bacillus* spp. *Phytopathology* 94(11), 1259–1266. doi:10.1094/PHYTO.2004.94.11.1259

Koulman, A., Lee, T.V., Fraser, K., Johnson, L., Arcus, V., *et al.* (2012) Identification of extracellular siderophores and a related peptide from the endophytic fungus *Epichloë festucae* in culture and endophyte-infected *Lolium perenne*. *Phytochemistry* 75, 128–139. doi:10.1016/J.PHYTOCHEM.2011.11.020

Kusari, S., Lamshöft, M., Zühlke, S. and Spiteller, M. (2008) An endophytic fungus from *Hypericum perforatum* that produces hypericin. *Journal of Natural Products* 71(2), 159–162. doi:10.1021/np070669k

Kusari, S., Zühlke, S. and Spiteller, M. (2009) An endophytic fungus from *Camptotheca acuminata* that produces camptothecin and analogues. *Journal of Natural Products* 72(1), 2–7. doi:10.1021/np800455b

Lane, G.A., Christensen, M.J. and Miles, C.O. (2000) Coevolution of fungal endophytes with grasses: The significance of secondary metabolites. In: Bacon, C.W. and White, J.F. (eds) *Microbial Endophytes*. Marcel Dekker, New York, pp. 341–388.

Leis, B., Angelov, A. and Liebl, W. (2013) Screening and expression of genes from metagenomes. *Advances in Applied Microbiology* 83, 1–68. doi:10.1016/B978-0-12-407678-5.00001-5

Lemanceau, P., Corberand, T., Gardan, L., Latour, X., Laguerre, G., Boeufgras, J. and Alabouvette, C. (1995) Effect of two plant species, flax (*Linum usitatissinum* L.) and tomato (*Lycopersicon esculentum* Mill.), on the diversity of soilborne populations of fluorescent pseudomonads. *Applied and Environmental Microbiology* 61(3), 1004–1012.

Liebl, W., Angelov, A., Juergensen, J., Chow, J., Loeschcke, A., *et al.* (2014) Alternative hosts for functional (meta)genome analysis. *Applied*

Microbiology and Biotechnology 98(19), 8099–8109. doi:10.1007/s00253-014-5961-7

Lindell, S.D., Pattenden, L.C. and Shannon, J. (2009) Combinatorial chemistry in the agrosciences. *Bioorganic & Medicinal Chemistry* 17(12), 4035–4046. doi:10.1016/J.BMC.2009.03.027

Liu, R., Li, X. and Lam, K.S. (2017) Combinatorial chemistry in drug discovery. *Current Opinion in Chemical Biology* 38, 117–126. doi:10.1016/J.CBPA.2017.03.017

Lorenzo, J.M., Munekata, P.E.S., Gómez, B., Barba, F.J., Mora, L., Pérez-Santaescolástica, C. and Toldrá, F. (2018) Bioactive peptides as natural antioxidants in food products – A review. *Trends in Food Science & Technology* 79, 136–147. doi:10.1016/J.TIFS.2018.07.003

Maria, G.L., Sridhar, K.R. and Raviraja, N.S. (2005) Antimicrobial and enzyme activity of mangrove endophytic fungi of southwest coast of India. *International Journal of Agricultural Technology* 1(1), 67–80.

Martinez-Klimova, E. and Rodríguez-Peña, K. (2017) Endophytes as sources of antibiotics. *Biochemical Pharmacology* 134, 1–17. doi:10.1016/J.BCP.2016.10.010

McChesney, J.D., Venkataraman, S.K. and Henri, J.T. (2007) Plant natural products: Back to the future or into extinction? *Phytochemistry* 68(14), 2015–2022. doi:10.1016/j.phytochem.2007.04.032

Medina-Franco, J.L. (2012) Interrogating novel areas of chemical space for drug discovery using chemoinformatics. *Drug Development Research* 73(7), 430–438. doi:10.1002/ddr.21034

Medina-Franco, J.L., Martinez-Mayorga, K., Giulianotti, M., Houghten, R. and Pinilla, C. (2008) Visualization of the chemical space in drug discovery. *Current Computer Aided-Drug Design* 4(4), 322–333. doi:10.2174/157340908786786010

Medina-Romero, Y.M., Roque-Flores, G. and Macías-Rubalcava, M.L. (2017) Volatile organic compounds from endophytic fungi as innovative postharvest control of *Fusarium oxysporum* in cherry tomato fruits. *Applied Microbiology and Biotechnology* 101(22), 8209–8222. doi:10.1007/s00253-017-8542-8

Mookherjee, A., Bera, P., Mitra, A. and Maiti, M.K. (2018) Characterization and synergistic effect of antifungal volatile organic compounds emitted by the *Geotrichum candidum* PF005, an endophytic fungus from the eggplant. *Microbial Ecology* 75(3), 647–661. doi:10.1007/s00248-017-1065-0

Müller, C.A., Obermeier, M.M. and Berg, G. (2016a) Bioprospecting plant-associated microbiomes. *Journal of Biotechnology* 235, 171–180. doi:10.1016/j.jbiotec.2016.03.033

Müller, C.A., Perz, V., Provasnek, C., Quartinello, F., Guebitz, G.M. and Berg, G. (2016b) Discovery of polyesterases from moss-associated microorganisms. *Applied and Environmental Microbiology* 83(4), e02641-16. doi:10.1128/AEM.02641-16

Nalini, M.S. and Prakash, H.S. (2017) Diversity and bioprospecting of actinomycete endophytes from the medicinal plants. *Letters in Applied Microbiology* 64(4), 261–270. doi:10.1111/lam.12718

Newman, D.J. and Cragg, G.M. (2015) Endophytic and epiphytic microbes as “sources” of bioactive agents. *Frontiers in Chemistry* 3, 34. doi:10.3389/fchem.2015.00034

Newman, D.J. and Cragg, G.M. (2016) Natural products as sources of new drugs from 1981 to 2014. *Journal of Natural Products* 79(3), 629–661. doi:10.1021/acs.jnatprod.5b01055

Nikolic, B., Schwab, H. and Sessitsch, A. (2011) Metagenomic analysis of the 1-aminocyclopropane-1-carboxylate deaminase gene (*acdS*) operon of an uncultured bacterial endophyte colonizing *Solanum tuberosum* L. *Archives of Microbiology* 193(9), 665–676. doi:10.1007/s00203-011-0703-z

Nikolouli, K. and Mossialos, D. (2012) Bioactive compounds synthesized by non-ribosomal peptide synthetases and type-I polyketide synthases discovered through genome-mining and metagenomics. *Biotechnology Letters* 34(8), 1393–1403. doi:10.1007/s10529-012-0919-2

Ortholand, J.-Y. and Ganesan, A. (2004) Natural products and combinatorial chemistry: Back to the future. *Current Opinion in Chemical Biology* 8(3), 271–280. doi:10.1016/J.CBPA.2004.04.011

Owen, J.G., Robins, K.J., Parachin, N.S. and Ackerley, D.F. (2012) A functional screen for recovery of 4′-phosphopantetheinyl transferase and associated natural product biosynthesis genes from metagenome libraries. *Environmental Microbiology* 14(5), 1198–1209. doi:10.1111/j.1462-2920.2012.02699.x

Pace, N.R., Stahl, D.A., Lane, D.J. and Olsen, G.J. (1986) *The Analysis of Natural Microbial Populations by Ribosomal RNA Sequences*. Springer, Boston, Massachusetts, pp. 1–55. doi:10.1007/978-1-4757-0611-6_1

Patridge, E., Gareiss, P., Kinch, M.S. and Hoyer, D. (2016) An analysis of FDA-approved drugs: Natural products and their derivatives. *Drug Discovery Today* 21(2), 204–207. doi:10.1016/J.DRUDIS.2015.01.009

Pleban, S., Chernin, L. and Chet, I. (1997) Chitinolytic activity of an endophytic strain of *Bacillus cereus*. *Letters in Applied Microbiology* 25(4), 284–288. doi:10.1046/j.1472-765X.1997.00224.x

Qin, S., Xing, K., Jiang, J.-H., Xu, L.-H. and Li, W.-J. (2010) Biodiversity, bioactive natural products and biotechnological potential of plant-associated endophytic actinobacteria. *Applied*

Microbiology and Biotechnology 89(3), 457–473. doi:10.1007/s00253-010-2923-6
Ravin, N.V., Mardanov, A.V. and Skryabin, K.G. (2015) Metagenomics as a tool for the investigation of uncultured microorganisms. *Russian Journal of Genetics* 51(5), 431–439. doi:10.1134/S1022795415050063
Reinhold-Hurek, B. and Hurek, T. (2011) Living inside plants: Bacterial endophytes. *Current Opinion in Plant Biology* 14(4), 435–443. doi:10.1016/J.PBI.2011.04.004
Reinhold-Hurek, B., Maes, T., Gemmer, S., Van Montagu, M. and Hurek, T. (2006) An endoglucanase is involved in infection of rice roots by the not-cellulose-metabolizing endophyte *Azoarcus* sp. strain BH72. *Molecular Plant-Microbe Interactions* 19(2), 181–188. doi:10.1094/MPMI-19-0181
Ribes, S., Fuentes, A., Talens, P. and Barat, J.M. (2018) Prevention of fungal spoilage in food products using natural compounds: A review. *Critical Reviews in Food Science and Nutrition* 58(12),2002–2016.doi:10.1080/10408398.2017.1295017
Rosconi, F., Davyt, D., Martínez, V., Martínez, M., Abin-Carriquiry, J.A., *et al.* (2013) Identification and structural characterization of serobactins, a suite of lipopeptide siderophores produced by the grass endophyte *Herbaspirillum seropedicae*. *Environmental Microbiology* 15(3), 916–927. doi:10.1111/1462-2920.12075
Rudrappa, T., Czymmek, K.J., Paré, P.W., and Bais, H.P. (2008) Root-secreted malic acid recruits beneficial soil bacteria. *Plant Physiology* 148(3), 1547–1556. doi:10.1104/pp.108.127613
Schaible, G.A., Strobel, G.A., Mends, M.T., Geary, B. and Sears, J. (2015) Characterization of an endophytic *Gloeosporium* sp. and its novel bioactivity with "synergistans". *Microbial Ecology* 70(1), 41–50. doi:10.1007/s00248-014-0542-y
Schmidt, T.M., DeLong, E.F. and Pace, N.R. (1991) Analysis of a marine picoplankton community by 16S rRNA gene cloning and sequencing. *Journal of Bacteriology* 173(14), 4371–4378. doi:10.1128/JB.173.14.4371-4378.1991
Schmitt, E.K., Hoepfner, D. and Krastel, P. (2016) Natural products as probes in pharmaceutical research. *Journal of Industrial Microbiology & Biotechnology* 43(2–3), 249–260. doi:10.1007/s10295-015-1691-9
Sears, J., Dirkse, E., Markworth, C. and Strobel, G.A. (2001) Volatile antimicrobials from *Muscodor albus*, a novel endophytic fungus. *Microbiology* 147(11), 2943–2950. doi:10.1099/00221287-147-11-2943
Sessitsch, A., Hardoim, P., Döring, J., Weilharter, A., Krause, A., *et al.* (2011) Functional characteristics of an endophyte community colonizing rice roots as revealed by metagenomic analysis. *Molecular Plant-Microbe Interactions* 25(1), 28–36. doi:10.1094/MPMI-08-11-0204
Shen, B. (2015) A new golden age of natural products drug discovery. *Cell* 163(6), 1297–1300. doi:10.1016/j.cell.2015.11.031
Sheoran, N., Valiya Nadakkakath, A., Munjal, V., Kundu, A., Subaharan, K., *et al.* (2015) Genetic analysis of plant endophytic *Pseudomonas putida* BP25 and chemo-profiling of its antimicrobial volatile organic compounds. *Microbiological Research* 173, 66–78. doi:10.1016/J.MICRES.2015.02.001
Shi, X., Liu, Q., Ma, J., Liao, H., Xiong, X., *et al.* (2015) An acid-stable bacterial laccase identified from the endophyte *Pantoea ananatis* Sd-1 genome exhibiting lignin degradation and dye decolorization abilities. *Biotechnology Letters* 37(11), 2279–2288. doi:10.1007/s10529-015-1914-1
Shi, Y., Yang, H., Zhang, T., Sun, J. and Lou, K. (2014) Illumina-based analysis of endophytic bacterial diversity and space-time dynamics in sugar beet on the north slope of Tianshan mountain. *Applied Microbiology and Biotechnology* 98(14), 6375–6385. doi:10.1007/s00253-014-5720-9
Simon, C. and Daniel, R. (2011) Metagenomic analyses: Past and future trends. *Applied and Environmental Microbiology* 77(4), 1153–1161. doi:10.1128/AEM.02345-10
Simon, C. and Daniel, R. (2017) *Construction of Small-Insert and Large-Insert Metagenomic Libraries*. Humana Press, New York, pp. 1–12. doi:10.1007/978-1-4939-6691-2_1
Singh Arora, D. and Kumar Sharma, R. (2010) Ligninolytic fungal laccases and their biotechnological applications. *Applied Biochemistry and Biotechnology* 160(6), 1760–1788. doi:10.1007/s12010-009-8676-y
Sparks, T.C., Hahn, D.R. and Garizi, N.V. (2017) Natural products, their derivatives, mimics and synthetic equivalents: Role in agrochemical discovery. *Pest Management Science* 73(4), 700–715. doi:10.1002/ps.4458
Stewart, E.J. (2012) Growing unculturable bacteria. *Journal of Bacteriology* 194(16), 4151–4160. doi:10.1128/JB.00345-12
Stierle, A., Strobel, G. and Stierle, D. (1993) Taxol and taxane production by *Taxomyces andreanae*, an endophytic fungus of Pacific yew. *Science* 260(5105), 214–216. doi:10.1126/SCIENCE.8097061
Strobel, G. and Daisy, B. (2003) Bioprospecting for microbial endophytes and their natural products. *Microbiology and Molecular Biology Reviews* 67(4), 491–502. doi:10.1128/MMBR.67.4.491-502.2003

Sunitha, V., Nirmala Devi, D. and Srinivas, C. (2013) Extracellular enzymatic activity of endophytic fungal strains isolated from medicinal plants. *World Journal of Agricultural Sciences* 9(1), 1–9.

Tan, R.X. and Zou, W.X. (2001) Endophytes: A rich source of functional metabolites (1987 to 2000). *Natural Product Reports* 18(4), 448–459. doi: 10.1039/b100918o

Tian, B.-Y., Cao, Y. and Zhang, K.-Q. (2015) Metagenomic insights into communities, functions of endophytes and their associates with infection by root-knot nematode, *Meloidogyne incognita*, in tomato roots. *Scientific Reports* 5(1), 17087. doi:10.1038/srep17087

Truyens, S., Weyens, N., Cuypers, A. and Vangronsveld, J. (2015) Bacterial seed endophytes: Genera, vertical transmission and interaction with plants. *Environmental Microbiology Reports* 7(1), 40–50. doi:10.1111/1758-2229.12181

Uchiyama, T., Abe, T., Ikemura, T. and Watanabe, K. (2005) Substrate-induced gene-expression screening of environmental metagenome libraries for isolation of catabolic genes. *Nature Biotechnology* 23(1), 88–93. doi:10.1038/nbt1048

Ulloa-Benítez, Á., Medina-Romero, Y.M., Sánchez-Fernández, R.E., Lappe-Oliveras, P., Roque-Flores, G., *et al.* (2016) Phytotoxic and antimicrobial activity of volatile and semi-volatile organic compounds from the endophyte *Hypoxylon anthochroum* strain Blaci isolated from *Bursera lancifolia* (Burseraceae). *Journal of Applied Microbiology* 121(2), 380–400. doi:10.1111/jam.13174

Wall, M.E., Wani, M.C., Cook, C.E., Palmer, K.H., McPhail, A.T. and Sim, G.A. (1966) Plant antitumor agents. I. The isolation and structure of camptothecin, a novel alkaloidal leukemia and tumor inhibitor from *Camptotheca acuminata*. *Journal of the American Chemical Society* 88(16), 3888–3890. doi:10.1021/ja00968a057

Wang, H.-X., Geng, Z.-L., Zeng, Y. and Shen, Y.-M. (2008) Enriching plant microbiota for a metagenomic library construction. *Environmental Microbiology* 10(10), 2684–2691. doi:10.1111/j.1462-2920.2008.01689.x

Weisskopf, L. (2013) The potential of bacterial volatiles for crop protection against phytophathogenic fungi. In: Méndez-Vilas, A. (ed.) *Microbial Pathogens and Strategies for Combating Them: Science, Technology and Education*. Formatex Research Center, Badajoz, Spain.

Williamson, L.L., Borlee, B.R., Schloss, P.D., Guan, C., Allen, H.K. and Handelsman, J. (2005) Intracellular screen to identify metagenomic clones that induce or inhibit a quorum-sensing biosensor. *Applied and Environmental Microbiology* 71(10), 6335–6344. doi:10.1128/AEM.71.10.6335-6344.2005

Yu, P., Cheng, S., Xiang, J., Yu, B., Zhang, M., Zhang, C. and Xu, X. (2015) Expectorant, antitussive, anti-inflammatory activities and compositional analysis of *Aster tataricus*. *Journal of Ethnopharmacology* 164, 328–333. doi:10.1016/J.JEP.2015.02.036

Zhang, H.W., Song, Y.C. and Tan, R.X. (2006) Biology and chemistry of endophytes. *Natural Product Reports* 23(5), 753. doi:10.1039/b609472b

Zhang, W., Zhao, B., Du, L. and Shen, Y. (2017) Cytotoxic polyketides with an oxygen-bridged cyclooctadiene core skeleton from the mangrove endophytic fungus *Phomosis* sp. A818. *Molecules* 22(9), 1547. doi:10.3390/molecules22091547

12 Interplay Between Endophyte and Host Plant in the Synthesis and Modification of Metabolites

Jutta Ludwig-Müller*

Institut für Botanik, Technische Universität Dresden, Dresden, Germany

Abstract

The interplay between plants and microbes in their contribution of secondary metabolite synthesis is still not well understood. While plant pathogens might contribute by the synthesis of toxic compounds and the host by making antimicrobial compounds, the synthesis of secondary metabolites in the interaction of plants with endophytic organisms might be much more sophisticated. One possibility is that the plant makes antimicrobial compounds that are either sequestered or metabolized by the endophytic organisms. Alternatively, the inhabitant makes compounds that are released into the plant or environment for its own benefit. Such a compound can also be altered biochemically by the host. Finally, the two partners can contribute to one pathway by using an intermediate synthesized by one and later made to another compound by the other partner. Such examples and the implication for future research to identify the organisms mainly responsible for the biosynthetic pathways are discussed in this chapter.

12.1 Short Introduction to the Levels of Interaction between Plants and Microbes

The interplay between plants and microbes can occur on many different levels. To understand how much the interaction as either pathogen, symbiont (both with visible phenotypes) or endophyte (not causing any visible phenotypes on the host) can influence the host metabolism, a brief summary on the levels of interactions is provided in the beginning, although this is also covered by other chapters. It is clear that many microbes cause disease in plants after inoculation. Such pathogens can be of prokaryotic (bacteria) and eukaryotic (protists, oomycetes, fungi) origin. They can be roughly divided into biotrophic, hemibiotrophic and necrotrophic pathogens. While necrotrophs readily kill their hosts and eventually feed on dead tissue (Zheng *et al.*, 2013), biotrophs rely on the living host, at least for the phase in their life cycle where proliferation occurs (Panstruga, 2003). Some have formed intricate relationships with their hosts and evolved an intimate interaction. In such phases, the pathogens might be considered as an endophyte as well (Rodriguez *et al.*, 2009). Many biotrophs cannot be cultivated outside their host due to this close dependence, for example, rust fungi or some oomycota (Kemen *et al.*, 2011). In general, endophytes are living inside their host, but upon isolation, they also might grow without their host in culture (Singh *et al.*, 2011; Jahn *et al.*, 2017a). Nevertheless, some are highly specialized, such as arbuscular mycorrhizal fungi, so that they also need the host plant for development and nutrition (Harrison, 1998). Others, such as *Serendipita indica* (formerly *Piriformospora indica*), interact with their host

* Corresponding author e-mail: Jutta.Ludwig-Mueller@tu-dresden.de

to promote growth and resistance against biotic and abiotic stress factors (see Chapter 5, this volume), but the fungus can also be cultivated on artificial media and it was isolated originally from soil (Varma *et al.*, 2012). Again, other endophytes do not cause any apparent phenotypes, but can protect the host against unfavourable environmental conditions (Auer and Ludwig-Müller, 2015). Such fungi have been isolated from plant tissues such as grass endophytes (see Chapter 7, this volume) or fungi from the genus *Acremonium*, *Trichoderma*, *Cyanodermella*, just to name a few (Pu *et al.*, 2013; van Nieuwenhuijzen *et al.*, 2016; Jahn *et al.*, 2017a). However, there is no specific way by which an organism's lifestyle with a plant can be attributed to it being beneficial or pathogenic. Endophytes can turn from beneficial into pathogenic under changing conditions and can be beneficial for one species and a pathogen for another (Herre *et al.*, 2007; Fesel and Zuccaro, 2016; Hiruma *et al.*, 2016). Despite such difficulties in definition, the endophyte–plant interaction with respect to their potential for the synthesis of secondary metabolites will be described in the following sections mainly for those endophytes that do not cause any symptoms in their hosts. Although bacterial endophytes produce secondary metabolites with relevant bioactivities in their interaction with a host (Trapp *et al.*, 2015), the work summarized here will deal with aspects of endophytic fungi.

12.2 Role of Secondary Metabolites in Plants and Fungi

Secondary metabolites are involved in many aspects of a plant's life. There are an enormous number of secondary metabolites with antimicrobial or insecticidal properties that are important in plant–pathogen or plant–insect interactions (reviewed in Pusztahely *et al.*, 2015). In the case of phytopathogens, the plant is supposed to synthesize the secondary metabolites in response to the attack solely by the induction of the respective host pathways (Cheong, 2000). Nevertheless, plant pathogens can also synthesize their own set of metabolites that could, for example, act as virulence factors. An example is coronatine, which is synthesized by *Pseudomonas syringae* and suppresses the plant's defence response by interacting with receptors (Zheng *et al.*, 2012). However, such metabolites also occur in host plants during the interaction with endophytes and symbionts in a more strict sense (Abdel-Lateif *et al.*, 2012). Saprophytic microbes can also produce a vast array of compounds with bactericidal and fungicidal properties (Panaccione and Coyle, 2005; Schafhauser *et al.*, 2016). Endophytes are able to produce secondary metabolites in the host plant or in culture as well (Nisa *et al.*, 2015; Jahn *et al.*, 2017b), but so far for growth-promoting species, the focus was more on the synthesis of plant hormones (Vadassery *et al.*, 2008). In endophytes, which do not cause any apparent phenotype, the secondary metabolites they produce inside the host plant could be beneficial to the plant by protecting against pathogens or insects. However, even though from a biological viewpoint these considerations are highly important, the major focus of this chapter will be on the synthesis of secondary metabolites, mainly through the interplay between endophytic fungi and host plant.

There is much evidence that we do not understand completely as yet which partner contributes to which extent to the biosynthesis of secondary metabolites in a close interaction such as the endophyte–plant one (Ludwig-Müller, 2015; Nicoletti and Fiorentino, 2015; Nisa *et al.*, 2015; Vasundhara *et al.*, 2016). The idea is that the synthesis of secondary metabolites by the endophytic fungus is beneficial for its host plant, but to what extent has not been investigated much (Owen and Hundley, 2004). For example, toxic metabolites from the fungus could protect the plant against insects or other herbivores, if the compounds are brought into the correct compartments or exuded via plant fluids (Koulman *et al.*, 2007). Such compartments are ideal places for storage and use against sucking insects, but the distinct localization is not necessary for defence against chewing herbivores. Other mechanisms of the involvement of endophytes in biological control are given in Chapters 2, 6 and 9.

In a tight interaction with plants and endophytes, as well as in looser associations, the question arises as to whether secondary metabolites are exclusively produced by one partner at a given time or to which extent both partners contribute to the synthesis of particular derivatives within a given class of compounds. In many cases, a given metabolite can be exchanged with the partner and then metabolized further by the receiving partner. Some examples for the metabolism of either a plant metabolite by fungi or vice versa will be given. Following the exchange of compounds between partners, the next question arises as to whether plants are able to metabolize secondary metabolites from endophytes. One example was found for the interaction of grasses with their endophytes. The typical alkaloid from the fungus was found in plant fluids, which points to an active mechanism for the secretion of such fungal metabolites from plants (Koulman *et al.*, 2007). It is important to trace the origin of such compounds found in plant compartments, since it is not always clear from which organism they arise.

12.3 Potential of Endophyte–Plant Interactions to Synthesize Secondary Metabolites

As reviewed by Nisa *et al.* (2015) and Nicoletti and Fiorentino (2015), the endophytic fungi can produce a vast amount of different chemical classes of metabolites including alkaloids, steroids, terpenoids, peptides, polyketones, flavonoids, quinols, phenols, xanthones, chinones, isocumarines, benzopyranones, tetralones, cytochalasins, amongst others. Many as yet undescribed compounds are among those as well. In Fig. 12.1, some selected structures of compounds from plant–endophyte interactions are shown. Many compounds are believed to be synthesized by both partners and are therefore claimed to be of fungal and plant origin (Kusari *et al.*, 2013). This might be true when the fungus and the host plant can be cultivated separately and microbe-free host plant variants are present. Otherwise, this categorization might be too simplified for some compounds, as detailed below. Recent evidence has placed more attention on the endophytes as actual producers of the 'plant compounds' (Ali *et al.*, 2016; Arora *et al.*, 2016; Zhang *et al.*, 2016; Vassaux *et al.*, 2019). Therefore, the exact origin of some natural products found in plants may need to be reinvestigated.

In former times, bioactive compounds were isolated mainly from plants with known beneficial properties, and especially medicinal plants were also known for the presence of endophytes (Strobel, 2002). The huge potential of naturally occurring endophytes can be estimated by the report that 130 endophytic fungi were isolated from 12 Chinese traditional medicinal plants collected in Southwest China (Li *et al.*, 2005). Since the discovery of the biosynthetic potential of endophytes, the production of the compounds in question by these organisms if they can be cultivated outside of the host is a very attractive alternative (Priti *et al.*, 2009). However, some endophytic fungi may lose their ability to produce a certain compound that is found in the interaction with the host plant over time in aseptic culture (Kusari *et al.*, 2009; Pu *et al.*, 2013), whereas others do not (Vassaux *et al.*, 2019). Another advantage, besides the cultivation in bioreactors as single or co-cultivation, is the genetic manipulation of such fungi, which is usually a better option. Through various mutagenesis protocols, several endophytic fungi have been successfully transformed with foreign DNA to alter their biosynthetic potential (Venugopalan and Srivastava, 2015).

The potential of endophytic fungi for the production of valuable compounds needs to be harnessed, but the vast array of compounds possibly synthesized only allows either a non-targeted approach or, nowadays, metabolome analyses (Tian *et al.*, 2014), or one can use the molecular information on genes for selected synthesis pathways to screen in a more targeted approach for the presence of pathways (Vasundhara *et al.*, 2016). The latter has been employed in the screening for skyrin synthesis by genome

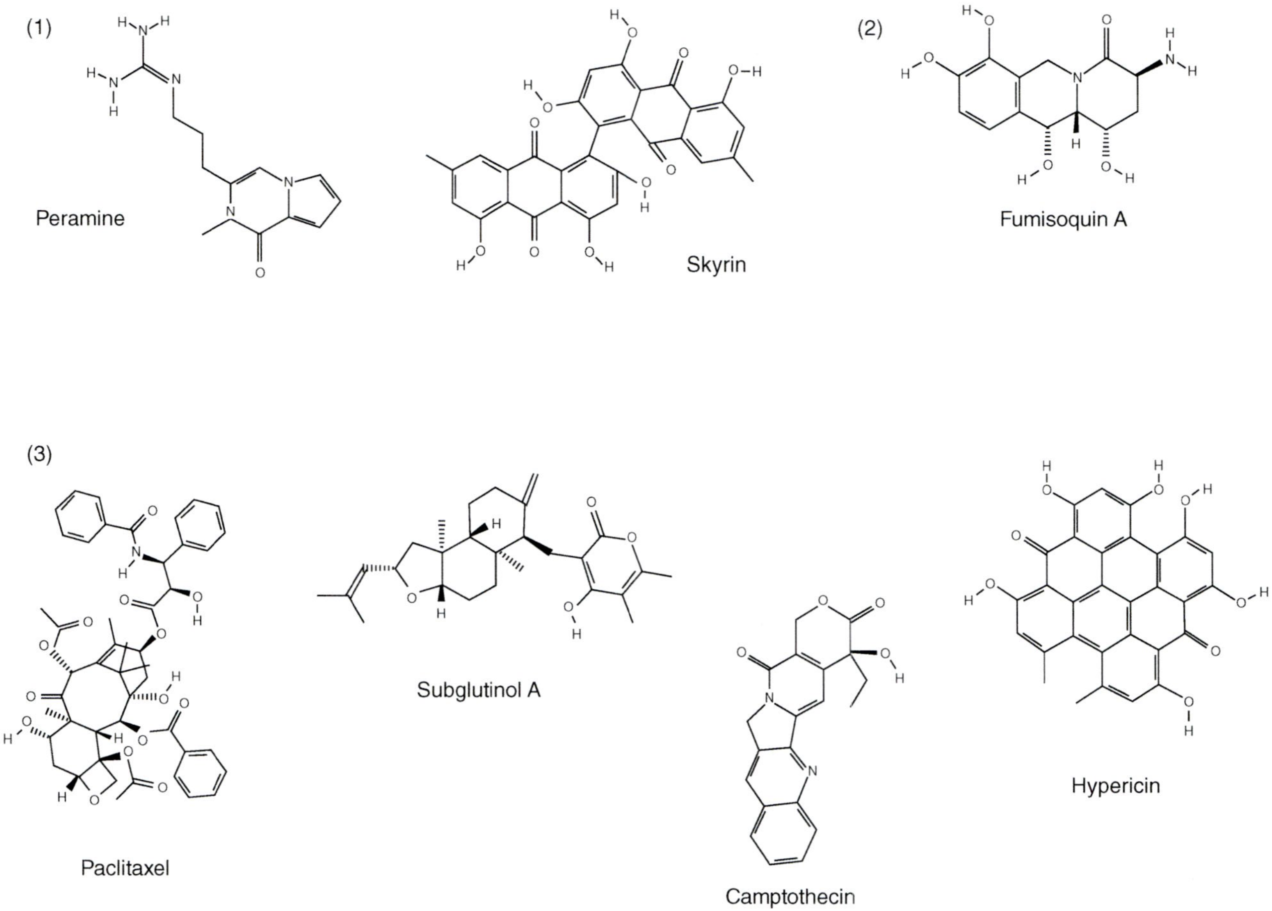

Fig. 12.1. Some examples for secondary metabolites in plant–endophyte interactions. Structures of secondary metabolites isolated (1) from fungi, (2) from fungi but with partial plant-like biosynthesis and (3) from both plant and endophyte. (The structures were taken from The PubChem Project (https://pubchem.ncbi.nlm.nih.gov).)

mining (Jahn *et al.*, 2017b) or for the anticancer compounds like paclitaxel, podophyllotoxin and camptothecin using molecular markers (Kusari *et al.*, 2009, 2013). Furthermore, innovative bioinformatics approaches, which involve the utilization of the various available databases on the Internet, can help identify novel pathways. For example, the analysis of the *Cyanodermella asteris* genome with the online tool antiSMASH (Weber *et al.*, 2015) revealed ample potential genes involved in secondary metabolite biosynthesis (Jahn *et al.*, 2017b).

Despite these observations, the potential of endophytes to synthesize bioactive metabolites is strongly exploited in biotechnology (Nisa *et al.*, 2015; see Chapter 10, this volume). This is also documented by the huge number of US patents that were granted in the years between 1993, a landmark in the discovery of plant secondary metabolites produced in fungi (Stierle *et al.*, 1993), and 2009, on endophytic fungi producing important metabolites and exhibiting biologically important activity (Priti *et al.*, 2009). What followed were many reports of the *in vitro* production of secondary metabolites by endophytic fungi (Venugopalan and Srivastava, 2015). The key here is the 'host-independent biosynthesis of plant secondary metabolites in endophytes', which are still termed plant metabolites as long as it is not clear whether both can produce them alike. Some of these plant metabolites can, however, now be attributed to the endophyte only (Vassaux *et al.*, 2019), whereas for others, like paclitaxel, it is not so clear (see section 12.3.3).

Some endophyte–plant combinations have shown to be sources of secondary metabolites with agricultural and/or pharmaceutical potential such as taxol, subglutinol A and B and the peptide leucinostatin A, where these metabolites can be produced by both the host and the respective endophytic fungus. However, if an endophyte is cultivated in a medium outside the host, the spectrum of secondary metabolites synthesized cannot directly be attributed to the interaction. For example, the aster endophyte *Cyanodermella asteris* characterized by a pink colour in culture, which is mainly due to the synthesis of the bisanthraquinone skyrin, is also discussed to possess anticancer potential (Jahn *et al.*, 2017b). Whether skyrins are synthesized also inside of the host has not yet been investigated.

Another problem is that, if no sterile host plant exists, it is difficult to judge whether the endophyte might exclusively synthesize the compound or whether the plant is also contributing. The ability of fungi to be cultivated without a host can be attributed to the biosynthetic potential of the secondary metabolites, but sometimes endophytes lose the ability for secondary product production after several rounds of culture, which was the case for camptothecin production (Kusari *et al.*, 2009; Pu *et al.*, 2013). Since this might be a species-specific effect, the biosynthetic potential of a given fungus outside of the plants needs to be monitored over time.

For the biosynthesis pathways, polyketide, isoprenoid or amino acid–dependent pathways have been described. However, there are biosynthetic pathways in saprophytes that partially resemble those of plants using the type of enzymes present in plants and not those that would be typical for fungi. For example, the biosynthesis of an isoquinoline alkaloid in *Aspergillus fumigatus* fits that bill (Baccile *et al.*, 2016). The respective genetic information was also found in sequences of endophytic *Fusarium* sp.

12.3.1 Synthesis in plant and metabolism by fungus

On the one hand, host metabolites can be exploited by endophytes to their own necessity (Tian *et al.*, 2014). On the other hand, sometimes host metabolites have antimicrobial activities to protect themselves against pathogens and could also be toxic to the endophyte. Therefore, detoxification of metabolites is also another strategy for a fungus (Zikmundova *et al.*, 2002; Saunders and Kohn, 2008).

Up to now, there are more examples for endophytic fungi that metabolize host compounds than vice versa (see section 12.3.2). The metabolism of host compounds by fungi concerns a wide range of different structures including flavanes, phenols, alkaloids and even volatiles (Nisa *et al.*, 2015). This also means a wide range of different enzymatic reactions that take place (Fig. 12.2), among them oxidation, acylation, reduction, hydrolysis, nitration, hydroxylation, epoxidation, deacetylation, deglycosylation, to name a few, and some of these reactions can also be carried out in a stereospecific manner (reviewed in Ludwig-Müller, 2015 and Nisa *et al.*, 2015). A summary of selected biotransformation reactions is given in Fig. 12.2 using the reactions on 2-hydroxy-1,4-benzoxazin-3-one as an example. Most experiments have been done *in vitro* using the endophytes in culture and adding the respective plant metabolite. Added to this, many reactions can be carried out in a stereoselective manner as well (see 'Potential of bioconversions by endophytic fungi for industrial applications' section).

In the case of biotransformation of the phytoanticipins 2-benzoxazolinone and 2-hydroxy-1,4-benzoxazin-3-one by the endophytic fungi by *Fusarium sambucinum*, *Plectosporium tabacinum*, *Gliocladium cibotii* and *Chaetosphaeria* sp., all the following reactions were observed: acylation, oxidation, reduction, hydrolysis and nitration (Zikmundova *et al.*, 2002). Since the experiments were carried out in culture and not in the host plant, it is not possible to deduce whether the biotransformation is important in the interaction. The enzymes catalyzing the biotransformation have been assumed to be *N*-acetyl- and *N*-malonyl-transferases

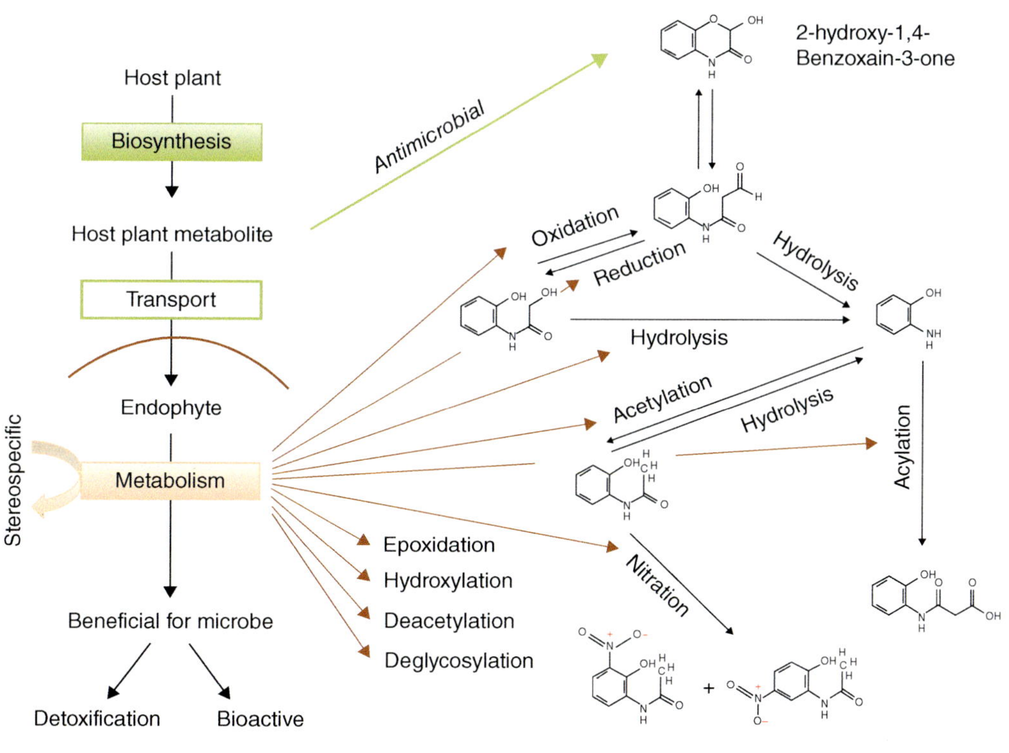

Fig. 12.2. Examples for host-derived compounds that are metabolized by an endophyte. Emphasis is given on the biotransformation of the fungitoxic 2-hydroxy-1,4-benzoxazin-3-one which has been derivatized by several different enzymatic reactions. Structures were created with the PubChem Sketcher function (Ihlenfeldt *et al.*, 2009).

as well as oxidases and reductases, e.g. cytochrome P-450 monooxygenases. The former are well-known detoxification enzymes for many xenobiotics in all organisms, whereas the latter act often as highly specific biocatalysts. Therefore, the authors discussed their potential for biotransformations in biotechnology. Since the compounds mentioned above are toxic to some fungi *in vitro*, a role of the biotransformation in tolerance for the fungi to live in their host can be hypothesized (Saunders and Kohn, 2008). Several endophytes that were found to be associated with maize in the genus *Fusarium* can indeed metabolize these phytoanticipins. The tolerance levels in ten species of *Fusarium* and in the maize endophytes *Nigrospora oryzae*, *Acremonium zeae* and *Periconia macrospinosa* varied in *in vitro* experiments. Interestingly, in the presence of the toxin, fungal species that are better in detoxifying their substrate can enhance the colonization rate of less tolerant fungi. This might be of ecological importance since the endophytes can be found not alone, but presumably co-localized in a given host.

There is also evidence for the bioconversion in planta by metabolome analyses (Tian *et al.*, 2014). In this study, the metabolome of host leaves of *Cephalotaxus harringtonia* together with the endophytic fungus *Paraconiothyrium variabile* was analysed and the changes caused by the fungus were monitored. Finally, the products that showed changes were identified. The result was a specific biotransformation of glycosylated flavonoids by the endophyte (Tian *et al.*, 2014). Interestingly, such deglycosylated products, in this case the aglycones apigenin and chrysoeriol, had a positive effect on hyphal growth, indicating indeed a possible function of this reaction for the endophyte besides detoxification reactions. In addition, the authors hypothesized that the flavonoid aglycones could also constitute beneficial compounds for the plant's defence (Tian *et al.*, 2014).

Resveratrol is a bioactive compound from grapes which acts as a phytoalexin for the plant. While its biosynthesis, also in microbes, has been elucidated to improve levels in plants which do not naturally possess (high) resveratrol levels, its metabolism has only marginally been studied (Mei *et al.*, 2015). There is one report on the cleavage of resveratrol by fungi using the so-called *Ustilago maydis* resveratrol cleavage oxygenase 1 (Brefort *et al.*, 2011), the homologs of which were also found in *Aspergillus* and other fungi.

Other biotransformations of host compounds by the endophytes include curcumin from turmeric plants (*Curcuma longa*) (Prana *et al.*, 2010), juglone (Prado *et al.*, 2013), huperzine A (Ying *et al.*, 2014), catechin/epicatechin (Agusta *et al.*, 2005), luteolin (Wang *et al.*, 2015) and phenanthrene (Fu *et al.*, 2018). Biotransformation was shown, but it is not clear what the benefit for the fungi would be.

Potential of bioconversions by endophytic fungi for industrial applications

In the area of producing antimicrobial compounds, biotransformation by endophytes is also applied as novel resource (Bianchini *et al.*, 2015). Therefore, for some biotransformations of plant metabolites by the fungi, mainly a commercial application was the aim of the study. For example, the biotransformation of protoberberine alkaloids by the endophytic fungus *Coelomycetes* sp. to a 7-*N*-oxide may not have a function for the fungus in the host plant. However, the product was shown to have a similar antimicrobial activity as the substrate of the bioconversion (Agusta *et al.*, 2014). Similarly, the endophyte *Xylaria* sp. isolated from *Cinchona pubescens* can transform some *Cinchona* alkaloids to the 1-*N*-oxide derivative. Here, the product had a better antimalaria activity due to lower toxicity than the native quinine (Shibuya *et al.*, 2003).

Volatiles and aroma compounds are also of interest to the food and cosmetics industry, so it is not surprising that endophytes were found that could use plant volatiles as substrates and convert them to interesting compounds (Abrahão *et al.*, 2013). Some examples are the conversion of α-pinene into verbenol via hydroxylation reactions by endophytic fungi from fruits of the plant *Dipteryx alata* (Molina *et al.*, 2012); the compound is of interest due to its camphor- and mint-like properties (Rottava *et al.*, 2010). However, this reaction was found more often also in other (endophytic) fungi. For the food industry, sometimes caffeine-free products are of interest, and therefore the decaffeination by biotransformation in endophytes isolated from *Coffee arabica* was investigated (Baker *et al.*, 2012). It was indeed possible to find microbes, in this case bacteria, that fit the demands because they degraded caffeine. One could hypothesize that the true function could indeed be the detoxification of the alkaloids or antimicrobial monoterpenes.

Other highly interesting reactions are stereospecific conversions that can also be catalyzed by endophytes (Borges *et al.*, 2009). Such stereospecific reactions include stereoselective hydroxylation, sulfoxidation, epoxidation, Baeyer-Villiger oxidation, deracemization and stereo- and enantioselective reduction.

12.3.2 Synthesis in fungus and metabolism by plant

The huge potential of naturally occurring endophytes has been explored in the isolation of many strains from different Chinese traditional medicinal plants (Li *et al.*, 2005). Based on the assumption that the plant–endophyte interaction thus might yield new compounds, endophytes were also isolated from medicinal plants to potentially isolate antimycobacterial compounds (Alvin *et al.*, 2014). Bioprospecting occurred for polyketides and small (cyclic) peptides that exhibited antituberculosis activity, and these classes of compounds seem to be very promising for therapy. Whether such compounds play a role during the host–fungus interaction is usually not clear, since the production of interesting bioactive secondary metabolites is mainly studied in culture, if the endophyte belongs to the culturable species.

Some endophytes synthesize compounds that may even be toxic for the host. For example, in culture, the endophyte *Acremonium* sp. isolated from *Taxus baccata* can make a phytotoxic peptide, leucinostatin A (Fig. 12.3; Strobel and Hess, 1997). The isolated peptide causes necrosis in non-host plants but no symptoms in the host. Therefore, it was hypothesized that the actual host can detoxify the peptide. Indeed, *T. baccata* and several other plants have a uridine diphosphate (UDP) glucose glucosyl transferase which transforms the toxic peptide to the non-toxic glycoside. It should be mentioned that the fungus could also make the glycosylated version of the peptide (Strobel and Hess, 1997). However, if the glycosylated version is the only one that occurs in the host, this would be a proof for the toxicity of the compound for the plant.

However, there are also some surprises waiting in the research on interesting secondary metabolites from (medicinal) plants. Plant parts from the species *Aster tataricus* have been used in Traditional Chinese Medicine (TCM) for its beneficial effects on human health (Morita *et al.*, 1995). Another work points to the possible cytotoxic characteristics of such plant extracts or individual compounds (Xu *et al.*, 2013). Some reports describe the bioactive potential of a compound class named astins (Morita *et al.*, 1995), cyclic peptides that have characteristics of a biosynthetic pathway via non-ribosomal peptide synthesis. Such cyclic peptides with modified amino acid residues are typically synthesized by non-ribosomal peptide synthetases (NRPSs), which constitute large multi-functional enzymes that assemble simple building blocks, the amino acids, into complex molecules (Hori *et al.*, 1989; Finking and Marahiel, 2004). Such NRPS genes have so far only been described from bacteria and fungi. Actually, plants also synthesize cyclic peptides, but these are made via ribosomal pathways and a cyclization step (Craik *et al.*, 1999). Recently, an endophytic fungus named *Cyanodermella asteris* was identified from inflorescences of *A. tataricus* and the fungus shown to be able to grow outside the plant (Jahn *et al.*, 2017a). The fungus is presumably able to produce a wealth of secondary metabolites based on its genome (Jahn *et al.*, 2017b), and one metabolite, skyrin (Fig. 12.1), was identified as part of the distinct pink

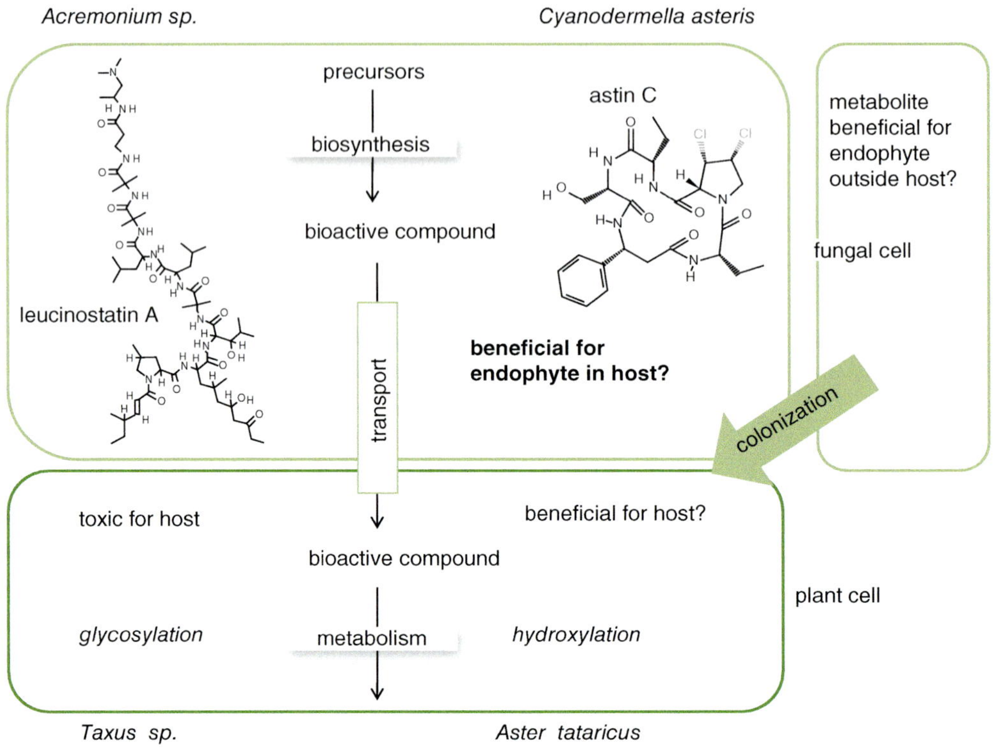

Fig. 12.3. Example for host metabolism of an endophyte-derived metabolite. Specific astin derivatives are synthesized in the endophytic fungus *Cyanodermella asteris* in culture (left), but there are more, especially hydroxylated, variants of astins found in the host plant. Therefore, a metabolism of astin C and a similar derivative after transport in the host plant is assumed. In the right part of the figure, another structure, the peptide leucinostatin A, for a compound produced by an endophytic fungus *Acremonium* sp. in *Taxus*, which is glycosylated by the host plant, is shown. It is possible that metabolism occurs to detoxify the compound in the host or to increase its bioactive potential. For leucinostatin A, the former possibility is the most likely one. For the metabolism of astin in the host this is not clear.

colour of the fungus (Jahn *et al.*, 2017a,b). While this was not a metagenomic approach per se, such investigations can be used to find novel metabolites (see Chapter 11). Follow-up work showed that the fungus *C. asteris* is actually the producer of a set of the bioactive astins because the production was shown outside of the plant (Vassaux *et al.*, 2019). However, some astins have only been detected in the plant tissues, namely hydroxylated ones, so it can be assumed that the plant can metabolize the astins derived from the endophyte (Fig. 12.3), even though there is no direct proof for this hypothesis so far. In the case of the occurrence of astins in *A. tataricus*, it is less likely that the non-hydroxylated compounds produced by the endophyte are toxic to the plant itself, since both hydroxylated and non-hydroxylated versions occur in the host.

Similar cyclic peptides are also synthesized by a saprophyte, *Talaromyces islandicus*, originally identified from spoiled rice, the compounds of which were named cyclochlorotines (Schafhauser *et al.*, 2016). Genetic evidence for the involvement of a non-ribosomal peptide synthetase (NRPS) was provided for the synthesis of cyclochlorotine, so it can be assumed that the same mechanism occurs in *C. asteris* for the synthesis of astins (Schafhauser *et al.*, 2016; Vassaux *et al.*, 2019). Other cyclic peptides with non-proteinogenic amino acids such as dihydroxyornithine or other dihydroxy derivatives of proteinogenic amino acids, the cryptocandin class, exhibit antifungal activity. *Cryptosporiopsis* cf. *quercina* is the endophyte producing the compound that is active against a series of different phytopathogenic fungi, thereby maybe giving its host an advantage (Strobel *et al.*, 1999). Notably, one should expect more 'plant-derived' metabolites that are actually made by an endophyte, whereas the grade of modification within the host is more difficult to understand due to missing genomes of such medicinal plants.

12.3.3 Sharing responsibilities: Plants and fungi contribute alike? The case of paclitaxel

In the following paragraphs, some aspects of the question asked above will be highlighted. While the question posed seems to implicate a co-ordinated synthesis between fungi and plants, this does not seem to be the case for the examples found so far and described below. For the quinoline alkaloid camptothecin (Fig. 12.1), which is produced in several plants with medicinal properties, the contribution of plant and endophyte is far from clear (Kumara *et al.*, 2014). For both plant and fungi, the biosynthesis is only partially characterized on the molecular level. Genes encoding strictosidine synthase, the enzyme for a key step in the pathway, were isolated from plants, but in an endophyte that was known to produce camptothecin, this crucial gene was not found. The authors of this study argued that the fungus might use the plant enzyme (Kusari *et al.*, 2011), but this would not explain how the fungus produces camptothecin in culture without host (Kumara *et al.*, 2014). The case of paclitaxel will be described in more detail below.

The search for endophytes producing taxol®, generic name paclitaxel (Fig. 12.1), has been conducted for quite some time. Since the demand of the anticancer drug paclitaxel is enormous, a focus is, of course, on alternative production possibilities due to the problem with the availability of the plant materials (Kusari *et al.*, 2014). Since the chemical synthesis is demanding, a semi-synthesis of taxol® and derivatives from an intermediate of the pathway, 10-deacetylbaccatin III, is often carried out (Jennewein and Croteau, 2001). The intermediate 10-deacetylbaccatin III has always been isolated from *Taxus* trees. Some of the endophytes discussed below are also able to synthesize the intermediate 10-deacetylbaccatin III (Zhang *et al.*, 2009) and might therefore be of use to be exploited for the semi-synthesis. Thus, efforts were underway to increase the availability of paclitaxel, for example, by *in vitro* plant cultures. The latter have to be optimized for production either by changing growth conditions and/or elicitation, which is also not trivial (Cusidó *et al.*, 1999; Cusido *et al.*, 2014; Bonfill *et al.*, 2007). More recent work also identified negative regulators for the jasmonate-induced paclitaxel production, so this can also be a strategy to circumvent low production if such regulators can be decreased (Lenka *et al.*, 2015).

Furthermore, expression of paclitaxel synthesis genes in prokaryotic as well as eukaryotic heterologous hosts has been attempted (Flores-Bustamante *et al.*, 2010). In *Escherichia coli*, it was possible to produce taxadiene in cell-free extracts of strains over-expressing three different early genes from the pathway (Huang *et al.*, 2001). In *Saccharomyces cerevisiae* the expression of eight genes from the paclitaxel pathway from *Taxus canadensis* was achieved (DeJong *et al.*, 2006). Genome shuffling to generate fungal strains with higher paclitaxel production was also used as an alternative approach (Zhao *et al.*, 2008). The authors could create a strain of *Nodulisporium sylviform* that had, after the alterations, a higher paclitaxel production than the initial strain. However, such information would need to be compared with production values from other high taxol producers, either plant or fungus, under optimum conditions.

In the beginning of intensive research on paclitaxel-producing plants, the focus was exclusively on the plant side and the compound was thought to be solely of plant origin, similar to the case of astins described above (see section 12.3.2). The genus *Taxus* has well been explored, but other plant species have not been well studied, rendering information on other paclitaxel-producing plant species scarce. However, there is one report on the occurrence of paclitaxel in a non-gymnosperm, the plant species *Corylus avellena*, belonging to the *Betulaceae* (Service, 2000; Qaderi *et al.*, 2012). This is of importance since from this plant material also an endophyte has been identified that produces paclitaxel (Yang *et al.*, 2014).

During the search for alternative sources, endophyte-containing plants came into focus, even though it was not clear as to what extent the microbes are involved in the synthesis of paclitaxel. After paclitaxel-producing endophytic fungi were identified, the discussion was on as to which extent these are contaminations or whether they are the sole producers. Nowadays it seems to be verified that both plant and fungi can produce the compound. However, the contribution of one partner to the whole synthesis scheme is still in question due to the incomplete microbial synthesis pathways found so far (Garyali *et al.*, 2013; Kusari *et al.*, 2014; Fig. 12.4). For example, in fungi harbouring only two pathway genes, no paclitaxel could be found (Garyali *et al.*, 2013).

Therefore, endophytic fungi from various sources were investigated for taxol® production. For example, *Fusarium oxysporum* was isolated as an endophyte from a mangrove plant and analysed for paclitaxel production (Elavasari *et al.*, 2012), but the compound was only tentatively identified by thin-layer chromatography and staining of terpene groups. Taxol®-producing fungal endophytes, for which tentative pathway genes have been known for quite some time, and

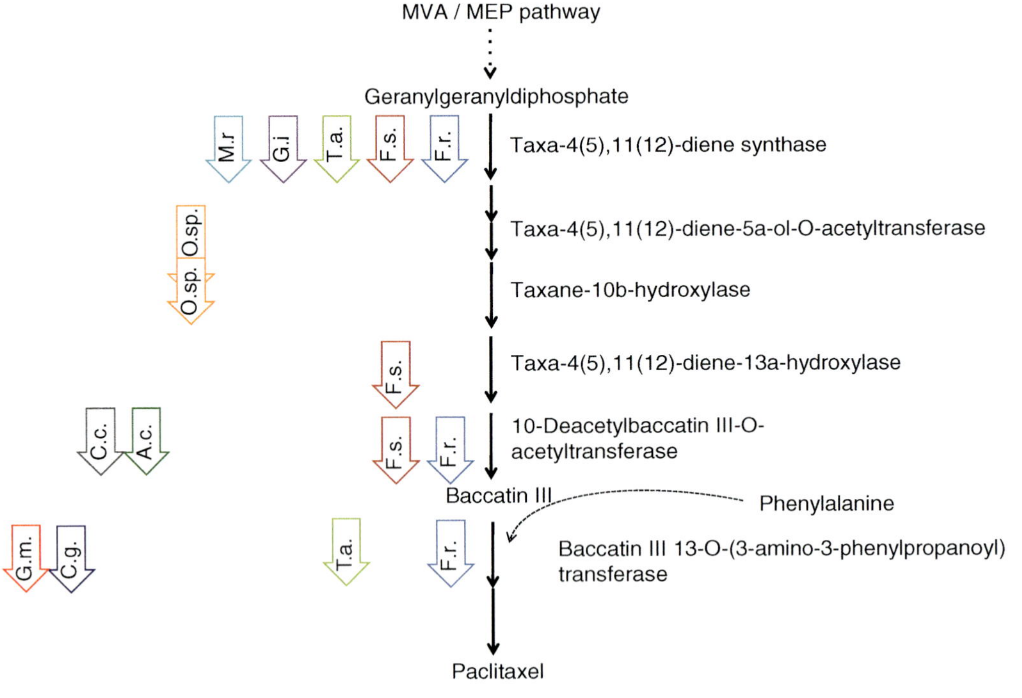

Fig. 12.4. Pathway to paclitaxel and respective genes that encode for putative enzymes in the pathway found in endophytic fungi (based on the compilation by Kusari *et al.*, 2014). Only the major metabolites are shown. Each solid arrow stands for one reaction; dashed arrows represent several reactions. Abbreviations for the respective fungi are as follows: A.c. = *Aspergillus candidus*; C.c. = *Cladosporium cladosporoides*; C.g. = *Colletotrichum gloeosporioides*; F.r. = *Fusarium redolens*; F.s. = *Fusarium solani*; G.i. = *Gibberella intermedia*; G.m. = *Guignardia mangiferae*; M.r. = *Mucor rouxianus*; O.sp. = *Ozonium* sp. BT2; T.a. = *Taxomyces andreanae*.

their hosts were compiled by Kusari *et al.* (2014). From one plant, as many as 80 endophytic fungi could be isolated and at least three of them were able to produce paclitaxel (Xiong *et al.*, 2013). To make screens for production easier, a potential gene from the biosynthetic pathway could also be used as marker (Garyali *et al.*, 2013).

So far, no complete pathway could be assigned to an endophyte (Fig. 12.4). Several organisms were found from which one of the following genes were isolated: taxa-4(5),11(12)-diene synthase (five fungi); taxa-4(5),11(12)-diene-5α-ol-O-acetyltranseferase (one fungus); taxane-10β-hydroxylase (one fungus); taxa-4(5),11(12)-diene-13α-hydroxylase (one fungus); 10-deacetylbaccatin III-O-acetyltransferase (four fungi); and baccatin III 13-O-(3-amino-3-phenylpropanoyl)transferase (four fungi) (reviewed in Kusari *et al.*, 2014). Taxadiene synthase is thought to be the committing step in the biosynthetic pathway (Walker and Croteau, 2001). For *Fusarium solani* and *F. redolens* three genes for putative paclitaxel pathway enzymes were annotated, which was the highest number in one species according to this compilation, probably indicating a most complete pathway. *Taxomyces andreanae* and an *Ozonium* species had two putative genes annotated, whereas all other fungi showed only one gene in the pathway, but that was always a different one. Whether in other fungi more genes of the pathway would be present, or whether such an incompleteness could be an indication for the contribution of plant and fungus alike cannot be stated from such data.

An endophytic fungus of hazel, *Penicillium aurantiogriseum*, was found to be able to produce paclitaxel, and in the genome of the fungus, several candidate genes that could encode for proteins in paclitaxel biosynthesis were identified by comparing the fungal genome with sequences known from *Taxus* species (Yang *et al.*, 2014). However, the function of these genes were not shown thus far. In addition, based on their results, the authors hypothesized that the genes in question evolved independently due to low sequence similarities and therefore ruling out horizontal gene transfer. Another argument against horizontal gene transfer is presented by Kusari *et al.* (2014), who noted the high number of seemingly unrelated fungal genera isolated from distant hosts which do not even all produce paclitaxel. Based on the latter observation that fungi produce the compound but the host plant does not, the assumption of horizontal gene transfer is difficult to explain. However, one could speculate that the gene transfer could also occur between fungi in the host or soil microbiome and not only between host and endophyte. At least, some evidence for a more complex interaction in the microbiome of *Taxus* species was provided by the analysis of Soliman and Raizada (2013), who showed that pyramiding several endogenous endophytes increased the paclitaxel synthesis as compared to single interactions. Their results suggest that fungi residing within a host plant can interact with one another to stimulate secondary metabolite biosynthesis, either directly or through their metabolites as signals or precursors.

Also other valuable bioactive compounds such as those from the harziane tetracyclic diterpene family were isolated from *Taxus* species or their endophytes, i.e. *Trichoderma atroviridae* (Adelin *et al.*, 2014). Some of these might be the result of the bioconversion of taxane derivatives by fungi (see section 12.3.1 for other examples); for example, hydrolyzation and epimerization of two taxoids was described by three fungi of the genera *Microsphaeropsis*, *Mucor* and *Alternaria* (Zhang *et al.*, 1998).

Despite the success of finding endophytes producing paclitaxel as early as 1993 (Stierle *et al.*, 1993), the fungi have not been used in biotechnological production so far. Maybe this can be attributed to the endophytic lifestyle that is not understood very well and the different conditions in culture compared to the natural habitat (discussed by Kusari *et al.*, 2014). In addition, fungi might lose their ability to produce the secondary metabolites over a long time in culture (Kusari *et al.*, 2009; Pu *et al.*, 2013). Nevertheless, Li *et al.* (2009) reported on a successful co-cultivation of *Taxus* cell cultures and its paclitaxel-producing endophytic fungus

Fusarium mairei in a 20-litre bioreactor with paclitaxel production that was significantly higher than the production without co-cultivation. However, from such experiments, it is not possible to judge whether this is a co-production by two organisms or an elicitation effect by the fungus. The latter idea is actually supported by earlier data from Yuan *et al.* (2001) who showed an increase in paclitaxel production in *Taxus chinensis* var. *mairei* induced by oligosaccharide from *Fusarium oxysporum*.

Another possible reason was brought up by Heinig *et al.* (2013) who could not find any evidence for paclitaxel production in fungi such as *Taxomyces andreanae*. Despite the many reports on fungal paclitaxel biosynthesis, they questioned the results based on several problems in other publications. One point of criticism was the amplification of fungal DNA from the host plant with primers based on the host DNA. The amplified fungal DNA corresponded quite closely to the plant sequences making an independent evolution of fungal and plant pathways questionable. They examined endophytes from *Taxus* and could not find high paclitaxel concentrations but found the precursors 10-deacetylbaccatin III and baccatin III. However, they could not isolate any plant-like sequences from DNA libraries when they used probes resembling plant genes for the biosynthetic pathway. After genome sequencing of endophytes again, they could not detect any genes with significant homology to *Taxus* genes involved in taxane biosynthesis (Heinig *et al.*, 2013). Even though the pathway might have been completely independently evolved, a low similarity of the respective fungal genes should be expected. Such experiments at least should be taken into account when future work on endophytes with paclitaxel-producing potential are studied. On the contrary, work from 2014 by Yang *et al.* on a genome sequencing approach identified in an endophyte of *Taxus brevifolia*, *Penicillium aurantiogriseum*, gene sequences that seem to encode for proteins involved in paclitaxel synthesis. Therefore, much more work is still needed to solve these puzzles.

12.4 Concluding Remarks

The interplay between fungi and their hosts can function on many levels. On the one hand, there are strong symbiotic or pathogenic interactions in which a wealth of secondary metabolites is produced. On the other hand, there are those endophytic interactions that lead to no host symptoms unless the plant is stressed or the fungus–plant homoeostasis is otherwise altered. The potential of such endophytes to produce secondary metabolites has only recently been unravelled. Both partners in such a relationship contribute to the synthesis by their own metabolism, maybe even by the induction of pathways upon signals from the other partner, i.e. the elicitation of synthesis pathways. Therefore, one should be careful with the term 'plant metabolites' in such interactions. Plants can sometimes metabolize the secondary metabolites from the endophytic fungus and vice versa – the fungi are capable of carrying out many enzymatic reactions on plant metabolites. Such strategies add to the number and complexity of compounds found in nature. While research is directed to identify novel compounds and their bioactive potential, the role of such substances within the plant–endophyte relationship is significantly less, if at all, understood. Here future work is needed, although this is often complicated by non-sequenced host plants, host plants that are not amenable to genetic manipulation or the lack of endophyte-free host plants. Such problems are also often encountered when studying the contribution from one or the other partner. While most endophytes can be grown in culture, it is not clear as to whether they are synthesizing the complete repertoire of compounds that is made in the plant, but it also cannot be analysed to which extent a fungus-free host plant would contribute or to which extent in the interaction both are needed for the full pathway or set of metabolites. Work on pyramiding microbes and getting better results for secondary metabolite production also indicates that the endophytes should be studied in their natural habitat. That leaves a lot to think about for future research.

References

Abdel-Lateif, K., Bogusz, D. and Hocher, V. (2012) The role of flavonoids in the establishment of plant roots endosymbioses with arbuscular mycorrhiza fungi, rhizobia and *Frankia* bacteria. *Plant Signaling & Behavior* 7, 636–641. doi: 10.4161/psb.20039

Abrahão, M.R.E., Molina, G. and Pastore, G.M. (2013) Endophytes: Recent developments in biotechnology and the potential for flavor production. *Food Research International* 52, 367–372. doi:10.1016/j.foodres.2013.03.007

Adelin, E., Servy, C., Martin, M.-T., Arcile, G., Iorga, B.I., *et al.* (2014) Bicyclic and tetracyclic diterpenes from a *Trichoderma* symbiont of *Taxus baccata*. *Phytochemistry* 97, 55–61. doi:10.1016/j.phytochem.2013.10.016

Agusta, A., Maehara, S., Ohashi, K., Simanjuntak, P. and Shibuya, H. (2005) Stereoselective oxidation at C-4 of flavans by the endophytic fungus *Diaporthe* sp. isolated from a tea plant. *Chemical and Pharmaceutical Bulletin* 53, 1565–1569. doi:10.1248/cpb.53.1565

Agusta, A., Wulansari, D., Praptiwi, Nurkanto, A. and Fathoni, A. (2014) Biotransformation of protoberberine alkaloids by the endophytic fungus *Coelomycetes* AFKR-3 isolated from Yellow Moonsheed Plant (*Archangelisia flava* (L.) Merr.). *Procedia Chemistry* 13, 38–43. doi:10.1016/j.proche.2014.12.004

Ali, L., Khan, A.L., Hussain, J., Al-Harrasi, A., Waqas, M., *et al.* (2016) Sorokiniol: A new enzymes inhibitory metabolite from fungal endophyte *Bipolaris sorokiniana* LK12. *BMC Microbiology* 16, 103. doi:10.1186/s12866-016-0722-7

Alvin, A., Miller, K.I. and Neilan, B.A. (2014) Exploring the potential of endophytes from medicinal plants as sources of antimycobacterial compounds. *Microbiological Research* 169, 483–495. doi:10.1016/j.micres.2013.12.009

Arora, P., Wani, Z.A., Nalli, Y., Ali, A. and Riyaz-Ul-Hassan, S. (2016) Antimicrobial potential of thiodiketopiperazine derivatives produced by *Phoma* sp., an endophyte of *Glycyrrhiza glabra* Linn. *Microbial Ecology* 72, 802–812. doi:10.1007/s00248-016-0805-x

Auer, S. and Ludwig-Müller, J. (2015) Biological control of clubroot (*Plasmodiophora brassicae*) by the endophytic fungus *Acremonium alternatum*. *Journal of Endocytobiosis and Cell Research* 26, 43–49. https://zs.thulb.uni-jena.de/receive/jportal_jparticle_00346200 (accessed 16 May 2019).

Baccile, J.A., Spraker, J.E., Le, H.H., Brandenburger, E., Gomez, C., *et al.* (2016) Plant-like biosynthesis of isoquinoline alkaloids in *Aspergillus fumigatus*. *Nature Chemical Biology* 12, 419–424. doi:10.1038/nchembio.2061

Baker, S., Sahana, S., Rakshith, D., Kavitha, H.U., Kavitha, K.S. and Satish, S. (2012) Biodecaffeination by endophytic *Pseudomonas* sp. isolated from *Coffea arabica* L. *Journal of Pharmacy Research* 5, 3654–3657.

Bianchini, L.F., Arruda, M.F.C., Vieira, S.R., Campelo, P.M.S., Grégio, A.M.T. and Rosa, E.A.R. (2015) Microbial biotransformation to obtain new antifungals. *Frontiers in Microbiology* 6, 1433. doi:10.3389/fmicb.2015.01433

Bonfill, M., Bentebibel, S., Moyano, E., Palazón, J., Cusidó, R.M., *et al.* (2007) Paclitaxel and baccatin III production induced by methyl jasmonate in free and immobilized cells of *Taxus baccata*. *Biologia Plantarum* 51, 647–652. doi:10.1007/s10535-007-0137-2

Borges, K.B., Borges, W.D.S., Durán-Patrón, R., Pupo, M.T., Bonato, P.S. and Collado, I.G. (2009) Stereoselective biotransformations using fungi as biocatalysts. *Tetrahedron: Asymmetry* 20, 385–397. doi:10.1016/j.tetasy.2009.02.009

Brefort, T., Scherzinger, D., Limón, M.C., Estrada, A.F., Trautmann, D., *et al.* (2011) Cleavage of resveratrol in fungi: Characterization of the enzyme Rco1 from *Ustilago maydis*. *Fungal Genetics and Biology* 48, 132–143. doi:10.1016/j.fgb.2010.10.009

Cheong, J.-J. (2000) Phytoalexin biosynthesis as a model system for biochemistry of plant-microbe interaction. *Plant Pathology Journal* 16, 70–78.

Craik, D.J., Daly, N.L., Bond, T. and Waine, C. (1999) Plant cyclotides: A unique family of cyclic and knotted proteins that defines the cyclic cystine knot structural motif. *Journal of Molecular Biology* 294, 1327–1336. doi:10.1006/jmbi.1999.3383

Cusidó, R.M., Palazón, J., Navia-Osorio, A., Bonfill, A.M.M., Morales, C. and Pinol, M.T. (1999) Production of Taxol® and baccatin III by a selected *Taxus baccata* callus line and its derived cell suspension culture. *Plant Science* 146, 101–107. doi:10.1016/S0168-9452(99)00093-X

Cusido, R.M., Sabater-Jara, A.B., Moyano, E., Bonfill, M., Goossens, A., *et al.* (2014) A rational approach to improving the biotechnological production of taxanes in plant cell cultures of *Taxus* spp. *Biotechnology Advances* 32, 1157–1167. doi:10.1016/j.biotechadv.2014.03.002

DeJong, J.M., Liu, Y., Bollon, A.P., Long, R.M., Jennewein, S., *et al.* (2006) Genetic engineering of taxol biosynthetic genes in *Saccharomyces cerevisiae*. *Biotechnology and Bioengineering* 93, 212–224. doi:10.1002/bit.20694

Elavarasi, A., Rathna, G.S. and Kalaiselvam, M. (2012) Taxol producing mangrove endophytic fungi *Fusarium oxysporum* from *Rhizophora annamalayana*. *Asian Pacific Journal of Tropical Biomedicine* 2, S1081–S1085.

Fesel, P.H. and Zuccaro, A. (2016) Dissecting endophytic lifestyle along the parasitism/mutualism continuum in *Arabidopsis*. *Current Opinion in Microbiology* 32, 103–112. doi:10.1016/j.mib.2016.05.008

Finking, R. and Marahiel, M.A. (2004) Biosynthesis of nonribosomal peptides. *Annual Reviews of Microbiology* 58, 453–488. doi:10.1146/annurev.micro.58.030603.123615

Flores-Bustamante, Z.R., Rivera-Orduna, F.N., Martinez-Cardenas, A. and Flores-Cotera, L.B. (2010) Microbial paclitaxel: Advances and perspectives. *The Journal of Antibiotics* 63, 460–467. doi:10.1038/ja.2010.83

Fu, W., Xu, M., Sun, K., Hu, L., Cao, W., *et al.* (2018) Biodegradation of phenanthrene by endophytic fungus *Phomopsis liquidambari* in vitro and in vivo. *Chemosphere* 203, 160–169. doi:10.1016/j.chemosphere.2018.03.164

Garyali, S., Kumar, A. and Reddy, M.S. (2013) Taxol production by an endophytic fungus, *Fusarium redolens*, isolated from Himalayan yew. *Journal of Microbiology and Biotechnology* 23, 1372–1380. doi:10.4014/jmb.1305.05070

Harrison, M. (1998) Development of the arbuscular mycorrhizal symbiosis. *Current Opinion in Plant Biology* 1, 360–365. doi:10.1146/annurev micro.58.030603.123749

Heinig, U., Scholz, S. and Jennewein, S. (2013) Getting to the bottom of Taxol biosynthesis by fungi. *Fungal Diversity* 60, 161–170. doi:10.1007/s13225-013-0228-7

Herre, E.A., Mejía, L.C., Kyllo, D., Rojas, E.I., Maynard, Z., *et al.* (2007) Ecological implications of anti-pathogen effects. *Ecology* 88, 550–558. doi:10.1890/05-1606

Hiruma, K., Gerlach, N., Sacristán, S., Nakano, R.T., Hacquard, S., *et al.* (2016) Root endophyte *Colletotrichum tofieldiae* confers plant fitness benefits that are phosphate status dependent. *Cell* 165, 464–474. doi:10.1016/j.cell.2016.02.028

Hori, K., Yamamoto, Y., Minetoki, T., Kurotsu, T., Kanda, M., *et al.* (1989) Molecular cloning and nucleotide sequence of the gramicidin S synthetase 1 gene. *Journal of Biochemistry* 106, 639–645. doi:10.1093/oxfordjournals.jbchem.a122909

Huang, Q.L., Roessner, C.A., Croteau, R. and Scott, A.I. (2001) Engineering *Escherichia coli* for the synthesis of taxadiene, a key intermediate in the biosynthesis of taxol. *Bioorganic & Medicinal Chemistry* 9, 2237–2242. doi:10.1016/S0968-0896(01)00072-4

Ihlenfeldt, W.D., Bolton, E.E. and Bryant, S.H. (2009) The PubChem chemical structure sketcher. *Journal of Cheminformatics* 1, 20. doi:10.1186/1758-2946-1-20

Jahn, L., Schafhauser, T., Pan, S., Weber, T., Wohlleben, W., *et al.* (2017a) *Cyanodermella asteris* sp. nov. (*Ostropales*) from the inflorescence axis of *Aster tataricus*. *Mycotaxon* 132, 107–123. doi:10.5248/132.107

Jahn, L., Schafhauser, T., Wibberg, D., Rückert, C., Winkler, A., *et al.* (2017b) Linking secondary metabolites to biosynthesis genes in the fungal endophyte *Cyanodermella asteris*: The anticancer bisanthraquinone skyrin. *Journal of Biotechnology* 257, 233–239. doi:10.1016/j.jbiotec.2017.06.410

Jennewein, S. and Croteau, R. (2001) Taxol: Biosynthesis, molecular genetics, and biotechnological applications. *Applied Microbiology and Biotechnology* 57, 13–19. doi:10.1007/s002530100757

Kemen, E., Gardiner, A., Schultz-Larsen, T., Kemen, A.C., Balmuth, A.L., *et al.* (2011) Gene gain and loss during evolution of obligate parasitism in the white rust pathogen of *Arabidopsis thaliana*. *PLOS Biology* 9, e1001094. doi:10.1371/journal.pbio.1001094

Koulman, A., Lane, G.A., Christensen, M.J., Fraser, K. and Tapper, B.A. (2007) Peramine and other fungal alkaloids are exuded in the guttation fluid of endophyte-infected grasses. *Phytochemistry* 68, 355–360. doi:10.1016/j.phytochem.2006.10.012

Kumara, P.M., Shweta, S., Vasanthakumari, M.M., Sachin, N. and Manjunatha, B.L. (2014) Endophytes and plant secondary metabolite synthesis: Molecular and evolutionary perspective. In: Verma, V.C. and Gange, A.C. (eds) *Advances in Endophytic Research*. Springer, New Delhi, pp. 177–190. doi:10.1007/978-81-322-1575-2_9

Kusari, S., Zühlke, S. and Spiteller, M. (2009) An endophytic fungus from *Camptotheca acuminata* that produces camptothecin and analogues. *Journal of Natural Products* 72, 2–7. doi:10.1021/np800455b

Kusari, S., Zühlke, S. and Spiteller, M. (2011) Effect of artificial reconstitution of the interaction between the plant *Camptotheca acuminata* and the fungal endophyte *Fusarium solani* on camptothecin biosynthesis. *Journal of Natural Products* 74, 764–775. doi:10.1021/np1008398

Kusari, S., Pandey, S.P. and Spiteller, M. (2013) Untapped mutualistic paradigms linking host plant and endophytic fungal production of similar bioactive secondary metabolites. *Phytochemistry* 91, 81–87. doi:10.1016/j.phytochem.2012.07.021

Kusari, S., Singh, S. and Jayabaskaran, C. (2014) Rethinking production of Taxol® (paclitaxel) using endophyte biotechnology. *Trends in Biotechnology* 32, 304–311. doi:10.1016/j.tibtech.2014.03.011

Lenka, S.K., Nims, N.E., Vongpaseuth, K., Boshar, R.A., Roberts, S.C. and Walker, E.L. (2015) Jasmonate-responsive expression of paclitaxel biosynthesis genes in *Taxus cuspidata* cultured cells is negatively regulated by the bHLH transcription factors TcJAMYC1, TcJAMYC2, and TcJAMYC4. *Frontiers in Plant Science* 6, 1–13. doi:10.3389/fpls.2015.00115

Li, H., Qing, C., Zhang, Y. and Zhao, Z. (2005) Screening for endophytic fungi with antitumour and antifungal activities from Chinese medicinal plants. *World Journal of Microbiology and Biotechnology* 21, 1515–1519. doi:10.1007/s11274-005-7381-4

Li, Y.-C., Tao, W.Y. and Cheng, L. (2009) Paclitaxel production using co-culture of *Taxus* suspension cells and paclitaxel-producing endophytic fungi in a co-bioreactor. *Applied Microbiology and Biotechnology* 83, 233–239. doi:10.1007/s00253-009-1856-4

Ludwig-Müller, J. (2015) Plants and endophytes – equal partners in secondary metabolite production? *Biotechnology Letters* 37, 1325–1334. doi:10.1007/s10529-015-1814-4

Mei, Y.Z., Liu, R.X., Wang, D.P., Wang, X. and Dai, C.C. (2015) Biocatalysis and biotransformation of resveratrol in microorganisms. *Biotechnology Letters* 37, 9–18. doi:10.1007/s10529-014-1651-x

Molina, G., Pimentel, M.R., Bertucci, T.C.P. and Pastore, G.M. (2012) Application of fungal endophytes in biotechnological processes. *Chemical Engineering Transactions* 27, 289–294. doi:10.3303/CET1227049

Morita, H., Nagashima, S., Takeya, K., Itokawa, H. and Iitaka, Y. (1995) Structures and conformation of antitumour cyclic pentapeptides, astins A, B and C, from *Aster tataricus*. *Tetrahedron* 51, 1121–1132. doi:10.1016/0040-4020(94)01000-P

Nicoletti, R. and Fiorentino, A. (2015) Plant bioactive metabolites and drugs produced by endophytic fungi of Spermatophyta. *Agriculture* 5, 918–970. doi:10.3390/agriculture5040918

Nisa, H., Kamili, A.N., Nawchoo, I.A., Shafi, S., Shameem, N. and Bandh, S.A. (2015) Fungal endophytes as prolific source of phytochemicals and other bioactive natural products: A review. *Microbial Pathogenesis* 82, 50–59. doi:10.1016/j.micpath.2015.04.001

Owen, N.L. and Hundley, N. (2004) Endophytes – the chemical synthesizers inside plants. *Science Progress*87,79–99.doi:10.3184/003685004783238553

Panaccione, D.G. and Coyle, C.M. (2005) Abundant respirable ergot alkaloids from the common airborne fungus *Aspergillus fumigatus*. *Applied and Environmental Microbiology* 71, 3106–3111. doi:10.1128/AEM.71.6.3106-3111.2005

Panstruga, R. (2003) Establishing compatibility between plants and obligate biotrophic pathogens. *Current Opinion in Plant Biology* 6, 320–326. doi:10.1016/S1369-5266(03)00043-8

Prado, S., Buisson, D., Ndoye, I., Vallet, M. and Nay, B. (2013) One-step enantioselective synthesis of (4S)-isosclerone through biotranformation of juglone by an endophytic fungus. *Tetrahedron Letters* 54, 1189–1191. doi:10.1016/j.tetlet.2012.12.038

Prana, T.K., Srikandace, J., Sumitro, E. and Wulandari, D. (2010) The potency of endophytic fungi of turmeric (*Curcuma longa* L.) in biotransformation of curcumin compounds in various media. *Research Journal of Microbiology* 5, 1189–1198. Available at: https://scialert.net/abstract/?doi=jm.2010.1189.1198 (accessed 16 May 2019). doi: 10.3923/jm.2010.1189.1198

Priti, V., Ramesha, B.T., Singh, S., Ravikanth, G., Ganeshaiah, K.N., *et al.* (2009) How promising are endophytic fungi as alternative sources of plant secondary metabolites? *Current Science* 97, 477–478. Available at: https://core.ac.uk/download/pdf/143845664.pdf (accessed 15 July 2019).

Pu, X., Qu, X., Chen, F., Bao, J., Zhang, G. and Luo, Y. (2013) Camptothecin-producing endophytic fungus *Trichoderma atroviride* LY357: Isolation, identification, and fermentation conditions optimization for camptothecin production. *Applied Microbiology and Biotechnology* 97, 9365–9375. doi:10.1007/s00253-013-5163-8

Pusztahelyi, T., Holb, I.J. and Pócsi, I. (2015) Secondary metabolites in fungus plant interactions. *Frontiers in Plant Science* 6, 573. doi:10.3389/fpls.2015.00573

Qaderi, A., Omidi, M., Etminan, A., Oladzad, A., Ebrahimi, C., *et al.* (2012) Hazel (*Corylus avellana* L.) as a new source of taxol and taxanes. *Journal of Medicinal Plants* 11, 66–77.

Rodriguez, R.J., White, J.F., Arnold, A.E. and Redman, R.S. (2009) Fungal endophytes: Diversity and functional roles. *The New Phytologist* 182, 314–330. doi:10.1111/j.1469-8137.2009.02773.x

Rottava, I., Cortina, P., Zanella, C., Cansian, R., Toniazzo, G., *et al.* (2010) Microbial oxidation of (–)-α-pinene to verbenol production by newly isolated strains. *Applied Biochemistry and Biotechnology* 162, 2221–2231. doi:10.1007/s12010-010-8996-y

Saunders, M. and Kohn, L.M. (2008) Host-synthesized secondary compounds influence the in vitro interactions between fungal endophytes of maize. *Applied and Environmental Microbiology* 74, 136–142. doi:10.1128/AEM.01538-07

Schafhauser, T., Kirchner, N., Kulik, A., Hujbers, M.M.E., Flor, L., *et al.* (2016) The cyclochlorotine mycotoxin is produced by the nonribosomal peptide synthetase CctN in *Talaromyces islandicus* (‘*Penicillium islandicum*’). *Environmental Microbiology* 18, 3728–3741. doi:10.1111/1462-2920.13294

Service, R. (2000) Hazel tree offers a new source of cancer drug. *Science* 288, 27–28. doi:10.1126/science.288.5463.27a

Shibuya, H., Kitamura, C., Maehara, S., Nagahata, M., Winarno, H., *et al.* (2003) Transformation of *Cinchona* alkaloids into 1-N-oxide derivatives by endophytic *Xylaria* sp. isolated from *Cinchona pubescens*. *Chemical and Pharmaceutical Bulletin* 51, 71–74. doi:10.1248/cpb.51.71

Singh, L.P., Gill, S.S. and Tuteja, N. (2011) Unraveling the role of fungal symbionts in plant abiotic stress tolerance. *Plant Signaling & Behavior* 6, 175–191. doi:10.4161/psb.6.2.14146

Soliman, S.S.M. and Raizada, M.N. (2013) Interactions between co-habitating fungi elicit synthesis of Taxol from an endophytic fungus in host *Taxus* plants. *Frontiers in Microbiology* 4, 3. doi:10.3389/fmicb.2013.00003

Stierle, A., Strobel, G.A. and Stierle, D. (1993) Taxol and taxane production by *Taxomyces andreanae*, an endophytic fungus of Pacific yew. *Science* 260, 214–216. doi:10.1126/science.8097061

Strobel, G.A. (2002) Microbial gifts from the rain forest. *Canadian Journal of Phytopathology* 24, 14–20. doi:10.1080/07060660109506965

Strobel, G.A. and Hess, W.M. (1997) Glucosylation of the peptide leucinostatin A, produced by an endophytic fungus of European yew, may protect the host from leucinostatin toxicity. *Chemistry & Biology* 4, 529–536. doi:10.1016/S1074-5521(97)90325-2

Strobel, G.A., Miller, R.V., Miller, C., Condron, M., Teplow, D.B. and Hess, W.M. (1999) Cryptocandin, a potent antimycotic from the endophytic fungus *Cryptosporiopsis* cf. *quercina*. *Microbiology* 145, 1919–1926. doi:10.1099/13500872-145-8-1919

Tian, Y., Amand, S., Buisson, D., Kunz, C., Hachette, F., *et al.* (2014) The fungal leaf endophyte *Paraconiothyrium variabile* specifically metabolizes the host-plant metabolome for its own benefit. *Phytochemistry* 108, 95–101. doi:10.1016/j.phytochem.2014.09.021

Trapp, M.A., Kai, M., Mithöfer, A. and Rodrigues-Filho, E. (2015) Antibiotic oxylipins from *Alternanthera brasiliana* and its endophytic bacteria. *Phytochemistry* 110, 72–82. doi:10.1016/j.phytochem.2014.11.005

Vadassery, J., Ritter, C., Venus, Y., Camehl, I., Varma, A., *et al.* (2008) The role of auxins and cytokinins in the mutualistic interaction between *Arabidopsis* and *Piriformospora indica*. *Molecular Plant-Microbe Interactions* 21, 1371–1383. doi:10.1094/MPMI-21-10-1371

van Nieuwenhuijzen, E., Miadlikowska, J., Houbraken, J., Adan, O., Lutzoni, F. and Samson, R. (2016) Wood staining fungi revealed taxonomic novelties in *Pezizomycotina*: New order *Superstratomycetales* and new species *Cyanodermella oleoligni*. *Studies in Mycology* 85, 107–124. doi:10.1016/j.simyco.2016.11.008

Varma, A., Sherameti, I., Tripathi, S., Prasad, R., Das, A., *et al.* (2012) The symbiotic fungus *Piriformospora indica*: Review. In: Hock, B. (ed.) *Fungal Associations. The Mycota IX*, 2nd edn. Springer, Berlin, pp. 231–254. doi:10.1007/978-3-642-30826-0_13

Vassaux, (2019) Astin C production by the endophytic fungus *Cyanodermella asteris* in planktonic and immobilized culture conditions. *Biotechnology Journal*. doi:10.1002/biot.201800624

Vasundhara, M., Kumar, A. and Reddy, M.S. (2016) Molecular approaches to screen bioactive compounds from endophytic fungi. *Frontiers in Microbiology* 7, 1774. doi:10.3389/fmicb.2016.01774

Venugopalan, A. and Srivastava, S. (2015) Endophytes as *in vitro* production platforms of high value plant secondary metabolites. *Biotechnology Advances* 33, 873–887. doi:10.1016/j.biotechadv.2015.07.004

Walker, K. and Croteau, R. (2001) Taxol biosynthetic genes. *Phytochemistry* 58, 1–7. doi:10.1016/S0031-9422(01)00160-1

Wang, H.W., Zhang, W., Su, C.L., Zhu, H. and Dai, C.C. (2015) Biodegradation of the phytoestrogen luteolin by the endophytic fungus *Phomopsis liquidambari*. *Biodegradation* 26, 197–210. doi:10.1007/s10532-015-9727-4

Weber, T., Blin, K., Duddela, S., Krug, D., Kim, H.U., *et al.* (2015) antiSMASH 3.0 – a comprehensive resource for the genome mining of biosynthetic gene clusters. *Nucleic Acids Research* 43, W237–W243. doi:10.1093/nar/gkv437

Xiong, Z.-Q., Yang, Y.-Y., Zhao, N. and Wang, Y. (2013) Diversity of endophytic fungi and screening of fungal paclitaxel producer from Anglojap yew, *Taxus × media*. *BMC Microbiology* 13, 71. http://www.biomedcentral.com/1471-2180/13/71

Xu, H.M., Zeng, G.Z., Zhou, W.B., He, W.J. and Tan, N.H. (2013) Astins K-P, six new chlorinated cyclopentapeptides from *Aster tataricus*. *Tetrahedron* 69, 7964–7969. doi:10.1016/j.tet.2013.07.006

Yang, Y., Zhao, H., Barrero, R.A., Zhang, B., Sun, G., *et al.* (2014) Genome sequencing and analysis of the paclitaxel-producing endophytic fungus *Penicillium aurantiogriseum* NRRL 62431.

BMC Genomics 15, 69. http://www.biomedcentral.com/1471-2164/15/69

Ying, Y.-M., Shan, W.-G. and Zhan, Z.-J. (2014) Biotransformation of huperzine A by a fungal endophyte of *Huperzia serrata* furnished sesquiterpenoid–alkaloid hybrids. *Journal of Natural Products* 77, 2054–2059. doi:10.1021/np500412f

Yuan, Y.-J., Li, C., Hu, Z.-D. and Wu, J.-C. (2001) Signal transduction pathway for oxidative burst and taxol production in suspension cultures of *Taxus chinensis* var. *mairei* induced by oligosaccharide from *Fusarium oxysprum*. *Enzyme and Microbial Technology* 29, 372–379. doi:10.1016/S0141-0229(01)00406-9

Zhang, J., Zhang, L., Wang, X., Qiu, D., Sun, D., *et al*. (1998) Microbial transformation of 10-deacetyl-7-epitaxol and 1β-hydroxybaccatin I by fungi from the inner bark of *Taxus yunnanensis*. *Journal of Natural Products* 61, 497–500. doi:10.1021/np960647q

Zhang, N., Zhang, C., Xiao, X., Zhang, Q. and Huang, B. (2016) New cytotoxic compounds of endophytic fungus *Alternaria* sp. isolated from *Broussonetia papyrifera* (L.) Vent. *Fitoterapia* 110, 173–180. doi:10.1016/j.fitote.2016.03.014

Zhang, P., Zhou, P.-P. and Yu, L.-J. (2009) An endophytic taxol-producing fungus from *Taxus media*, *Cladosporium cladosporioides* MD2. *Current Microbiology* 59, 227–232. doi:10.1007/s00284-008-9270-1

Zhao, K., Ping, W.X., Zhang, L.N., Liu, J., Lin, Y., *et al*. (2008) Screening and breeding of high taxol producing fungi by genome shuffling. *Science in China Series C: Life Sciences* 51, 222–231. Available at: https://link.springer.com/article/10.1007/s11427-008-0037-5 (accessed 15 July 2019).

Zheng, A., Lin, R., Zhang, D., Qin, P., Xu, L., *et al*. (2013) The evolution and pathogenic mechanisms of the rice sheath blight pathogen. *Nature Communications* 4, 1424. doi:10.1038/ncomms2427

Zheng, X.Y., Spivey, N.W., Zeng, W., Liu, P.P., Fu, Z.Q., *et al*. (2012) Coronatine promotes *Pseudomonas syringae* virulence in plants by activating a signaling cascade that inhibits salicylic acid accumulation. *Cell Host & Microbe* 11, 587–596. doi:10.1080/10810730902873927

Zikmundova, M., Drandarov, K., Bigler, L., Hesse, M. and Werner, C. (2002) Biotransformation of 2-benzoxazolinone and 2-hydroxy-1,4-benzoxazin-3-one by endophytic fungi isolated from *Aphelandra tetragona*. *Applied and Environmental Microbiology* 68, 4863–4870. doi:10.1128/AEM.68.10.4863-4870.2002

Index